ICM-90 Satellite Conference Proceedings

M. Kashiwara · T. Miwa (Eds.)

Special Functions

Proceedings of the Hayashibara Forum 1990
held in Fujisaki Institute, Okayama, Japan
August 16-20, 1990

Springer-Verlag
Tokyo Berlin Heidelberg
New York London Paris
Hong Kong Barcelona

Editors
Masaki Kashiwara and Tetsuji Miwa
RIMS, Kyoto University
Sakyoku, Kyoto, 606 Japan

Mathematics Subject Classification (1980): 33-06, 33A30, 33A35, 33A70, 33A75, 33A99

ISBN-13: 978-4-431-70085-2 e-ISBN-13: 978-4-431-68170-0
DOI: 10.1007/978-4-431-68170-0

PREFACE

The 1990 Hayashibara Forum, "the International Conference on Special Functions", was held at Fujisaki Institute, Hayashibara Biochemical Laboratories, Inc., Okayama, Japan for five days (August 16–20, 1990). This volume is the proceedings for that meeting.

On January 14,1985, Heisuke Hironaka and Ken Hayashibara, the president of Chairman, Board of Trustees, Hayashibara Foundation, met and decided to have an international conference on mathematics in the summer of 1990. This was pushed forward by Kiyosi Itô, who proposed "Special functions" as the theme of the conference. He also asked the present editors to join in the organizing committee of the Hayashibara Forum, 1990.

On May 13, 1989 the organizing committee sent letters to major Japanese mathematical institutions asking their members to give suggestions about whom it should invite. Receiving the replies, the organizing committee decided the invited speakers, and sent invitation letters to them, in which it was written that

> "Special functions have been created and explored to describe scientific and mathematical phenomena. Trigonometric functions give the relation of angle to length. Riemann's zeta function was invented in order to describe the prime number distribution. Legendre's spherical functions and Bessel's functions were born in connection with the eigenvalue problems for partial differential equations. On the other hand, progress in mathematics was often made through attempts to understand special functions more deeply. The role of hypergeometric functions in the theory of generalized functions are notable in this respect. Sometimes, developments in mathematics shed light on previously known special functions from a completely different angle. The re-discovery of Painlevé transcendents as correlation functions in statistical mechanical models is one of the best examples of this sort. Several approaches were pursued to handle various special functions in a unified manner, for example, by means of differential equations, or by group representations. However, the nature of special functions is not yet fully explored. To make further progress, entirely new standpoints are to be sought. The theory of quantum groups applied to functions in q-analogue and the Hodge theory as an abstract theory of algebraic integrals are successful examples of recent innovations.
>
> In this symposium we would like to invite people from various branches of mathematics, hoping that special functions are discussed from broader viewpoints than usual."

Most of the invitations were accepted. The organizing committee then decided on a few Japanese speakers and other participants of the conference. After this period, the preparation of the conference was done mainly by the present editors with the support of Kazuya Masaki of Manager, Administration & Public Relations, Hayashibara Company, Limited and Takeo Shiozaki of Japan Association for Mathematical Sciences, and Yasuko Shichida and Kazuko Kôno, secretaries of RIMS.

In Okayama, 20 talks and poster sessions were given, and the programme is shown at the end of this volume.

We thank all the participants of the conference for their active participation, and extend our gratitude to those who supported this conference and helped the activities, in particular, to Ken Hayashibara, Heisuke Hironaka, Kiyosi Itô, Kazuya Masaki, Takeo Shiozaki, Takaya Miyamoto, Masashi Kurimoto, Yasuko Shichida and Kazuko Kôno.

Masaki Kashiwara
Tetsuji Miwa
RIMS, Kyoto University
May, 1991

TABLE OF CONTENTS

The Bender-Wu Analysis and the Voros Theory

— To the memory of the late Professor K. Yosida —

by

Takashi AOKI

Department of Mathematics and Physics,

Faculty of Science and Technology

Kinki University

Higashi-Osaka 577, Japan

Takahiro KAWAI

Research Institute for Mathematical Sciences

Kyoto University

Kyoto 606, Japan

and

Yoshitsugu TAKEI

Department of Mathematics

Faculty of Science

Kyoto University

Kyoto 606, Japan

§0. Introduction.

In their pioneering work [BW], Bender and Wu presented the secular equation for anharmonic oscillators ([BW], (F.56)~(F.58) in p.1257), which was supported by their ingenious WKB analysis. As we shall discuss in our forthcoming article, we can validate their conjecture by Voros' epoch-making article [V]. At the same time several ideas contained in [BW] can be effectively employed to understand the Voros theory from the viewpoint of (micro)differential operators, and this is what we report here.

To be more precise, we will show the following:

In §2 we give another proof of the Voros connection formula ([V], §6) in the simplest non-trivial case with the aid of microdifferential operators ([SKK], Chap.II; see also [P1], where the importance of microdifferential operators in alien calculus was first emphasized).

We hope our proof based on the Gauss connection formula applied to the Borel-transformed WKB solution of the canonical equation elucidates the analytic background of the geometric reasoning of Voros.

In §3 we discuss a more global situation than in §1. The main result in this section (Theorem 3.1) was motivated by the ingenious idea of Bender and Wu to the effect that the Weber function can be used in some region as an approximate wave function for the Schrödinger operator for anharmonic oscillators. Our reasoning will make manifest the relevance of period integrals to WKB solutions. We also plan to use it in our subsequent article to discuss "connection automorphism" (=automorphisme de passage) for WKB solutions.

In ending this introduction, we sincerely thank Professor M. Sato for having patiently led us to this subject.

§1. Preliminaries.

We first fix the notations and terminologies to be used in what follows. We mainly follow [V] except that we use the symbol S_{odd} as a substitute of the symbol u in [V]. (Cf. (1.6)ff.)

The central object we are concerned with in this article is an equation of the following form:

$$\left(\frac{d^2}{dq^2} - x^2 Q(q)\right)\psi(q,x) = 0, \tag{1.1}$$

where $Q(q)$ is a holomorphic function of $q \in \mathbf{C}$ and x is a complex (large) parameter. A WKB solution of (1.1) is, by definition, a formal solution of (1.1) which is of the following form:

$$\psi(q,x) = \exp\left(\int^q S(q,x)dq\right), \tag{1.2}$$

where

$$S(q,x) = \sum_{j \geqq -1} x^{-j} S_j(q) \tag{1.3}$$

formally (i.e., as a formal power series in x^{-1} with holomorphic coefficients) satisfies the following Riccati equation:

$$\frac{\partial S}{\partial q} + S^2 = x^2 Q(q). \tag{1.4}$$

Let us note some important properties of a WKB solution. First of all, $S_j(q)$ in the formal expansion in (1.3) is uniquely fixed once we fix $S_{-1}(q) = \pm\sqrt{Q(q)}$. Actually $S_j(q)$ is determined by the following recursive relation:

$$2S_{-1}S_{j+1} + \sum_{\substack{k+\ell=j \\ k,\ell \geqq 0}} S_k S_\ell + \frac{dS_j}{dq} = 0 \quad (j \geqq -1). \tag{1.5}$$

Hence the ambiguity of a WKB solution solely stems from the choice of the initial end-point of the integral in (1.2). Thus a WKB solution bears some resemblance to a holonomic function, the central subject of algebraic analysis. This is the main reason why we are interested in WKB analysis.

Second, in spite of the above mentioned interesting property, a WKB solution suffers from a notorious divergence problem, which is common in singular perturbations. But, the important discovery of Voros [V] is that the systematic use of the Borel transformations rescues us. This is also the viewpoint of Ecalle [E]. As is emphasized by Pham [P1], the Borel transformation is closely related to microdifferential operators, and hence, microlocal analysis. Actually we surmise that the notion of resurgence introduced by Ecalle would be important in the non-symmetry between x and ξ in microlocal analysis in that ξ-variables enjoy some homogeneity property.

Third, a WKB solution of the equation has the following interesting property (cf. [Z] for example):

Let S_{odd} (resp., S_{even}) denote

$$\sum_{j=2m-1} x^{-j} S_j \quad (\text{resp.}, \sum_{j=2m} x^{-j} S_j), \tag{1.6}$$

where m ranges over natural numbers; $0, 1, 2, \cdots$. Then

$$S_{\text{even}} = -\frac{1}{2}\frac{\partial}{\partial q} \log S_{\text{odd}} \tag{1.7}$$

holds. This is the reason why most of the literature in physics (including [V], where the symbol u is used to denote S_{odd} [with some difference of a factor $\pm i$]) prefers to use S_{odd}. Note, however, the relation (1.7) is peculiar to the equation (1.1); if we consider the case where the potential Q depends not only on q but also on x^{-1}, the relation (1.7) should be modified, or, in general, abandoned. In this article we consider the case where Q depends only on q, and we freely use the relation (1.7).

As we emphasized above, the exact (i.e., not merely asymptotic) analysis of a WKB solution relies on the Borel transformation. Since the definition of the Borel transform is not uniform in literature, we follow Voros [V] in fixing the definition and notations as follows:

Definition 1.1. Let $f(x)$ be a formal series of descending powers of (a large variable) x multiplied by an exponential factor, i.e.,

$$f(x) = e^{\xi_0 x}\Big(\sum_{\lambda} f_\lambda x^{-\lambda}\Big), \tag{1.8}$$

where λ ranges over $\Lambda_+(\alpha) \underset{\text{def}}{=} \{\alpha + n; \alpha \text{ is a real number, and } n = 0, 1, 2, \ldots\}$.

Then its Borel transform $f_B(\xi)$ is, by definition,

$$\sum_{\lambda} \frac{f_\lambda}{\Gamma(1+\lambda)}(\xi + \xi_0)^\lambda.$$

Remark 1.2. The definition of $f_B(\xi)$ is the same as in [V], but differs from the definition in most literature, where $\sum_\lambda f_\lambda(\xi + \xi_0)^{\lambda-1}/\Gamma(\lambda)$ is used as the definition of $f_B(\xi)$.

What we really need is the Borel transform of a formal series depending on a holomorphic parameter, that is, the Borel transform $f_B(q, \xi)$ of a formal series

$$f(q, x) = e^{\xi_0(q) x}\Big(\sum_{\lambda \in \Lambda_+(\alpha)} f_\lambda(q) x^{-\lambda}\Big), \tag{1.10}$$

where q ranges over an open subset U of $\mathbf{C}$ and $\xi_0(q)$ and $f_\lambda(q)$ are holomorphic on U. (Cf. [V] and [P1].) The Borel transform $f_B(q, \xi)$ of $f(q, x)$ is, again by definition,

$$\sum_{\lambda} \frac{f_\lambda(q)}{\Gamma(1+\lambda)}(\xi + \xi_0(q))^\lambda. \tag{1.11}$$

If, for each compact set K in U, there exist constants A_K and C_K for which

$$\sup_K |f_\lambda(q)| < A_K C_K^\lambda \Gamma(1+\lambda) \tag{1.12}$$

holds, then $f_B(q, \xi)$ is holomorphic on a neighborhood of $(q, -\xi_0(q))$ in $\mathbf{C}^2$. Most of the formal series we encounter in this article satisfy this condition, and if it is the case, we say that $f(q, x)$ is a pre-Borel-summable series with exponent $\xi_0(q)$. Note that, if a formal series $f(q, x)$ is pre-Borel-summable and $\xi_0(q) = 0$, then it is a symbol of a microdifferential operator. (Cf. [SKK] Chap. II, Definition 1.4.6 and [P1]) Note that the Borel sum of the formal series $f(q, x)$ is defined through analytic properties of $f_B(q, \xi)$. (Cf. [V], for example.)

§2. A connection formula at a simple turning point.

The purpose of this section is to give a (micro)differential operator theoretic proof of a basic connection formula (Theorem 2.1 below) of Voros. Another approach with more emphasis on its relevance to resurgence theory can be found in [P2] and [CNP].

The situation we shall discuss is as follows: Let $Q(q)$ be a holomorphic function defined on a neighborhood U of the origin 0 of $\mathbf{C}$, and suppose that the origin is a unique simple turning point of the equation (1.1) in U, that is,

(2.1) $$Q(0) = 0,$$

(2.2) $$(dQ/dq)(0) \neq 0,$$

(2.3) $$Q(q) \neq 0 \quad \text{on} \quad U - \{0\}.$$

We introduce a cut $\{q \in U; Q(q) \leqq 0\}$, and denote by $\sqrt{Q(q)}$ the principal branch, i.e., $\sqrt{Q(q)} > 0$ if $Q(q) > 0$. Let $\xi(q)$ denote the following integral considered on this cut-plane:

(2.4) $$\int_0^q \sqrt{Q(q')}dq'.$$

Let us now denote by $\psi_\pm(q,x)$ the WKB solution of (1.1) with $S_{-1}(q)$ being $\pm\sqrt{Q(q)}$ and the integration in (1.2) is performed from 0 to q. Note that, as $S_j(q)$ has the form $q^{-\frac{3}{2}j-1}R_j(q)$ with a holomorphic function $R_j(q)$ near the origin, there arises no ambiguity in defining the integral $\int_0^q S_j(q')dq'$ when j is odd. Hence

(2.5) $$\frac{1}{\sqrt{S_{\rm odd}(q,x)}}\exp\left(\int_0^q S_{\rm odd}(q',x)dq'\right)$$

is well-defined. We can readily verify that $\psi_\pm(q,x)$ is pre-Borel-summable (on the cut-plane introduced above) with exponent $\pm\xi(q)$ respectively. (See Appendix §A.1.) Their Borel transforms shall be denoted respectively by $\psi_{\pm,B}(q,\xi)$. Then we have the following theorem. (Cf. [V], §6.)

Theorem 2.1. *Let the origin 0 of* $\mathbf{C}$ *be a simple turning point of the equation (1.1), and let* $\psi_{\pm,B}(q,\xi)$ *denote the Borel transform of the WKB solutions of (1.1), which are normalized as above. Then, for a sufficiently small neighborhood* W *of the origin of* $\mathbf{C}^2(=\mathbf{C}_q \times \mathbf{C}_\xi)$, *both* $\psi_{+,B}(q,\xi)$ *and* $\psi_{-,B}(q,\xi)$ *have their singularities in* W *only along* $\{(q,\xi) \in W; \xi = \pm\xi(q)\}$. *Furthermore the discontinuity* $\Delta_{\xi(q)}\psi_{+,B}(q,\xi)$ *(resp.,* $\Delta_{-\xi(q)}\psi_{-,B}(q,\xi)$*) of* $\psi_{+,B}(q,\xi)$ *(resp.,* $\psi_{-,B}(q,\xi)$*) along the cut* $\{(q,\xi) \in W; \mathrm{Im}\xi = \mathrm{Im}\xi(q),\ \mathrm{Re}\xi \geqq \mathrm{Re}\xi(q)\}$ *(resp.,* $\{(q,\xi) \in W; \mathrm{Im}\xi = \mathrm{Im}(-\xi(q)), \mathrm{Re}\xi \geqq \mathrm{Re}(-\xi(q))\}$*) coincides with* $i\psi_{-,B}(q,\xi)$ *(resp.,* $-i\psi_{+,B}(q,\xi)$*) along the cut.*

Our strategy of the proof of this theorem is as follows: We first construct a transformation that reduces $Q(q)$ to q near the origin; (Lemma 2.2 and Proposition 2.3) We next compute the WKB solution of (1.1) when $Q(q) = q$. Its Borel transform can be explicitly expressed with the aid of a hypergeometric function, and the assertion of Theorem 2.1 in this special case is an immediate consequence of the classical connection formula of Gauss. (Lemma 2.4.) The (micro)local character in (q,ξ)-space of the transformation we shall use then entails the assertion in general, completing the proof of Theorem2.1.

Now we state preparatory results concerning the transformation we shall use.

Lemma 2.2. *Let* $Q(\tilde{q})$ *be a holomorphic function defined on a neighborhood of the origin 0 of* $\mathbf{C}$. *Suppose* $Q(0) = 0$ *and* $(dQ/d\tilde{q})(0) \neq 0$. *Then we can find a pre-Borel-summable series (with exponent 0)*

$$q(\tilde{q},x) = q_0(\tilde{q}) + q_1(\tilde{q})x^{-1} + q_2(\tilde{q})x^{-2} + \cdots . \tag{2.6}$$

with a holomorphic parameter $\tilde{q}$ *so that it satisfies*

$$x^2Q(\tilde{q}) = x^2\left(\frac{\partial q(\tilde{q},x)}{\partial \tilde{q}}\right)^2 q(\tilde{q},x) - \frac{1}{2}\{q(\tilde{q},x);\tilde{q}\} \tag{2.7}$$

and

$$(dq_0/d\tilde{q})(0) \neq 0. \tag{2.8}$$

Here $\{q(\tilde{q},x);\tilde{q}\}$ denotes the Schwarzian derivative, i.e.,

$$\frac{\frac{\partial^3 q}{\partial \tilde{q}^3}}{\frac{\partial q}{\partial \tilde{q}}} - \frac{3}{2}\left(\frac{\frac{\partial^2 q}{\partial \tilde{q}^2}}{\frac{\partial q}{\partial \tilde{q}}}\right)^2. \tag{2.9}$$

Propositiion 2.3. *Let $Q(\tilde{q})$ be a holomorphic function defined on a neighborhood U of the origin 0 of $\mathbf{C}$, and suppose that $Q(0)=0$ and that $(dQ/d\tilde{q})(0)\neq 0$. Let $\tilde{\psi}(\tilde{q},x)$ be a pre-Borel-summable series with a holomorphic parameter $\tilde{q}$ ranging over U, and suppose that it satisfies the following equation:*

$$\frac{d^2\tilde{\psi}(\tilde{q},x)}{d\tilde{q}^2} - x^2 Q(\tilde{q})\tilde{\psi}(\tilde{q},x) = 0. \tag{2.10}$$

Let $q(\tilde{q},x)$ be the series given in Lemma 2.2. Then there exists a pre-Borel-summable series $\psi(q,x)$ which satisfies the following two conditions:

$$\left(\frac{d^2}{dq^2} - x^2 q\right)\psi(q,x) = 0 \tag{2.11}$$

$$\tilde{\psi}(\tilde{q},x) = \left(\frac{\partial q(\tilde{q},x)}{\partial \tilde{q}}\right)^{-1/2} \psi(q(\tilde{q},x),x). \tag{2.12}$$

Furthermore, if $\tilde{\psi}(\tilde{q},x)$ is a WKB solution of (2.10), then $\psi(q,x)$ is a WKB solution of (2.11).

The proof of these results shall be given in Appendix, §A. 2. We note that Lemma 2.2, which was motivated by [BW], is a refined version of the Liouville transformation in classical WKB analysis. An important point in our results is that $q(\tilde{q},x)$ is pre-Borel-summable. As we shall see later, this is crucially important in our reasoning. We also note that $q_j(\tilde{q})$ in (2.6) is identically 0 for j odd (cf. §A.2); this is a consequence of a special feature of the "potential" Q in that it does not depend on x^{-1}.

We now prove Theorem 2.1 in a special case where $Q(q)=q$, to which the general case shall be reduced eventually by the preceding proposition.

Lemma 2.4. *Theorem 2.1 holds if $Q(q) = q$.*

Proof. If $Q(q) = q$, then we can verify by the induction that $S_j(q)$ in (1.3) has the form

$$c_j q^{-3j/2-1}, \tag{2.13}$$

where c_j is a complex number. Hence $\int_0^q S_{\mathrm{odd}} dq$ is a function of $xq^{3/2}$. Therefore $\psi_{\pm,B}(q,\xi)$ has the form $\sqrt{q}h_\pm(\xi q^{-3/2})$. Letting t denote $\xi q^{-3/2}$, we obtain

$$(\frac{9}{4}t^2 - 1)\frac{d^2}{dt^2}h_\pm + \frac{9}{4}t\frac{d}{dt}h_\pm - \frac{1}{4}h_\pm = 0 \tag{2.14}$$

from the Balian-Bloch equation that $\psi_{\pm,B}$ solves, i.e.,

$$(\frac{\partial^2}{\partial q^2} - q\frac{\partial^2}{\partial \xi^2})\psi_{\pm,B}(q,\xi) = 0. \tag{2.15}$$

Since the exponent of $\psi_\pm(q,x)$ is $\pm\frac{2}{3}q^{3/2}$, we conclude from (2.14) the following:

$$\psi_{+,B} = \frac{4}{\sqrt{3\pi}}\sqrt{q}s^{1/2}F(\frac{5}{6},\frac{1}{6};\frac{3}{2};s)\Big|_{s=\frac{3\xi}{4q^{3/2}}+\frac{1}{2}} \quad \text{near} \quad \xi = -\frac{2}{3}q^{3/2} \tag{2.16}$$

and

$$\psi_{-,B} = \frac{4i}{\sqrt{3\pi}}\sqrt{q}(1-s)^{1/2}F(\frac{1}{6},\frac{5}{6};\frac{3}{2};1-s)\Big|_{s=\frac{3\xi}{4q^{3/2}}+\frac{1}{2}} \quad \text{near} \quad \xi = \frac{2}{3}q^{3/2}. \tag{2.17}$$

Here we have adjusted the constant factors by observing

$$\left[\frac{1}{\sqrt{S_{\mathrm{odd}}}}\exp(\pm\int_0^q S_{\mathrm{odd}}\,dq)\right]_B = \frac{1}{\Gamma(3/2)}q^{-1/4}(\xi \pm \frac{2}{3}q^{3/2})^{1/2}(1+\cdots). \tag{2.18}$$

Let us now compare the singular part of $\psi_{+,B}$ at $\xi = \frac{2}{3}q^{3/2}$ with $\psi_{-,B}$ considered there. Since the point $\xi = \frac{2}{3}q^{3/2}$ correspondens to $s = 1$ in the expression (2.16), we apply Gauss' connection formula to find that the singular part of $\psi_{+,B}$ at $\xi = \frac{2}{3}q^{3/2}$ is

$$\frac{4}{\sqrt{3\pi}}\sqrt{q}s^{1/2}\frac{\Gamma(\frac{3}{2})\Gamma(-\frac{1}{2})}{\Gamma(\frac{5}{6})\Gamma(\frac{1}{6})}(1-s)^{1/2}F(\frac{2}{3},\frac{4}{3};\frac{3}{2};1-s)\Big|_{s=\frac{3\xi}{4q^{3/2}}+\frac{1}{2}}. \tag{2.19}$$

Then Kummer's relation for hypergeometric functions entails that it is equal to

$$\frac{4}{\sqrt{3\pi}}\sqrt{q}(-\frac{1}{2})(1-s)^{1/2}F(\frac{5}{6},\frac{1}{6};\frac{3}{2};1-s)\Big|_{s=\frac{3\xi}{4q^{3/2}}+\frac{1}{2}}, \tag{2.20}$$

which is nothing but $\frac{i}{2}\psi_{-,B}$ considered near $\xi = \frac{2}{3}q^{3/2}$. Therefore as the singular part of $\psi_{+,B}$ at $\xi = \frac{2}{3}q^{3/2}$ is of square-root type, its discontinuity there is $i\psi_{-,B}$. The same reasoning goes well if we replace the role of $\psi_{+,B}$ and $\psi_{-,B}$. This completes the proof of Lemma 2.4.

Q.E.D.

Let us complete the proof of Theorem 2.1 by reducing the general case to the case where the "potential" $Q(q)$ is q.

Let $q(\tilde{q},x) = q_0(\tilde{q}) + \sum_{j\geqq 1} q_j(\tilde{q})x^{-j}$ with $q_1 = 0$ (cf. §A.2) be the formal series given in Lemma 2.2, and let $\psi_{\pm}(q,x)$ be the WKB solutions studied in Lemma 2.4, i.e., suitably normalized WKB solutions in the case where the "potential" $Q(q)$ is q. Then we find the following relation among formal series in x:

$$\psi_+(q(\tilde{q},x);x) = \sum_{n=0}^{\infty}\left[\frac{1}{n!}(\sum_{j=1}^{\infty} q_j(\tilde{q})x^{-j})^n \frac{\partial^n}{\partial q^n}\psi_+(q,x)\right]\Big|_{q=q_0(q)}. \tag{2.21}$$

Note that the right-hand side of (2.21) formally makes sense; in fact, $(\partial/\partial q)^n\psi_+(q;x) = S^n + L_{n-1}(S, \partial S/\partial q, \cdots, \partial^{n-1}S/\partial q^{n-1})$ holds formally for $S = S_{\rm odd} - \frac{1}{2}(\partial/\partial q)\log S_{\rm odd}$ (cf. (1.7)) and a polynomial L_{n-1} of degree at most $n-1$, and $(\sum_{j=2}^{\infty} q_j(\tilde{q})x^{-j})^n(S^n + L_{n-1})$ has the form $q_2(\tilde{q})^n S_{-1}^n x^{-n} + \cdots$ (a formal power seires of x^{-1}), showing that the right-hand side of (2.21) is of the form (1.10). Let us consider the Borel transform of the right-hand side of (2.21), first by replacing $(\partial/\partial q)^n\psi_+(q,x)$ with $(S^n + L_{n-1})\psi_+$ as above and then by recovering $(\partial/\partial q)^n\psi_{+,B}(q,\xi)|_{q=q_0(\tilde{q})}$ from $((\underbrace{S_B * \cdots * S_B}_{n} + L_{n-1}^*) * \psi_{+,B}(q,\xi))|_{q=q_0(\tilde{q})}$, where L_{n-1}^* is the polynomial L_{n-1} with the product operation given by the convolution product, and $\psi_{+,B}$ denotes the Borel tranform of ψ_+. Thus the Borel transform of the right-hand side of (2.21) can be written as

$$\sum_{n=0}^{\infty}\left[\frac{1}{n!}(\sum_{j=2}^{\infty} q_j(\tilde{q})(\frac{\partial}{\partial\xi})^{-j})^n \frac{\partial^n}{\partial q^n}\psi_{+,B}(q,\xi)\right]\Big|_{q=q_0(\tilde{q})}. \tag{2.22}$$

As $(dq_0/d\tilde{q})(0) \neq 0$, we can rewrite $q_j(\tilde{q}) = \tilde{q}_j(q)$ for some holomorphic function $\tilde{q}_j$ of q, and further verify

(2.23) $$B(q, \partial/\partial q, \partial/\partial\xi) = \sum_{n=0}^{\infty} \frac{1}{n!} (\sum_{j=2}^{\infty} \tilde{q}_j (\frac{\partial}{\partial\xi})^{-j})^n \frac{\partial^n}{\partial q^n}$$

is a well-defined microdifferential operator (near the codirection $d\xi$ at $q = 0$). (Cf. [AKK] for example.) Since $q = q_0(\tilde{q})$ determines a coordinate transformation near $q = 0$, the operator B can be rewritten as $\tilde{B}(\tilde{q}, \partial/\partial\tilde{q}, \partial/\partial\xi)$. Then by Borel transforming the both sides of (2.12) we obtain

(2.24) $$\tilde{\psi}_{+,B}(\tilde{q}, \xi) = \left(\frac{\partial q(\tilde{q}, \partial_\xi)}{\partial \tilde{q}}\right)^{-1/2} \tilde{B}(\tilde{q}, \partial/\partial\tilde{q}, \partial/\partial\xi)\psi_{+,B}(q_0(\tilde{q}), \xi),$$

where $\tilde{\psi}_{+,B}(\tilde{q}, \xi)$ is the Borel transform of the WKB solution $\tilde{\psi}_+(\tilde{q}, x)$ of (2.10) normalized as is required in Theorem 2.1. Since $(\partial q_0/\partial\tilde{q})(0) \neq 0, \partial q(\tilde{q}, \partial_\xi)/\partial\tilde{q}$ is a microdifferential operator with a non-vanishing principal symbol near $\tilde{q} = 0$. Hence $(\partial q(\tilde{q}, \partial_\xi)/\partial\tilde{q})^{-1/2}$ is a well-defined microdifferential operator; we denote its composition with $\tilde{B}$ as A. Although we have obtained the operator A by dealing with $\psi_{+,B}$, the same operator A appears if we deal with $\psi_{-,B}$. The point is that the operator B is written down with $\partial/\partial q$, not with $S_{B}*$ etc.(Cf. (2.23))

Hence we find

(2.25) $$\psi_{\pm,B}(\tilde{q}, \xi) = A(\tilde{q}, \partial/\partial\tilde{q}, \partial/\partial\xi)\psi_{\pm,B}(q_0(\tilde{q}), \xi),$$

or, if we represent $A(\tilde{q}, \partial/\partial\tilde{q}, \partial/\partial\xi)$ as an integral operator, we find, on a sufficiently small neighborhood of $(\tilde{q}, \xi) = (0, 0)$,

(2.26) $$\tilde{\psi}_{\pm,B}(\tilde{q}, \xi) = A(\tilde{q}, \partial/\partial\tilde{q}, \partial/\partial\xi)\psi_{\pm,B}(q_0(\tilde{q}), \xi) + h_\pm(\tilde{q}, \xi)$$

with some holomorphic functions $h_\pm(\tilde{q}, \xi)$.

Let us now study the singular part of $\tilde{\psi}_{-,B}(\tilde{q}, \xi)$ near $\xi = -\frac{2}{3}q_0(\tilde{q})^{3/2}$ $(\tilde{q} \neq 0, |\tilde{q}| \ll 1)$. It then follows from Lemma 2.4 and (2.26) that the singular part of $\tilde{\psi}_{-,B}(\tilde{q}, \xi)$ consedered there has the form $\frac{i}{2}A(\tilde{q}, \partial/\partial\tilde{q}, \partial/\partial\xi)\psi_{+,B}(q_0(\tilde{q}), \xi)$ modulo holomorphic functions there. Hence it coincides with $i\tilde{\psi}_{+,B}(\tilde{q}, \xi)$ modulo holomorphic function near the point. As $\tilde{\psi}_{+,B}(\tilde{q}, \xi)$ has a square root singularity near $\xi = -\frac{2}{3}q_0(a)^{3/2}$, this proves Theorem 2.1.

§3. A canonical form with two turning points.

As we have seen in §2, we can find a canonical form of the equation (1.1) considered near a simple turning point. A natural question that follows will be what if we consider two simple turning points simultaneously. This is not a question of mere curiosity; as we shall show in what follows, the canonical form in this case contains a formal power series whose coefficients are related to period integrals determined by the potential $Q(q)$. Although we have not yet proved, we conjecture that the formal series should be resurgent if $Q(q)$ is a rational function.

Our main result in this section is summarized as follows:

Theorem 3.1. *Let U be an open neighborhood of the origin 0 of $\mathbf{C}$, and let $Q(\tilde{q})$ be a holomorphic function defined there. Suppose that the equation*

$$(3.1)\qquad \left(\frac{d^2}{d\tilde{q}^2} - x^2 Q(\tilde{q})\right)\tilde{\psi}(\tilde{q},x) = 0$$

has exactly two distinct simple turning points p_0 and p_1 in U, and suppose further that p_0 and p_1 are connected by a curve in U defined by

$$(3.2)\qquad \operatorname{Im}\int_{p_0}^{\tilde{q}} \sqrt{Q(\tilde{q})}\,d\tilde{q} = 0.$$

Then we can find pre-Borel-summable series

$$(3.3)\qquad q(\tilde{q},x) = \sum_{j\geqq 0} q_j(\tilde{q})x^{-j}$$

and

$$(3.4)\qquad E(x) = \sum_{j\geqq 0} E_j x^{-j}$$

so that

$$(3.5)\qquad q_0(p_0) = -2\sqrt{E_0} \quad \text{and} \quad q_0(p_1) = 2\sqrt{E_0}$$

hold and $\tilde{\psi}(\tilde{q},x)$ can be expressed as

$$(3.6)\qquad \tilde{\psi}(\tilde{q},x) = \left(\frac{\partial q(\tilde{q},x)}{\partial \tilde{q}}\right)^{-1/2} \psi(q(\tilde{q},x),x)$$

where $\psi(q,x)$ satisfies

$$\left(\frac{d^2}{dq^2} - x^2(E(x) - \frac{1}{4}q^2)\right)\psi(q,x) = 0. \tag{3.7}$$

In particular, E_0 is a real positive constant given by

$$E_0 = \frac{1}{\pi}\int_{p_0}^{p_1} \sqrt{Q(\tilde{q})}d\tilde{q}. \tag{3.8}$$

Remark 3.2. In the definition of the curve (3.2) we may of course interchange p_0 and p_1. This curve (3.2) is one of the so-called Stokes lines. Stokes lines play an essential role in the Voros theory, especially in relation to his connection formula. (Cf. [V],§6).

Remark 3.3. The equality (3.8) is a typical example of the relationships between the canonical form (3.7) and period integrals determined by the potential $Q(\tilde{q})$.

Theorem 3.1 is an immediate consequence of the following Lemma 3.4. (Cf. Appendix §A.2, where Proposition 2.3 is deduced from Lemma 2.2.)

Lemma 3.4. *Let $Q(\tilde{q})$ be a holomorphic function defined on a neighborhood V of the origin 0 of $\mathbf{C}$. Suppose that $Q(\tilde{q})$ has exactly two distinct simple zeros p_0 and p_1 in V and that these two zeros are connected by a curve in U defined by (3.2). Then we can find pre-Borel-summable series $q(\tilde{q},x)$ and $E(x)$ in (3.3) and (3.4) which enjoy the following properties:*

(i) $q_j(\tilde{q})$ ($j \geq 0$) is holomorphic on an open subset $V(\subset U)$ containing p_0 and p_1,

(ii) $q_0(\tilde{q})$ is a biholomorphic mapping on V,

(iii) $q_0(p_0) = -2\sqrt{E_0}$ and $q_0(p_1) = 2\sqrt{E_0}$,

(iv) E_0 is a real positive constant and given by

$$E_0 = \frac{1}{\pi}\int_{p_0}^{p_1} \sqrt{Q(\tilde{q})}d\tilde{q},$$

and

(v) the following equation is formally satisfied:

$$x^2 Q(\tilde{q}) = x^2\left(\frac{\partial q(\tilde{q},x)}{\partial \tilde{q}}\right)^2 (E(x) - \frac{1}{4}q^2) - \frac{1}{2}\{q(\tilde{q},x);\tilde{q}\}, \tag{3.9}$$

where $\{q(\tilde{q},x);\tilde{q}\}$ is the Schwarzian derivative defined by (2.9).

Now let us prove Lemma 3.4. Comparing the coefficients of $x^{-n+2}(n=0,1,2,\cdots)$ in (3.9), we have the following:

(3.10.0) $$Q(\tilde{q}) = q_0'^2(E_0 - \frac{1}{4}q_0^2)$$

(3.10.1) $$(E_0 - \frac{1}{4}q_0^2)q_1' - \frac{1}{4}q_0'q_0q_1 = -\frac{1}{2}q_0'E_1$$

(3.10.n)
$$(E_0 - \frac{1}{4}q_0^2)q_n' - \frac{1}{4}q_0'q_0q_n = -\frac{1}{2}q_0'E_n + R_n(q_{j_1}^{(\alpha)}, E_{j_2}; 0 \leqq \alpha \leqq 3, 0 \leqq j_1, j_2 \leqq n-1)(n \geqq 2),$$

where we denote $dq_j/d\tilde{q}$ etc. by q_j' or $q_j^{(1)}$ etc. and $R_n(n = 2,3,\cdots)$ is given by the following formula:

(3.11)
$$\begin{aligned} &R_n(q_{j_1}^{(\alpha)}, E_{j_2}; 0 \leqq \alpha \leqq 3, 0 \leqq j_1, j_2 \leqq n-1) \\ &= -\frac{1}{2} \sum_{\substack{k_1+k_2+\ell=n \\ k_1,k_2,\ell \leqq n-1}} \frac{q_{k_1}'q_{k_2}'}{q_0'}E_\ell + \frac{1}{8} \sum_{\substack{k_1+k_2+\ell_1+\ell_2=n \\ k_1,k_2,\ell_1,\ell_2 \leqq n-1}} \frac{q_{k_1}'q_{k_2}'}{q_0'}q_{\ell_1}q_{\ell_2} \\ &+ \frac{1}{4} \sum_{k+\ell+\mu=n-2} \sum_{\mu_1+\cdots+\mu_\ell=\mu} (-1)^\ell \frac{q_k'''}{(q_0')^{\ell+2}} q_{\mu_1+1}' \cdots q_{\mu_\ell+1}' \\ &- \frac{3}{8} \sum_{k_1+k_2+\ell+\mu=n-2} \sum_{\mu_1+\cdots+\mu_\ell=\mu} (-1)^\ell(\ell+1) \frac{q_{k_1}''q_{k_2}''}{(q_0')^{\ell+3}} q_{\mu_1+1}' \cdots q_{\mu_\ell+1}'. \end{aligned}$$

(Cf. (A.2.2)-(A.2.5) in the Appendix §A.2.).

Let us first show that there exists a solution $(q_0(\tilde{q}), E_0)$ of (3.10.0) such that $q_0(\tilde{q})$ defines a biholomorphic mapping on an open subset containing p_0 and p_1 and that (3.5) and (3.8) are satisfied. Taking the square root in both sides of (3.10.0), we have

(3.12) $$\sqrt{Q(\tilde{q})} = q_0'\sqrt{E_0 - \frac{1}{4}q_0^2}.$$

Note that there are two choices in determining the branch of $\sqrt{Q(\tilde{q})}$; we shall fix it later. Equation (3.12) can be integrated easily. In fact, if there exists a solution $q_0(\tilde{q})$ satisfying the above conditions, we find

(3.13) $$E_0 = \frac{1}{\pi}\int_{p_0}^{p_1} \sqrt{Q(\tilde{q})}d\tilde{q},$$

which determines the constant E_0. Since the right-hand side of (3.13) is a real number by the assumption, we may assume that E_0 is positive by choosing the appropriate branch of $\sqrt{Q(\tilde{q})}$. In discussing the existence of $q_0(\tilde{q})$, we also make use of the integrated form of (3.12). That is, by integrating (3.12) we find that, if a holomorphic function $z(\tilde{q})$ satisfies

(3.14) $$z - \sin z = \frac{2}{E_0}\int_{p_0}^{\tilde{q}} \sqrt{Q(\tilde{q})}d\tilde{q}$$

and

(3.15) $$z(p_0) = 0 \quad \text{and} \quad z(p_1) = 2\pi,$$

then

(3.16) $$q_0(\tilde{q}) = -2\sqrt{E_0}\cos(\frac{1}{2}z(\tilde{q}))$$

actually becomes a solution of (3.10.0) satisfying (3.5). Hence what we have to do is to show the existence of a solution $z(\tilde{q})$ of (3.14) and (3.15).

To show the existence of the required $z(\tilde{q})$, let us consider the following two holomorphic mappings:

$$w_1 = w_1(z) \underset{\text{def}}{=} z - \sin z,$$
$$w_2 = w_2(\tilde{q}) \underset{\text{def}}{=} \frac{2}{E_0}\int_{p_0}^{\tilde{q}} \sqrt{Q(\tilde{q})}d\tilde{q}.$$

We can easily verify that $w_1(z)$ maps the closed interval $[0, 2\pi](\subset \mathbf{R} \subset \mathbf{C})$ onto itself bijectively and that its derivative never vanishes on the interior $(0, 2\pi)$. Similarly $w_2(\tilde{q})$ maps the curve (3.2), which connects p_0 and p_1 by the assumption, onto $[0, 2\pi]$ bijectively, and its derivative never vanishes on the interior of the curve (3.2). These facts imply that $z(\tilde{q}) = w_1^{-1}(w_2(\tilde{q}))$ is a well-defined biholomorphic mapping on a neighborhood of the interior of the curve (3.2) and that

$$\lim_{\tilde{q}\to p_0} z(\tilde{q}) = 0, \quad \lim_{\tilde{q}\to p_1} z(\tilde{q}) = 2\pi.$$

By the definition of w_1 and $w_2, z(\tilde{q})$ clearly satisfies the equation (3.14). Thus we have obtained a solution $z(\tilde{q})$ of (3.14) and (3.15). Furthermore $z(\tilde{q})$ has the following expansion at $\tilde{q} = p_0$:

$$z(\tilde{q}) = C_0(\tilde{q} - p_0)^{1/2}\{1 + C_1(\tilde{q} - p_0) + C_2(\tilde{q} - p_0)^2 + \cdots\}$$

where C_0 is a non-zero real constant. Substituting this expansion into (3.16), we find that $q_0(\tilde{q})$ is holomorphic even at $\tilde{q} = p_0$ and $(dq_0/d\tilde{q})(p_0) \neq 0$. The same reasoning applies equally to $\tilde{q} = p_1$. Therefore $q_0(\tilde{q})$ thus obtained defines a biholomorphic mapping on an open subset V containing p_0 and p_1. This completes the proof of the existence of a solution $(q_0(\tilde{q}), E_0)$ of (3.10.0) with the required properties.

Next let us consider $(q_1(\tilde{q}), E_1)$ and $(q_n(\tilde{q}), E_n)(n \geqq 2)$. Because (3.10.1) and (3.10.n) are of the same form if we define $R_1 = 0$ as a convention, we discuss $(q_1(\tilde{q}), E_1)$ and $(q_n(\tilde{q}), E_n)(n \geqq 2)$ simultaneously. Our task is to show that the first order ordinary differential equation

$$(E_0 - \frac{1}{4}q_0^2)\frac{df}{d\tilde{q}} - \frac{1}{4}(q_0' q_0)f = -\frac{1}{2}q_0' E + R(\tilde{q}) \tag{3.17}$$

has a holomorphic solution $f(\tilde{q})$ on V for an appropriately chosen constant E, where $R(\tilde{q})$ is a given holomorphic function on V. Let us decompose (3.17) into the following two equations:

$$(E_0 - \frac{1}{4}q_0^2)\frac{du}{d\tilde{q}} - \frac{1}{4}(q_0' q_0)u = -\frac{1}{2}q_0' \tag{3.18}$$

$$(E_0 - \frac{1}{4}q_0^2)\frac{dv}{d\tilde{q}} - \frac{1}{4}(q_0' q_0)v = R. \tag{3.19}$$

Since

$$(E_0 - \frac{1}{4}q_0^2) = Q(\tilde{q})(q_0')^{-2}$$

holds by (3.10.0), equations (3.18) and (3.19) have singularities only at p_0 and p_1 in V, which are regular singular points of the equations in question. Furthermore the exponents at p_0 and p_1 are not integers; actually both of them are equal to $-\frac{1}{2}$. Hence the equation (3.18) [resp., (3.19)] has a unique holomorphic solution at $\tilde{q} = p_j (j = 0, 1)$, which we denote by $u^j(\tilde{q})$[resp., $v^j(\tilde{q})$]$(j = 0, 1)$. In particular, $f(\tilde{q}) = Eu^0(\tilde{q}) + v^0(\tilde{q})$ gives a holomorphic solution of (3.17) at $\tilde{q} = p_0$. Next we consider the analytic continuation of $f(\tilde{q})$ to $\tilde{q} = p_1$. The analytic continuations of u^0 and v^0 can be expressed as

$$\begin{cases} u^0(\tilde{q}) = u^1(\tilde{q}) + \lambda H(\tilde{q}) \\ v^0(\tilde{q}) = v^1(\tilde{q}) + \tilde{\lambda} H(\tilde{q}) \end{cases} \tag{3.20}$$

with some constants λ and $\tilde{\lambda}$, where $H(\tilde{q})$ is a solution of the homogeneous equation

$$(E_0 - \frac{1}{4}q_0^2)\frac{dH}{d\tilde{q}} - \frac{1}{4}(q_0' q_0)H = 0.$$

We shall first finish the proof of the existence of the required solution, assuming that λ and H are non-zero. The non-zero property of λ and H shall be shown later.

In order that

$$\begin{aligned} f(\tilde{q}) &= Eu^0(\tilde{q}) + v^0(\tilde{q}) \\ &= (Eu^1(\tilde{q}) + v^1(\tilde{q})) + (E\lambda + \tilde{\lambda})H(\tilde{q}) \end{aligned}$$

may be holomorphic also at $\tilde{q} = p_1, E\lambda + \tilde{\lambda}$ must be equal to zero since $H(\tilde{q})$ is not holomorphic at p_1. Conversely if we define E by $E = -\lambda^{-1}\tilde{\lambda}, f(\tilde{q}) = Eu^0 + v^0$ actually defines a holomorphic solution of (3.17) on V. Thus we have proved that (3.17) has a holomorphic solution on V for some constant E. We should remark that, since $R_1 = 0, q_1(\tilde{q})$ as well as E_1 is equal to zero just as in the case of the discussion about a canonical form near one simple turning point. (Cf. the proof of Lemma 2.2 in the Appendix).

It remains to prove the constant λ and H are different from zero. Since we are considering in the open set V, we may use $t = q_0(\tilde{q})$ as a holomorphic coordinate in V. With this new coordinate system the equation (3.18) is expressed as

$$(E_0 - \frac{1}{4}t^2)\frac{du}{dt} - \frac{1}{4}tu = -\frac{1}{2}. \tag{3.21}$$

Hence we can write down explicitly the holomorphic solution u^0 of (3.21) at $t_0 = q_0(p_0) = -2\sqrt{E_0}$. That is,

$$u^0 = -\frac{1}{2}(E_0 - \frac{1}{4}t^2)^{-1/2}\int_{-2\sqrt{E_0}}^{t}(E_0 - \frac{1}{4}s^2)^{-1/2}ds.$$

The analytic continuation of u^0 to $t_1 = q_0(p_1) = 2\sqrt{E_0}$ is given by

$$u^0 = -\frac{1}{2}(E_0 - \frac{1}{4}t^2)^{-1/2}\left[\int_{-2\sqrt{E_0}}^{2\sqrt{E_0}}(E_0 - \frac{1}{4}s^2)^{-1/2}ds - \int_{t}^{2\sqrt{E_0}}(E_0 - \frac{1}{4}s^2)^{-1/2}ds\right],$$

where

$$u^1 = \frac{1}{2}(E_0 - \frac{1}{4}t^2)^{-1/2}\left[\int_t^{2\sqrt{E_0}}(E_0 - \frac{1}{4}s^2)^{-1/2}ds\right],$$

is the unique holomorphic solution of (3.21) at t_1. Therefore we find

$$\begin{aligned}\lambda H(\tilde{q})\big|_{\tilde{q}=q_0^{-1}(t)} &= u^0 - u^1 \\ &= -\frac{1}{2}(E_0 - \frac{1}{4}t^2)^{-1/2}\int_{-2\sqrt{E_0}}^{2\sqrt{E_0}}(E_0 - \frac{1}{4}s^2)^{-1/2}ds\end{aligned}$$

which shows that λ and H are different from zero.

Thus we have verified the existence of the required formal series $q(\tilde{q}, x)$ and $E(x)$. Their pre-Borel-summability can be proved in the same way as in the proof of the pre-Borel-summability of $q(\tilde{q}, x)$ in one simple turning point case (Lemma 2.2), and we do not repeat it here.

We have thus finished the proof of Lemma 3.4.

Although we assumed in Theorem 3.1 that the turning points p_0 and p_1 are connected by a Stokes line, the required series $q(\tilde{q}, x)$ and $E(x)$ do exist even if we perturb the equation (3.1) slightly so that p_0 and p_1 are no longer connected by a Stokes line.(See Theorem 3.5 below). It should be noteworthy, however, that a reduction of $Q(\tilde{q})$ into $E - \frac{1}{4}q^2$ cannot be expected in general; for example when a Stokes line runs across the interval between two turning points p_0 and p_1 so that they are in opposite regions with respect to the Stokes line, we cannot expect that the pre-Borel-summable series $q(\tilde{q}, x)$ and $E(x)$ required in Theorem 3.1 should exist. The above mentioned condition in Theorem 3.1 that two turning points are connected by a Stokes line actually excludes such a possibility.

Now we summarize the perturbed version of Theorem 3.1 as follows:

Theorem 3.5. *Let U and Ω be open neighborhoods of the origin 0 of* **C**, *and let $Q(\tilde{q}, \varepsilon)$ be a holomorphic function on $U \times \Omega$. Suppose that the equation*

$$\left(\frac{d^2}{d\tilde{q}^2} - x^2 Q(\tilde{q}, \varepsilon)\right)\tilde{\psi}(\tilde{q}, \varepsilon, x) = 0$$

has exactly two distinct simple turning points $p_0(\varepsilon)$ and $p_1(\varepsilon)$ in U for any fixed ε in Ω. Suppose further that, when $\varepsilon = 0$, $p_0(0)$ and $p_1(0)$ are connected by a curve in U defined by

$$\operatorname{Im} \int_{p_0(0)}^{\tilde{q}} \sqrt{Q(\tilde{q}, 0)} d\tilde{q} = 0,$$

and that

$$\operatorname{Im} \int_{p_0(\varepsilon)}^{p_1(\varepsilon)} \sqrt{Q(\tilde{q}, \varepsilon)} d\tilde{q}$$

changes its sign according as ε is positive or negative. Then we can find pre-Borel-summable series

$$q(\tilde{q}, \varepsilon, x) = \sum_{j \geq 0} q_j(\tilde{q}, \varepsilon) x^{-j}$$

and

$$E(\varepsilon, x) = \sum_{j \geq 0} E_j(\varepsilon) x^{-j}.$$

with holomorphic parameters $\tilde{q}$ and ε so that

$$q_0(p_0(\varepsilon), \varepsilon) = -2\sqrt{E_0(\varepsilon)} \quad \text{and} \quad q_0(p_1(\varepsilon), \varepsilon) = 2\sqrt{E_0(\varepsilon)}$$

hold and that $\tilde{\psi}(\tilde{q}, \varepsilon, x)$ may be expressed as

$$\tilde{\psi}(\tilde{q}, \varepsilon, x) = \left(\frac{\partial q(\tilde{q}, \varepsilon, x)}{\partial \tilde{q}} \right)^{-1/2} \psi(q(\tilde{q}, \varepsilon, x), \varepsilon, x),$$

with a solution $\psi(q, \varepsilon, x)$ of the following equation:

$$\left(\frac{d^2}{dq^2} - x^2 (E(\varepsilon, x) - \frac{1}{4} q^2) \right) \psi(q, \varepsilon, x) = 0.$$

In particular, $E_0(\varepsilon)$ is given by the following formula

$$E_0(\varepsilon) = \frac{1}{\pi} \int_{p_0(\varepsilon)}^{p_1(\varepsilon)} \sqrt{Q(\tilde{q}, \varepsilon)} d\tilde{q},$$

where the branch of $\sqrt{Q(\tilde{q}, \varepsilon)}$ is chosen so that $E_0(0)$ may become a real positive constant.

Appendix

§A.1. Pre-Borel-summability of a WKB solution.

Although pre-Borel-summability of a WKB solution is well-known (cf. [V] p.252), we give its proof in view of its importance.

Proposition A.1.1. *Let U be the domain of analyticity of $Q(q)$ in (1.1). Then any WKB solution of (1.1) is pre-Borel-summable on each compact subset of $\{q \in U; Q(q) \neq 0\}$.*

Proof. In view of a well-known result for the manipulation of microdifferential operators ([SKK], Chap.II, Proposition 2.1.2), it suffices to prove the pre-Borel-summability of a solution S of the Riccati equation (1.4). Since the situation is local in q, we show the pre-Borel-summability on each small disk in $U_0 = \{q \in U; Q(q) \neq 0\}$. Let q_0 be a point in U_0. We denote by $D(r)$ a closed disk of rodius r centered at q_0. Let r_0 be an arbitrary positive number such that $D(r_0) \subset U_0$. Then there exists a positive number M for which

$$M^{-1} \leqq |S_{-1}(q)| \leqq M \tag{A.1.1}$$

holds in $D(r_0)$.

Let us now prove the following by the induction on j:

(A.1.2) There exist positive constants A and C for which

$$\sup_{D(r_1)} |S_j(q)| \leqq (j+1)! A C^j (r_0 - r_1)^{-j-1} \qquad (j = 0, 1, 2, \cdots)$$

holds for any $r_1 < r_0$.

Since $S_0 = -\frac{1}{2S_{-1}} \cdot \frac{\partial S_{-1}}{\partial q}$, the estimaation for S_0 is true if we choose $A = \frac{1}{2}M^2$. Let us suppose that the estimation for S_j holds up to j ($j \geqq 0$). For each $r < r_0$, we use the assumption of the induction by setting $r_1 = r + \frac{1}{j+2}(r_0 - r)$.

By using the Cauchy formula, we find

$$\begin{aligned} \sup_{D(r)} \left|\frac{\partial S_j}{\partial q}\right| &\leqq (j+1)! A C^j (r_0 - r)^{-j-1} \left(1 - \frac{1}{j+2}\right)^{-j-1} \frac{j+2}{r_0 - r} \\ &\leqq (j+2)! A C^j (r_0 - r)^{-j-2} e. \end{aligned}$$

Furthermore an application of the induction hypothesis immediately entails

$$\sup_{D(r)} \sum_{\substack{k+\ell=j \\ k,\ell \geqq 0}} |S_k S_\ell| \leqq \sum_{\substack{k+\ell=j \\ k,\ell \geqq 0}} (k+1)!(\ell+1)! A^2 C^{k+\ell} (r_0 - r)^{-k-\ell-2}.$$

The right-hand side is dominated by

$$(j+2)! A^2 C^j (r_0 - r)^{-j-2},$$

as we have the following inequality:

$$\sum_{\substack{k+\ell=j \\ k,\ell \geqq 0}} (k+1)!(\ell+1)! \leqq (j+2)!.$$

Since

$$S_{j+1} = -\frac{1}{2S_{-1}} \left(\frac{\partial S_j}{\partial q} + \sum_{\substack{k+\ell=j \\ k,\ell \geqq 0}} S_k S_\ell \right),$$

we thus obtain

$$\sup_{D(r)} |S_{j+1}| \leqq \frac{M}{2}(e+A) \cdot (j+2)! A C^j (r_0 - r)^{-j-2}.$$

Then, choosing C so that $\frac{M}{2}(e+A) \leqq C$, we finally get the required estimate for S_{j+1}:

$$\sup_{D(r)} |S_{j+1}| \leqq (j+2)! A C^{j+1} (r_0 - r)^{-j-2}.$$

Thus the induction proceeds, completing the proof.

§A.2. Proof of Lemma 2.2 and Proposition 2.3.

Let us first prove Lemma 2.2. We may assume without loss of generality that

(A.2.1) $$Q(\tilde{q}) = \tilde{q}(1 + \sum_{j \geq 1} a_j \tilde{q}^j)$$

We shall write down the relations which $q_j(\tilde{q})$ in (2.6) should satisfy formally. Denoting $dq_j/d\tilde{q}$ etc. by q_j' etc., we observe the following:

(A.2.2)
$$\frac{\frac{\partial^3 q(\tilde{q},x)}{\partial \tilde{q}^3}}{\frac{\partial q(\tilde{q};x)}{\partial \tilde{q}}} = \frac{1}{q_0'}\Big(\sum_{k=0}^{\infty} x^{-k} q_k'''\Big)\Big(\sum_{\ell=0}^{\infty}(-1)^{\ell}\Big(\sum_{\mu=1}^{\infty} x^{-\mu}\frac{q_\mu'}{q_0'}\Big)^{\ell}\Big)$$
$$= \frac{1}{q_0'}\Big(\sum_{n=0}^{\infty} x^{-n}\Big(\sum_{k+\ell+\mu=n}(-1)^{\ell}\sum_{\mu_1+\cdots+\mu_\ell=\mu}\frac{q_k'''}{(q_0')^{\ell}} q_{\mu_1+1}'\cdots q_{\mu_\ell+1}'\Big)\Big),$$

(A.2.3)
$$\left(\frac{\frac{\partial^2 q(\tilde{q},x)}{\partial \tilde{q}^2}}{\frac{\partial q(\tilde{q},x)}{\partial \tilde{q}}}\right)^2 = \frac{1}{(q_0')^2}\Big(\sum_{k_1,k_2=0}^{\infty} x^{-k_1-k_2} q_{k_1}'' q_{k_2}''\Big)\Big(\sum_{\ell=0}^{\infty}(-1)^{\ell}(\ell+1)\Big(\sum_{\mu=0}^{\infty} x^{-\mu-1}\frac{q_{\mu+1}'}{q_0'}\Big)^{\ell}\Big)$$
$$= \frac{1}{(q_0')^2}\sum_{n=0}^{\infty} x^{-n}\sum_{k_1+k_2+\ell+\mu=n}\ \sum_{\mu_1+\cdots\mu_\ell=\mu}\frac{(-1)^{\ell}(\ell+1)}{(q_0')^{\ell}} q_{k_1}'' q_{k_2}'' q_{\mu_1+1}'\cdots q_{\mu_\ell+1}'.$$

(A.2.4) $$\Big(\frac{\partial q(\tilde{q},x)}{\partial \tilde{q}}\Big)^2 q(\tilde{q},x) = \sum_{n=0}^{\infty} x^{-n}\Big(\sum_{k_1+k_2+\ell=n} q_{k_1}' q_{k_2}' q_\ell\Big).$$

Then the comparison of the coefficients of x^{-n+2} in (2.7) entails the following:

(A.2.5.0) $$Q(\tilde{q}) = q_0'^2 q_0,$$

(A.2.5.1) $$2q_0 q_1' + q_0' q_1 = 0,$$

(A.2.5.n + 2)
$$\sum_{k_1+k_2+\ell=n+2} q_{k_1}' q_{k_2}' q_\ell$$
$$= \frac{1}{2q_0'}\Big(\sum_{k+\ell+\mu=n}\ \sum_{\mu_1+\cdots+\mu_\ell=\mu}\frac{(-1)^{\ell} q_k'''}{(q_0')^{\ell}} q_{\mu_1+1}'\cdots q_{\mu_\ell+1}'\Big)$$
$$- \frac{3}{4(q_0')^2}\Big(\sum_{k_1+k_2+\ell+\mu=n}\ \sum_{\mu_1+\cdots+\mu_\ell=\mu}(-1)^{\ell}(\ell+1) q_{k_1}'' q_{k_2}'' q_{\mu_1+1}'\cdots q_{\mu_\ell+1}'(q_0')^{-\ell}\Big).$$

Let us first show that there exist an analytic solution $q_0(\tilde{q})$ of (A.2.5.0) and that it is essentially unique in the sense that its arbitrariness comes from the arbitrariness in the choice of some constant c which satisfies $c^3 = 1$. In fact, taking square root in both sides of (A.2.5.0), we can readily integrate it to find

$$(\text{A.2.5.0}') \qquad q_0(\tilde{q}) = \left(\frac{3}{2}\int_0^{\tilde{q}} \sqrt{Q(\tilde{q})}d\tilde{q}\right)^{2/3}$$

It follows from the simplicity assumption on the turning point 0, $\frac{3}{2}\int_0^{\tilde{q}} \sqrt{Q(\tilde{q})}d\tilde{q}$ has the form $\tilde{q}^{3/2}(1+\cdots)$, where $\cdots$ denotes a convergent power series of $\tilde{q}$ near the origin. Hence $q_0(\tilde{q}) = c\tilde{q}(1+\cdots)$ is holomorphic near 0, where c is a constant satisfying $c^3 = 1$. In what follows, c shall be chosen to be 1.

Substituting $q_0(\tilde{q}) = \tilde{q}(1+\cdots)$ in (A.2.5.1), we find q_1 should be 0. Hence (A.2.5.n) entail the following:

$$\begin{aligned}
(\text{A.2.7})\, q_0'\left(2q_0\frac{d}{d\tilde{q}} + q_0'\right)q_{n+2}(\tilde{q}) = \frac{1}{2q_0'}\Big\{ &\sum_{\ell+\mu=n} \sum_{\substack{\mu_1+\cdots+\mu_\ell=\mu \\ \mu_1,\cdots,\mu_\ell \geqq 1}} (-1)^\ell q_0''' q'_{\mu_1+1}\cdots q'_{\mu_\ell+1}(q_0')^{-\ell} \\
&+ \sum_{\substack{k+\ell+\mu=n \\ k>1}} \sum_{\substack{\mu_1+\cdots+\mu_\ell=\mu \\ \mu_1,\cdots,\mu_\ell \geqq 1}} (-1)^\ell q_k''' q'_{\mu_1+1}\cdots q'_{\mu_\ell+1}(q_0')^{-\ell}\Big\} \\
&- \frac{3}{4q_0'^2}\Big(\sum_{k_1+k_2+\ell+\mu=n} \sum_{\substack{\mu_1+\cdots+\mu_\ell=\mu \\ \mu_1\cdots,\mu_\ell \geqq 1}} (-1)^\ell(\ell+1) q''_{k_1} q''_{k_2} q'_{\mu_1+1}\cdots q'_{\mu_\ell+1}(q_0')^{-\ell}\Big) \\
&- \Big(\sum_{\substack{k_1+k_2+\ell=n+2 \\ k_1,k_2,\ell<n+2}} q'_{k_1} q'_{k_2} q_\ell \Big).
\end{aligned}$$

Before estimating $q_j(\tilde{q})$, we note that $q_j(\tilde{q})$ vanishes identically for odd j; this is intuitively obvious, as the potential $Q(\tilde{q})$ does not depend on x^{-1}. A rigorous proof of this fact can be readily obtained by the following two observations: First, if n is odd, then each term in the right-hand side of (A.2.7) contains q_j or its derivatives with $j(\leq n)$ odd. Second, if the right-hand side of (A.2.7) is 0, then, again by the specific form of $q_0(\tilde{q}), q_{n+2}(\tilde{q})$ should be 0. Since $q_1(\tilde{q})$ is 0, the induction on n combined with the above observations shows that $q_j(\tilde{q}) = 0$ for odd j.

Now, let us embark on the estimation of $q_j(\tilde{q})$. Since $q_0'(0) = 1$, we can find positive constants C_1 and r so that

(A.2.8.0.i) $$|q_0(\tilde{q})|,\ |q_0'(\tilde{q})|,\ 1/|q_0'(\tilde{q})| \leqq C_1$$

hold on $D(r) = \{\tilde{q}; |\tilde{q}| \leqq r\}$. Then, for each $\varepsilon > 0$, we have

(A.2.8.0.ii) $$|q_0''(\tilde{q})| \leqq C_1 \varepsilon^{-1},$$

(A.2.8.0.iii) $$|q_0'''(\tilde{q})| \leqq 2C_1 \varepsilon^{-2}$$

on $D(r-\varepsilon)$. In parallel with the reasoning in §A.1, we want to prove the following by the induction on $n : n = 1, 2, \cdots$

(A.2.8.n) There exist positive constants C for which the following holds for each $\varepsilon > 0$:

(n.i) $$\sup_{|\tilde{q}| \leqq r-\varepsilon} |q_n(\tilde{q})| \leqq n! C^{n-1} \varepsilon^{-n},$$

(n.ii) $$\sup_{|\tilde{q}| \leqq r-\varepsilon} |q_n'(\tilde{q})| \leqq n! C^{n-1} \varepsilon^{-n}.$$

Note that (A.2.8.1) trivially holds, as q_1 is identically 0.

To make the induction procedure run smoothly, we prepare the following sublemma. We would like to thank Dr. Y. Ohyama for suggesting us to use the Schwarz lemma in its proof.

Sublemma A.2.1. *Le $v(t)$ be a holomorphic function on $\{t \in \mathbf{C}; |t| < R\}$, and consider the followng differential equation:*

(A.2.9) $$(t\frac{d}{dt} + \frac{1}{2})u(t) = v(t).$$

Then there exists a unique holomorphic solution $u(t)$ on $\{t \in \mathbf{C}; |t| < R\}$, and it satisfies the following inequalities for any positive r which is smaller than R:

(A.2.10) $$\sup_{|t| \leqq r} |u(t)| \leqq 2 \sup_{|t| \leqq r} |v(t)|$$

(A.2.11) $$\sup_{|t| \leqq r} \left|\frac{du(t)}{dt}\right| \leqq \frac{2}{r} \sup_{|t| \leqq r} |v(t)|$$

Proof. It is easy to verify that

$$u(t)=\int_0^1 s^{-1/2}v(st)ds \tag{A.2.12}$$

is the unique holomorphic solution of (A.2.9). Hence we find

$$\begin{aligned}|u(t)|&\leqq \sup_{0\leqq s\leqq 1}|v(st)|\int_0^1 s^{-1/2}ds\\ &=2\sup_{0\leqq s\leqq 1}|v(st)|.\end{aligned} \tag{A.2.13}$$

This immediately implies (A.2.10). Since $u(t)$ satisfies (A.2.9), we find

$$t\frac{d}{dt}u=v-\frac{1}{2}u \tag{A.2.14}$$

and hence

$$\sup_{|t|\leqq r}|t\frac{du(t)}{dt}|\leq \sup_{|t|\leqq r}|v(t)|+\frac{1}{2}\sup_{|t|\leqq r}|u(t)|. \tag{A.2.15}$$

Therefore (A.2.10) entails

$$\sup_{|t|\leqq r}|t\frac{du(t)}{dt}|\leqq 2\sup_{|t|\leqq r}|v(t)| \tag{A.2.16}$$

Since

$$v(0)-\frac{1}{2}u(0)=0$$

holds by the definition of $u(t)$, (A.2.16) combined with the Schwarz lemma entails

$$\sup_{|t|\leqq r}|\frac{du(t)}{dt}|\leqq\frac{2}{r}\sup_{|t|\leqq r}|v(t)|. \tag{A.2.17}$$

This completes the proof of the sublemma.

Now we resume the proof of (A.2.8.n). It follows from Sublemma A.2.1 that, if holomorphic funcitons u and v defined on a neighborhood of $\{\tilde{q};|\tilde{q}|\leqq r'\}$ satisfy

$$q_0'(2q_0\frac{d}{d\tilde{q}}+q_0')u=v,$$

then u and u' can be estimated as follows:

there exists a positive constant C_2 such that

$$\sup_{|\tilde{q}|\leqq r'} |u(\tilde{q})|\leqq C_2 \sup_{|\tilde{q}|\leqq r'} |v(\tilde{q})|$$

and

$$\sup_{|\tilde{q}|\leqq r'} |u'(\tilde{q})|\leqq C_2 \sup_{|\tilde{q}|\leqq r'} |v(\tilde{q})|.$$

Using this fact, we estimate q_{n+2} under the induction hypothesis that (A.2.8.k) holds for $k\leqq n$. Note that the logical contents of the inductioin for $n = 2p+1$ are the same as those for $n = 2p$, because $q_{2p+1}(\tilde{q})$ is 0 as we observed before. To simplify the notations, we introduce symbols *I*,*II* and *III* as in (A.2.18) below so that the right-hand side of (A.2.7) may be expressed as *I* - *II*- *III*:

(A.2.18)

$$I = \frac{1}{2q_0'} \sum_{k+\ell+\mu=n} \sum_{\mu_1+\cdots+\mu_\ell=\mu} (-1)^\ell q_k''' q_{\mu_1+1}' \cdots q_{\mu_\ell+1}' (q_0')^{-\ell},$$

$$II = \frac{3}{4(q_0')^2} \sum_{k_1+k_2+\ell+\mu=n} \sum_{\mu_1+\cdots+\mu_\ell=\mu} (-1)^\ell (\ell+1) q_{k_1}'' q_{k_2}'' q_{\mu_1+1}' \cdots q_{\mu_\ell+1}' (q_0')^{-\ell}$$

$$III = \sum_{\substack{k_1+k_2+\ell=n+2 \\ k_1,k_2,\ell<n+2}} q_{k_1}' q_{k_2}' q_\ell.$$

By (A.2.8.k) ($k = 1, 2, \cdots, n$), we have

(A.2.19)
$$\sup_{|\tilde{q}|\leqq r-\varepsilon} |q_k''(\tilde{q})|\leqq (k+1)! C^{k-1} e \varepsilon^{-k-1},$$

(A.2.20)
$$\sup_{|\tilde{q}|\leqq r-\varepsilon} |q_k'''(\tilde{q})|\leqq (k+2)! C^{k-1} e^2 \varepsilon^{-k-2}.$$

To have estimate for *I*,*II* and *III* in $\{\tilde{q};\, |\tilde{q}|\leqq r-\varepsilon\}$, we use the following fact:

Sublemma A.2.2.

$$\sum_{\substack{n_1+\cdots+n_\ell=n \\ n_1,\cdots,n_\ell\geqq 1}} n_1! \cdot \cdots \cdot n_\ell! \leqq n!$$

This can be easily shown by the induction on ℓ.

1^0 *Estimation of I* : We write $I = I_1 + I_2$ with

$$I_1 = \frac{1}{2q_0'} \sum_{\ell+\mu=n} \sum_{\mu_1+\cdots+\mu_\ell=\mu} (-1)^\ell q_0''' q_{\mu_1+1}' \cdots\cdot q_{\mu_\ell+1}'(q_0')^{-\ell}$$

$$I_2 = \frac{1}{2q_0'} \sum_{\substack{k+\ell+\mu=n \\ k\geqq 1}} \sum_{\mu_1+\cdots+\mu_\ell=\mu} (-1)^\ell q_k''' q_{\mu_1+1}' \cdots q_{\mu_\ell+1}'(q_0')^{-\ell}.$$

First we estimate I_2. By (A.2.8.k) ($k = 1, 2, \cdots, n$), (A.2.8.0)'s, and (A.2.19), (A.2.20), we find

$$|I_2| \leqq \frac{1}{2} C^{n-1} \varepsilon^{-n-2} C_1 e^2 \sum_{\substack{k+\ell+\mu=n \\ k>1}} \sum_{\substack{\mu_1+\cdots+\mu_\ell=\mu \\ \mu_j\geqq 1}} (C_1 C^{-1})^\ell (k+2)!(\mu_1+1)!\cdots(\mu_\ell+1)!$$

It then follows from Sublemma A.2.2 that the right-hand side of this inequality is dominated by

$$\frac{1}{2} C^{n-1} \varepsilon^{-n-2} C_1 e^2 \sum_{\ell=0}^{n} (C_1 C^{-1})^\ell \sum_{k=1}^{n-\ell} (k+2)!(n-k)!$$

Since

$$\sum_{k=1}^{n-\ell} (k+2)!(n-k)! \leqq 2(n+2)!$$

holds, and since we may assme $C > C_1$, we thus obtain

$$|I_2| \leqq (n+2)! C^{n+1} \varepsilon^{-n-2} (C_1 e^2 (1 - C_1 C^{-1})^{-1} \cdot C^{-2})$$

Similarly, we find

$$|I_1| \leqq n! C^{n+1} \varepsilon^{-n-2} (C_1^2 (1 - C_1 C^{-1})^{-1} \cdot C^{-1}).$$

2^0 *Estimation of II* : We write $II = II_1 + II_2 + II_3 + II_4$ with

$$II_1 = \frac{3}{4q_0'^2} \sum_{k_1=k_2=0}{}' \sum_{\mu_1+\cdots+\mu_\ell=\mu} (-1)^\ell (\ell+1) q_{k_1}'' q_{k_2}'' q_{\mu_1+1}' \cdots q_{\mu_\ell+1}' (q_0')^{-\ell},$$

$$II_2 = \frac{3}{4q_0'^2} \sum_{k_1=0, k_2>0}{}' \sum_{\mu_1+\cdots+\mu_\ell=\mu} (-1)^\ell (\ell+1) q_{k_1}'' q_{k_2}'' q_{\mu_1+1}' \cdots q_{\mu_\ell+1}' (q_0')^{-\ell},$$

$$II_3 = \frac{3}{4q_0'^2} \sum_{k_1>0, k_2=0}{}' \sum_{\mu_1+\cdots+\mu_\ell=\mu} (-1)^\ell (\ell+1) q_{k_1}'' q_{k_2}'' q_{\mu_1+1}' \cdots q_{\mu_\ell+1}' (q_0')^{-\ell},$$

$$II_4 = \frac{3}{4q_0'^2} \sum_{k_1>0, k_2>0}{}' \sum_{\mu_1+\cdots+\mu_\ell=\mu} (-1)^\ell (\ell+1) q_{k_1}'' q_{k_2}'' q_{\mu_1+1}' \cdots q_{\mu_\ell+1}' (q_0')^{-\ell},$$

Here $\sum'$ denotes the sum over $k_1 + k_2 + \ell + \mu = n$.

By using (A.2.8.k) ($k = 1, 2, \cdots, n$), (A.2.8.0)'s, (A.2.19), (A.2.20) and Sublemma A.2.2, we find

$$|II_4| \leqq \frac{3}{4} C^{n-2} \varepsilon^{-n-2} C_1^2 e^2 \sum_{\substack{k_1+k_2+\ell+\mu=n \\ k_1,k_2 \geqq 1}} (\ell+1)(C_1 C^{-1})^{\ell} (k_1+1)!(k_2+1)! \times$$

$$\times \Big(\sum_{\substack{\mu_1+\cdots+\mu_\ell=\mu \\ \mu_1,\cdots,\mu_\ell \geqq 1}} (\mu_1+1)! \cdots (\mu_2+1)! \Big)$$

$$\leqq \frac{3}{4}(n+2)! C^{n+1} \varepsilon^{-n-2} (C_1^2 e^2 (1 - C_1 C^{-1})^{-2} C^{-3}).$$

Similarly, we get

$$|II_1| \leqq \frac{3}{4} n! C^{n+1} \varepsilon^{-n-2} (C_1^4 (1 - C_1 C^{-1})^{-2} C^{-1})$$

and

$$|II_2|, |II_3| \leqq \frac{3}{4}(n+1)! C^{n+1} \varepsilon^{-n-2} (C_1^3 e (1 - C_1 C^{-1})^{-2} C^{-2})$$

3^0 *Estimation of III* : A reasoning given similar to those above shows

$$|III| \leqq (n+2)! C^{n+1} \varepsilon^{-n-1} (C^{-2} + 3C_1 \cdot C^{-1}).$$

Summing up the estimations $1^0, 2^0$ and 3^0, we conclude that the right-hand side of (A.2.7) is dominated by

$$(n+2)! C^{n+1} \varepsilon^{-n-1} A,$$

where

$$A = C_1 (1 - C_1 C^{-1})^{-1} (e^2 C^{-1} + C_1) C^{-1} + \frac{3}{4} C_1^2 (1 - C_1 C^{-1})^{-2} (e C^{-1} + C_1)^2 C^{-1} + (C^{-1} + 3C_1) C^{-1}$$

Hence, if we choose C so that $C_2 A \leqq 1$, the induction proceeds. This completes the proof of Lemma 2.2.

Next let us prove Proposition 2.3. First we note the following fact: By an implicit function theorem for pre-Borel-summable series ([GG], p.176) we can find a pre-Borel-summable series $f(q,x) = f_0(q) + f_1(q) x^{-1} + \cdots$ that satisfies

(A.2.21) $$q(f(q,x),x) = q.$$

Since

(A.2.22) $$\{q; \tilde{q}\} = -\Big(\frac{\partial q}{\partial \tilde{q}}\Big)^2 \{\tilde{q}; q\}$$

holds, the relation (2.7) may be converted to

$$x^2\Big(\frac{\partial f(q,x)}{\partial q}\Big)^2 Q(f(q,x)) - \frac{1}{2}\{f;q\} = x^2 q. \tag{A.2.23}$$

Hence the role of $Q(\tilde{q})$ and q in Lemma 2.2 is symmetric, that is, any pair of potentials $Q(q)$ and $\tilde{Q}(\tilde{q})$ can be used in place of q and $Q(\tilde{q})$, on the condition that both $Q(q)$ and $\tilde{Q}(\tilde{q})$ have a simple turning point at the origin.

Now, for the series $f(q,x)$ given above, let $\psi(q,x)$ denote

$$\Big(\frac{\partial f(q,x)}{\partial q}\Big)^{-1/2}\tilde{\psi}(f(q,x),x).$$

Then $\psi(q,x)$ is pre-Borel-summable. (Cf. [SKK], Chap. II, Proposition 2.1.2.) Furthermore a straightforward computation based on (A.2.23) shows

$$\Big(\frac{d^2}{dq^2} - x^2 q\Big)\psi(q,x) = 0. \tag{A.2.24}$$

It also follows from the definition of $f(q,x)$ that

$$\tilde{\psi}(\tilde{q},x) = \Big(\frac{\partial q(\tilde{q},x)}{\partial \tilde{q}}\Big)^{-1/2}\psi(q(\tilde{q},x),x)$$

holds. Hence $\psi(q,x)$ thus defined is the required solution of (2.11).

To claim the correspondence between WKB solutions, again by the symmetry between the role of $Q(\tilde{q})$ and q, it suffices to verify that

$$\tilde{S}(\tilde{q},x) \underset{\text{def}}{=} \frac{\partial}{\partial \tilde{q}}\log\Big[\Big(\frac{\partial q(\tilde{q},x)}{\partial \tilde{q}}\Big)^{-1/2}\psi(q(\tilde{q},x),x)\Big]$$

satisfies

$$\frac{\partial \tilde{S}}{\partial \tilde{q}} + \tilde{S}^2 = x^2 Q(\tilde{q}) \tag{A.2.25}$$

on the condition that

$$S(q,x) \underset{\text{def}}{=} \frac{\partial}{\partial q}\log\psi(q,x)$$

satisfies

$$\frac{\partial S}{\partial q} + S^2 = x^2 q. \tag{A.2.26}$$

This can be readily confirmed by (2.7).

Finally, let us also note that, if the normalization fo the WKB solution $\tilde{\psi}(\tilde{q}, x)$ is done so that the initial end-point of the integral in (1.2) is 0, then the normalization of the corresponding WKB solution $\psi(q, x)$ of (2.11) is done in the same manner. In particular, the exponent of $\psi(q, x)$ is $\int_0^{q_0(\tilde{q})} \sqrt{q}dq$, which coincides with $\int_0^{\tilde{q}} \sqrt{Q(\tilde{q})}d\tilde{q}$ by (A.2.5.0′).

References

[AKK] Aoki, T., M. Kashiwara and T. Kawai: On a class of linear differential operators of infinite order with finite index. Adv. in Math., **62** (1986), 155-168.

[BW] Bender, C.M. and T.T. Wu: Anharmonic oscillator. Phys. Rev., **184** (1969), 1231-1260.

[CNP] Candelpergher, B., C. Nosmas and F. Pham: Résurgence et développements semi-classiques. To appear.

[E] Ecalle, J.: Cinq applications des fonctions résurgentes. Prépublications d'Orsay, 84T62, Univ. Paris-Sud, 1984.

[GG] Gérard, C. and A. Grigis: Precise estimates of tunneling and eigenvalues near a potential barrier. J. Diff. Eq., **72** (1988), 149-177.

[P1] Pham, F.: Resurgence, quantized canonical transformations, and multi-instanton expansions. Algebraic Analysis, vol. II, Academic Press, 1988, pp.699-726.

[P2] _____: A talk at Hayashibara Forum, 1990 (Okayama). In this volume.

[SKK] Sato, M., T. Kawai and M. Kashiwara: Microfunctions and pseudo-differential equations. Lecture Notes in Math., **287**, Springer, 1973, pp.265-529.

[V] Voros, A.: The return of the quartic oscillator. The complex WKB method. Ann. Inst. Henri Poincaré, **39** (1983), 211-338.

[Z] Zinn-Justin, J.: Instantons in quantum mechanics: Numerical evidence for a conjecture. J. Math. Phys., **25** (1984), 549-555.

A q-analogue of de Rham cohomology associated with Jackson integrals

Kazuhiko AOMOTO[1] and Yoshifumi KATO[2]

In this note we shall give a new formulation of Jackson integrals involved in basic hypergeometric functions through the classical Barnes representations. We define a q-analogue of de Rham cohomology which can be described by means of q-version of Sato's b-functions and derive an associated holonomic q-difference system. The evaluation of its multiplicity will be given as the number of different asymptotic behaviours of Jackson integrals.

The authors are indebted to Professors Tadao Oda and Mutsuo Oka for useful informations and discussions.

1. Structure of b-functions.

We take an elliptic modulus $q = e^{2\pi i\tau}$, $\mathrm{Im}\,\tau > 0$. Let X be an n dimensional integer lattice $\simeq Z^n$ with a basis $\chi_1, \chi_2, \dots, \chi_n$. An arbitrary element $\chi \in X$ can be uniquely written by $\chi = \sum_{j=1}^n \nu_j \chi_j$, $\nu_j \in Z$. We put $\overline{X} = X \otimes_q C^* = X \otimes_q (C/\frac{2\pi i}{\log q}Z)$, which is twisted by q. Then $\overline{X}$ is isomorphic to an n dimensional algebraic torus $(C^*)^n$. The inclusion $X \subset \overline{X}$ can be obtained by identifying χ_j with the

[1]Dept. of Mathematics, School of Science, Nagoya University, Nagoya, 464-01 Japan

[2]Faculty of Science and Technology, Meijo University, Nagoya, 468 Japan

element $t = (1,\dots,1,\underset{j\text{-th}}{q},1,\dots,1) \in (\mathbf{C}^*)^n$. We denote by Q_j the shift operator acting on functions $f(t), t=(t_1,t_2,\dots,t_n)$, on $\overline{X}$ by the rule $Q_j f(t)= f(\chi_j\cdot t)$. We put $Q^\chi = Q_1^{\nu_1}\cdots Q_n^{\nu_n}$ and consider the following q-difference equations

$$(1.1)\qquad Q^\chi\Phi(t) = b_\chi(t)\Phi(t) \quad \text{for } \chi \in X \text{ and } t \in \overline{X},$$

for a set of some rational functions $\{b_\chi(t)\}_{\chi\in X}$, $b_\chi(t)\not\equiv 0$, on $\overline{X}$. The set $\{b_\chi(t)\}_{\chi\in X}$ satisfies the compatibility condition

$$(1.2)\qquad b_{\chi+\chi'}(t) = b_\chi(t)\cdot Q^\chi b_{\chi'}(t).$$

We denote by $R(\overline{X})$ the field of rational functions on $\overline{X}$ and by $R^\times(\overline{X})$ the multiplicative group $R(\overline{X})-\{0\}$. (1.2) means that $\{b_\chi(t)\}_{\chi\in X}$ defines a 1-cocycle on X with values in $R^\times(\overline{X})$. A set $\{b_\chi(t)\}_{\chi\in X}$ is said to be a 1-coboundary if and only if we can write as $b_\chi(t) = Q^\chi\varphi(t)/\varphi(t)$, $\chi\in X$, for some $\varphi \in R^\times(\overline{X})$. $H^1(X,R^\times(\overline{X})) \simeq$ {1-cocycles}/{1-coboundaries} is the first cohomology group. We put $(a)_\infty = \prod_{\nu=0}^\infty (1- a q^\nu)$ and $(a)_n = (a)_\infty/(aq^n)_\infty$ for $n \in \mathbf{Z}$. Then the following important result holds.

Proposition 1. <u>For an arbitrary cohomology class in</u> $H^1(X,R^\times(\overline{X}))$, <u>we can choose such a representative cocycle</u> $\{b_\chi(t)\}_{\chi\in X}$ <u>that is determined by the rule</u> (1.1) <u>where</u> Φ <u>denotes a special</u> q-<u>multiplicative function</u>

$$(1.3)\qquad \Phi = \Pi_{i=1}^{n} t_i^{\alpha_i} \Pi_{j=1}^{m} \frac{(a_j' t^{\mu_j})_\infty}{(a_j t^{\mu_j})_\infty}$$

for $m \in Z^+$ and α_j, a_j, $a_j' \in C$. The elements μ_j belong to the dual $\check{X} = \mathrm{Hom}(X,Z)$ of X and t^{μ_j} denotes the monomial $t_1^{\mu_j(\chi_1)} \cdots t_n^{\mu_j(\chi_n)}$.

This is a q-version of Sato's theorem in [S1] and can be proved in the same way as in its appendix. If we put $u_j = q^{\alpha_j}$ then $u = (u_1,\ldots,u_n)$ can be considered as an element of the dual algebraic torus $\check{X} \otimes_q C^*$. We denote by $\mathcal{L} = C[t_1,t_1^{-1},\ldots, t_n,t_n^{-1}]$ the Laurent polynomial ring in t . The functions $b_\chi(t)$ can be expressed by $b_\chi(t) = u^\chi \frac{b_\chi^+(t)}{b_\chi^-(t)}$ for $u^\chi = u_1^{\nu_1}\cdots u_n^{\nu_n}$ and Laurent polynomials

$$(1.4)\qquad b_\chi^+(t) = \prod_{\substack{1\le j\le m\\ \text{s.t. } \mu_j(\chi)>0}} (a_j t^{\mu_j})_{\mu_j(\chi)} \times \prod_{\substack{1\le j\le m\\ \text{s.t. } \mu_j(\chi)<0}} (a_j' q^{\mu_j(\chi)} t^{\mu_j})_{-\mu_j(\chi)},$$

$$(1.5)\qquad b_\chi^-(t) = \prod_{\substack{1\le j\le m\\ \text{s.t. } \mu_j(\chi)>0}} (a_j' t^{\mu_j})_{\mu_j(\chi)} \times \prod_{\substack{1\le j\le m\\ \text{s.t. } \mu_j(\chi)<0}} (a_j q^{\mu_j(\chi)} t^{\mu_j})_{-\mu_j(\chi)}.$$

We shall impose the following condition.

Assp 1. None of a_j and a_j' vanish so that Φ is of regular singularity.

We represent $a_j = q^{s_j}$ and $a_j' = q^{s_j'}$. We introduce the transformations T_j, $1 \le j \le m$, which replace μ_j, a_j and a_j' by $-\mu_j$, $qa_j'^{-1}$ and qa_j^{-1}

respectively and multiply by $t^{(s_j-s_j')\mu_j}$. We apply T_j to Φ then

$$(1.6)\qquad T_j\Phi(t) = U_j(t)\Phi(t)$$

where

$$(1.7)\qquad U_j(t) = t^{(s_j-s_j')\mu_j}\,\frac{(qa_j^{-1}t^{-\mu_j})_\infty(a_jt^{\mu_j})_\infty}{(a_j't^{\mu_j})_\infty(q^{-1}a_j'^{-1}t^{-\mu_j})_\infty}.$$

Since $Q^\chi U_j = U_j$, $T_j\Phi(t)$ satisfies the same equation (1.1). <u>It is admissible and convenient to put</u> $\mu_{-j} = -\mu_j$, $a_{-j} = qa_j^{-1}$, $a'_{-j} = qa_j^{-1}$ <u>for</u> $1 \le j \le m$.

2. <u>Jackson integral and a</u> q-<u>analogue of de Rham cohomology</u>.

We denote by $\varpi = \frac{d_qt_1}{t_1}\wedge\cdots\wedge\frac{d_qt_n}{t_n}$ the canonical n-form on $\overline{X}$ which is invariant under the action of X. We denote by d_qt_j the dual of q-difference operators which satisfy $\langle d_qt_j,(1-Q_k)\rangle = \delta_{j,k}(1-q)t_j$. For a function f on $\overline{X}$, we can consider the Jackson integral over an orbit $X\cdot\xi$, $\xi \in \overline{X}$, as follows

$$(2.1)\qquad \tilde{f} = \tilde{f}(u;\xi) = \int_{X\cdot\xi} f\,\varpi = (1-q)^n\sum_{\chi\in X} Q^\chi f(\xi),$$

when it is summable. By definition $\tilde{f} = \widetilde{Q^\chi f}$ holds for any choice of $\xi \in \overline{X}$. If we denote $\langle\varphi\rangle = \langle\varphi;\xi\rangle = \widetilde{\Phi\varphi}$ for $\varphi \in R(\overline{X})$, we have

$$(2.2)\qquad \langle\varphi\rangle = (1-q)^n\,\Phi(\xi)\sum_{\chi\in X} b_\chi(\xi)\,Q^\chi\varphi(\xi),$$

$$(2.3)\qquad \langle\varphi - b_\chi\cdot Q^\chi\varphi\rangle = 0, \qquad \chi\in X.$$

In particular,

$$(2.4)\qquad \langle\varphi - b_{\chi_j}\cdot Q_j\varphi\rangle = 0, \qquad 1\le j\le n.$$

Actually the equalities (2.4) lead to (2.3).

Definition 1. The operators $\nabla_j = 1 - b_{\chi_j}Q_j$, $1\le j\le n$, acting on functions on $\overline{X}$, define a q-covariant differenciation ∇ (see also Definition 3). They commute each other

$$(2.5)\qquad \nabla_j\,\nabla_k = \nabla_k\,\nabla_j\ , \qquad 1\le j,k\le n,$$

because of the compatibility condition (1.2) for $\{b_\chi(t)\}_{\chi\in X}$.

It should be noted that this gives a q-analogue version of ordinary integrable covariant differentiations investigated in [D3] and [K1]. See also [A2] and [K4] which deal with them from the viewpoint of complex analytic integrals.

Definition 2. We denote by $\tilde{Q}_j^{\pm1}$, $Q_{a_j}^{\pm1}$ and $Q_{a_j'}^{\pm1}$ the operators acting on functions of u_j, a_j and a_j' induced by the displacements $u_j\to u_jq^{\pm1}$, $a_j\to a_jq^{\pm1}$ and $a_j'\to a_j'q^{\pm1}$ respectively. They satisfy

$$Q_j^{\pm1}\Phi = t_j^{\pm1}\Phi\ , \qquad 1\le j\le n.$$

$$(2.6)\qquad Q_{a_j}\Phi = (1-a_j t^{\mu_j})\Phi\ ,\qquad Q_{a_j}^{-1}\Phi = (1-a_j q^{-1} t^{\mu_j})^{-1}\Phi,$$

$$Q_{a_j'}\Phi = (1-a_j' t^{\mu_j})^{-1}\Phi,\qquad Q_{a_j'}^{-1}\Phi = (1-a_j' q^{-1} t^{\mu_j})\Phi\ ,$$

for $1 \le j \le m$.

Let $\mathcal{A}$ be the commutative algebra of operators generated over $\mathbb{C}$ by $\tilde{Q}_j^{\pm 1}$, $Q_{a_j}^{\pm 1}$ and $Q_{a_j'}^{\pm 1}$. We define a subspace V of $R(\overline{X})$ by

$$(2.7)\qquad V = \{\ A\Phi/\Phi\ \mid\ A \in \mathcal{A}\ \}\ .$$

The space $\mathcal{A}\Phi = \Phi\cdot V$ is invariant under the action of $\mathcal{A}$. With a sequence $(L;L')=(\ell_1,\dots,\ell_n;\ell_1',\dots,\ell_n')$ of non-negative integers, we associate a subspace of V

$$(2.8)\qquad V_{(L;L')} = \{\ \varphi = \frac{\overline{\varphi}}{\prod_{j=1}^m (a_j' t^{\mu_j})_{\ell_j'}\cdot \prod_{j=1}^m (a_j q^{-\ell_j} t^{\mu_j})_{\ell_j}}\ \mid\ \overline{\varphi} \in \mathcal{L}\ \}$$

We put $V_\ell = \sum_{|L|+|L'|=\ell} V_{(L;L')}$ for $\ell \in \mathbb{Z}^+ \cup \{0\}$.

Lemma 2.1. <u>We have</u> (i) $V_\ell \subset V_{\ell+1}$, $V_0 = \mathcal{L}$ <u>and</u> $V = \bigcup_{\ell=0}^\infty V_\ell$, (ii) $\nabla^\chi V \subset V$ <u>where</u> $\chi \in X$ <u>and</u> $\nabla^\chi = 1 - b_\chi Q^\chi$.

<u>Proof</u>. Since (i) is obvious, we prove (ii). Let φ be an arbitrary element of V. There exists an $A \in \mathcal{A}$ such that $\varphi = A\Phi/\Phi$. Then $b_\chi Q^\chi\varphi = Q^\chi(\Phi\varphi)/\Phi = Q^\chi(A\Phi)/\Phi = A(Q^\chi(\Phi))/\Phi = A(b_\chi\Phi)/\Phi$. Since $b_\chi = u^\chi \dfrac{b_\chi^+}{b_\chi^-} \in V$, we have $b_\chi\Phi \in \mathcal{A}\Phi$ and $A(b_\chi\Phi) \in \mathcal{A}\Phi$. This implies $b_\chi Q^\chi\varphi \in V$ and

$\nabla^{\chi}\varphi = (1-b_{\chi}Q^{\chi})\varphi \in V.$

Let us define the following Koszul complex .

Definition 3. (q-analogue of de Rham complex). We put $\Omega^{\cdot} = \sum_{r=0}^{n} \Omega^{r}$ for $\Omega^{r} = \Lambda^{r} \check{X} \otimes V$. Let $e_1, e_2, \ldots, e_n$ be the dual basis of $\check{X}$ to $\chi_1, \chi_2, \ldots, \chi_n$ and $e_{i_1} \wedge e_{i_2} \wedge \cdots \wedge e_{i_r}$, $1 \le i_1 < i_2 < \ldots < i_r \le n$, be the basis of $\Lambda^{r}\check{X}$. An arbitrary element $\sum (e_{i_1} \wedge \cdots \wedge e_{i_r}) \otimes \varphi_{i_1 \cdots i_r}$ of Ω^{r} can be represented by a set of functions $(\varphi_{i_1 \cdots i_r})_{i_1 < \cdots < i_r}$ of V. The boundary operators ∇ from Ω^{r} into Ω^{r+1}, $0 \le r \le n$, are given by

$$(2.9)\quad (\nabla\varphi)_{i_1,\ldots,i_{r+1}} = \sum_{\nu=1}^{r+1} (-1)^{\nu-1} \nabla_{i_\nu} \varphi_{i_1,\ldots,i_{\nu-1},i_{\nu+1},\ldots,i_{r+1}}$$

From (2,4) $\nabla^2 = 0$ and hence $(\Omega^{\cdot}, \nabla)$ becomes a complex.

We denote its total cohomology group by $H^{*}(\Omega^{\cdot},\nabla) = \sum_{r=0}^{n} H^{r}(\Omega^{\cdot},\nabla)$. The n-th cohomology group $H^{n}(\Omega^{\cdot},\nabla)$ is isomorphic to

$$(2.10)\quad V/ \sum_{j=1}^{n} \nabla_j V = V/ \sum_{\chi \in X} \nabla^{\chi} V.$$

This identity follows from Lemma 2.1. It is important to note that $\langle\varphi\rangle$ vanishes for $\varphi \in \nabla\Omega^{n-1}$ by (2.3).

Remark 1. We can show that $H^{*}(\Omega^{\cdot},\nabla)$ is independent, up to isomorphism, of the choice of basis $\{\chi_1,\ldots,\chi_n\}$. In the sequel we shall mainly be concerned about $H^{n}(\Omega^{\cdot},\nabla)$.

3. Critical points and stable q-cycles.

Through the saddle point method, the notion of stable cycles have played a central part in asymptotic analysis of complex analytic integrals(see [A2]). In this section we define stable q-cycles attached to the Jackson integral (2.2) and show that they also play a crucial role in search of their asymptotic behaviours.

<u>We assume for simplicity that</u> q <u>is real and</u> $0 < q < 1$. We put $\alpha = N\eta + \alpha'$, where $\eta \in \check{X}$ and $\alpha' \in \mathbf{C}^n$ being fixed, and study the asymptotic behaviour of Jackson integrals $\langle\varphi\rangle$, $\varphi \in V$, for $N \to +\infty$.

Since $\Phi(t) = (t^\eta)^N \cdot t^{\alpha'} \cdot \Pi_{j=1}^m \dfrac{(a_j' t^{\mu_j})_\infty}{(a_j t^{\mu_j})_\infty}$, the major part of the absolute value $|\Phi|$ is played by $|t^\eta|$ when N approches to $+\infty$. $|t^\eta|$ attains a maximum on a fixed subset of $\overline{X}$ if and only if the level function $L_\eta(\lambda) = \mathrm{Re}(\eta,\lambda)$, $\lambda = \log_q t$, takes a minimum on it. Here (η,λ) denotes $\eta(\lambda)$ for $\lambda = (\lambda_1,\ldots,\lambda_n) \in X_{\mathbf{C}} = X \otimes \mathbf{C}$. Remark that for any t, λ is determined up to $\frac{2\pi i}{\log q}X \simeq \frac{2\pi i}{\log q}\mathbf{Z}^n$ but this ambiguity does not affects $L_\eta(\lambda)$. For convenience we write

$$(3.1)\qquad a \equiv b \mod \frac{2\pi i}{\log q}\mathbf{Z} \quad \text{for } a,\ b \in \mathbf{C}\ ,$$

$$(3.2)\qquad a \geq b \mod \frac{2\pi i}{\log q}\mathbf{Z} \quad \text{for } a,\ b \in \mathbf{C}\ ,$$

when $\mathrm{Re}(a-b) = 0$, $\mathrm{Im}(a-b) \equiv 0 \mod \frac{2\pi i}{\log q}\mathbf{Z}$, or when $\mathrm{Re}(a-b) \geq 0$, $\mathrm{Im}(a-b) \equiv 0 \mod \frac{2\pi i}{\log q}\mathbf{Z}$.

Let $J=\{j_1,j_2,\ldots,j_n\}$ be such an unordered subset of $\{\pm1,\pm2,\ldots,\pm m\}$ that μ_{j_ν}, $1 \leq \nu \leq n$, are linearly independent in $\check{X}_{\mathbf{R}}$. We denote by $\overline{X}_J$ and $\overline{X}_J^+$ the subsets of points $t = q^\lambda \in \overline{X}$ which satisfy the properties

(3.3) and (3.4) respectively.

$$(3.3)\qquad \log_q a_j' + (\mu_j,\lambda) \equiv Z \mod \frac{2\pi i}{\log q} Z\ , \text{ for } j \in J,$$

$$(3.4)\qquad \log_q a_j' + (\mu_j,\lambda) \equiv Z^+ \mod \frac{2\pi i}{\log q} Z\ , \text{ for } j \in J\ .$$

$\overline{X}_J^+$ is divided into a family of X-orbits $\overline{X}_J^+ \cap X\cdot\xi$ for some $\xi \in \overline{X}$.

<u>Definition</u> 4. We say that a point $t = q^\lambda \in \overline{X}_J^+ \cap X\cdot\xi$ is <u>critical</u> with respect to the level function L_η if L_η attains a finite minimum at $t = q^\lambda$ in $\overline{X}_J^+ \cap X\cdot\xi$. Namely $|t^\eta|$ attains a finite maximum.

This is a special case of linear programming problems investigated in [D2] and [S3]. We denote by $Cr_J = Cr_J(L_\eta)$ the set of all critical points in $\overline{X}_J^+$. Note that for each J the set Cr_J is finite or empty. Let us impose the following assumptions of genericity.

Assp 2. $L_\eta(\lambda) \neq L_\eta(\lambda')$ <u>for any pair</u> $t=q^\lambda, t'=q^{\lambda'} \in Cr_J$, $t \neq t'$,

Assp 3. $\log_q a_k' + (\mu_k,\lambda) \not\equiv 0 \mod \frac{2\pi i}{\log q} Z$ <u>for any</u> $k \in \{\pm1,\dots,\pm m\} - J$ <u>and</u> $t=q^\lambda \in Cr_J$.

Then we see that for each subset $J = \{j_1,j_2,\dots,j_n\}$, Cr_J is not empty if and only if η lies in the convex hull of $\mu_{j_1},\dots,\mu_{j_n}$. Among 2^n subsets $J_\varepsilon = \{\varepsilon_1 j_1,\varepsilon_2 j_2,\dots,\varepsilon_n j_n\}$, $\varepsilon_i=\pm1$, there exists a unique J such that $Cr_J \neq \phi$. In this case each X-orbit $\overline{X}_J^+ \cap X\cdot\xi$ has a critical point. From now on we shall restrict ourselves treating such J's which satisfy this condition and put $Cr = \bigcup_J Cr_J$. Let κ_J denote the number of elements in Cr_J.

Definition 5. For any critical point $\xi_J \in Cr_J$, we call its associated X-orbit $\mathscr{C}(\xi_J) = \overline{X}_J^+ \cap X\cdot\xi_J$ in $\overline{X}_J^+$ a stable q-cycle.

From the choice of Φ, when summable, we have

$$(3.6) \qquad \langle\varphi;\xi_J\rangle = \int_{X\cdot\xi_J} \Phi\varphi\,\overline{\omega} = \int_{\mathscr{C}(\xi_j)} \Phi\varphi\,\overline{\omega}\,.$$

Lemma 3.1. (1) Each X-orbit $\overline{X}_J^+ \cap X\cdot\xi$ coincides with unique stable q-cycle $\mathscr{C}(\xi_J)$ for some $\xi_J \in Cr_J$. (2) $Cr_J \cap Cr_{J'} = \phi$, for $J \neq J'$.

Proof. (1) is obvious. (2) follows from the genericity condition Assp 2.

We denote by g_J the morphism from $\overline{X}_J$ to $Z^n \bmod \frac{2\pi i}{\log q} Z^n$ as follows

$$(3.7) \qquad g_J(t) = g_J(q^\lambda) = (\log_q a'_{j_1} + (\mu_{j_1},\lambda), \ldots, \log_q a'_{j_n} + (\mu_{j_n},\lambda))$$
$$\in Z^n \bmod \frac{2\pi i}{\log q} Z^n.$$

Then $g_J(\overline{X}_J^+) \subset (Z^+)^n \bmod \frac{2\pi i}{\log q} Z^n$. The image $g_J(\mathscr{C}(\xi_J))$ of a stable q-cycle $\mathscr{C}(\xi_J)$ consists of one point.

Lemma 3.2. If we let $\mathscr{C}(\xi_J)$ run over all stable q-cycles in $\overline{X}_J^+$, the number of images $g_J(\mathscr{C}(\xi_J))$ in $Z^n \bmod \frac{2\pi i}{\log q} Z^n$ is equal to the absolute value of $[\mu_{j_1},\ldots,\mu_{j_n}]$. Here $[\mu_{j_1},\ldots,\mu_{j_n}]$ denotes the determinant $\det((\mu_{j_r}(\chi_s)))_{1\le r,s\le n}$.

Proof. We denote by Γ the subgroup of Z^n consisting of the vectors $((\mu_{j_1},\chi),\ldots,(\mu_{j_n},\chi))$, $\chi \in X$. If $\lambda' = \lambda + \chi$, then

$$g_J(q^{\lambda'}) = g_J(q^{\lambda}) + ((\mu_{j_1},\chi),\ldots,(\mu_{j_n},\chi)). \tag{3.8}$$

This shows that $g_J(\overline{X}_J)/g_J(X)$ is isomorphic to Z^n/Γ and hence its order is $|[\mu_{j_1},\ldots,\mu_{j_n}]|$. Since the image of $\overline{X}_J^+$ in Z^n/Γ coincides with that of $\overline{X}_J$, Lemma 3.2 holds.

Lemma 3.3. <u>For given</u> $(\ell_1,\ldots,\ell_n) \in Z^n$, <u>the linear equations in</u> λ mod $\frac{2\pi i}{\log q}X$

$$\log_q a'_{j_r} + (\mu_{j_r},\lambda) \equiv \ell_r \mod \frac{2\pi i}{\log q} Z , \tag{3.9}$$

<u>have</u> $|[\mu_{j_1},\ldots,\mu_{j_n}]|$ X-<u>incongruent solutions</u>.

<u>Proof</u>. We apply such a unimodular transformation $\{\chi_r\}_{r=1}^{n} \to \{\tilde{\chi}_s\}_{s=1}^{n}$ that the matrix $(\mu_{j_r}(\tilde{\chi}_s))_{1\le r,s\le n}$ becomes triangular. Then (3.9) can be solved successively and have $|\mu_{j_1}(\tilde{\chi}_1)\cdots\mu_{j_n}(\tilde{\chi}_n)| = |[\mu_{j_1},\ldots,\mu_{j_n}]|$ solutions. This proves Lemma 3.3.

From the above lemmas, we see that the number of different stable q-cycles $\mathscr{C}(\xi_J)$ in $\overline{X}_J^+$ is equal to $\kappa_J = [\mu_{j_1},\ldots,\mu_{j_n}]^2$. By Summing up we have

Proposition 2. $\overline{X}_J^+$ <u>is divided into</u> κ_J <u>stable</u> q-<u>cycles</u> $\mathscr{C}(\xi_J)$ $1 \le r \le \kappa_J$. <u>The total number of stable</u> q-<u>cycles in</u> $\overline{X}$ <u>is equal to that of all critical points.</u> <u>Namely it is</u>

$$\kappa = \sum_J \kappa_J = \sum_J [\mu_{j_1},\ldots,\mu_{j_n}]^2 \tag{3.10}$$

$$= \det \left(\left(\sum_{j=1}^{m} \mu_j(\chi_r)\mu_j(\chi_s) \right) \right)_{1\le r,s\le n} .$$

The following is a crucial proposition.

Proposition 3. <u>A point</u> $t = q^{\lambda}$ <u>is critical,</u> i.e., $t \in Cr(L_{\eta})$, <u>if and only if</u>

(3.11) $$b_{\chi}^{-}(q^{\lambda-\chi}) = 0$$

<u>for any element</u> χ <u>in</u> X <u>such that</u> $(\eta,\chi) > 0$.

<u>Proof</u>. Let $t=q^{\lambda}$ be an element of Cr_J. Since η lies in the convex cone spanned by $\mu_{i_1},\dots,\mu_{i_n}$, for each $\chi \in X$ satisfying $(\eta,\chi) > 0$ there exists such a non-empty subset K of J that $\mu_j(\chi) > 0$ for $j \in K$ and $\mu_j(\chi) \le 0$ for $j \in J - K$. We prove that $(a_j' t^{\mu_j} q^{-\mu_j(\chi)})_{\mu_j(\chi)}$ vanishes for some $j \in K$. Indeed, suppose otherwise $\log_q a_j' + (\mu_j,\lambda) \ge \mu_j(\chi) + 1 \bmod \frac{2\pi i}{\log q} Z$ for all $j \in K$. Since $\log_q a_j' + (\mu_j,\lambda) \in Z^{+} \bmod \frac{2\pi i}{\log q} Z$ for $j \in J - K$ and $1 \le \log_q a_j' + (\mu_j,\lambda-\chi) \bmod \frac{2\pi i}{\log q} Z$ for all $j \in J$, $\lambda-\chi$ also lies in $\overline{X}_J^{+}$. Since $L_{\eta}(\lambda) > L_{\eta}(\lambda-\chi)$, this contradicts the minimality of L_{η} at $t=q^{\lambda}$. Hence there exists some $j \in K$ such that $(a_j' q^{-\mu_j(\chi)} t^{\mu_j})_{\mu_j(\chi)} = 0$ which implies $b_{\chi}^{-}(q^{\lambda-\chi}) = 0$.

Let us prove the converse. We can take an element $\chi^{(1)} \in X$ which satisfies $(\eta,\chi^{(1)}) > 0$ and $b_{\chi^{(1)}}^{-}(q^{\lambda-\chi^{(1)}}) = 0$. We show that there exists $j_1 \in \{\pm1,\pm2,\dots,\pm m\}$ such that

(3.12) $$\log_q a_{j_1}' + (\mu_{j_1},\lambda) \equiv 1,2,\dots,\mu_{j_1}(\chi^{(1)}) \bmod \frac{2\pi i}{\log q} Z.$$

Indeed from the expression (1.5) of b_χ^- for $\chi = \chi^{(1)}$, we see that there exists $1 \le j \le m$ such that

$$(3.13) \qquad (a_j' \, q^{-\mu_j(\chi^{(1)})} t^{\mu_j})_{\mu_j(\chi^{(1)})} = 0, \quad \text{for} \quad \mu_j(\chi^{(1)}) > 0 \text{ , or}$$

$$(3.14) \qquad (a_j t^{\mu_j})_{-\mu_j(\chi^{(1)})} = 0, \quad \text{for } \mu_j(\chi^{(1)}) < 0.$$

Suppose first $\mu_j(\chi^{(1)}) > 0$ for j. We put $j_1 = j$ then (3.12) follows from (3.13). Suppose now $\mu_j(\chi^{(1)}) < 0$ then $\log_q a_j + (\mu_j, \lambda) \equiv 0, -1, \ldots, \mu_j(\chi^{(1)})+1 \bmod \frac{2\pi i}{\log q}\mathbb{Z}$. Since we have defined $a'_{-j} = q a_j^{-1}$, we can choose $-j$ as j_1 and then (3.9) holds. Next we take $\chi^{(2)} \in X$ such that $(\eta, \chi^{(2)}) > 0$ and $(\mu_{j_1}, \chi^{(2)}) = 0$. Then in the same manner as above, we can show that there exists $j_2 \in \{\pm 1, \pm 2, \ldots, \pm m\}$ such that

$$(3.15) \qquad \log_q a'_{j_2} + (\mu_{j_2}, \lambda) \equiv 1, 2, \ldots, \mu_{j_2}(\chi^{(2)}) \bmod \frac{2\pi i}{\log q}\mathbb{Z}.$$

Continuing this procedure, we can prove by induction that there exist $j_1, \ldots, j_n \in \{\pm 1, \pm 2, \ldots, \pm m\}$ and $\chi^{(1)}, \ldots, \chi^{(n)} \in X$ such that

$$(3.16) \qquad (\eta, \chi^{(r)}) > 0, \qquad 1 \le r \le n \text{ ,}$$

$$(3.17) \qquad (\mu_{j_s}, \chi^{(r)}) = 0, \qquad 1 \le s < r \le n,$$

$$(3.18)\qquad \log_q a'_{j_r} + (\mu_{j_r},\lambda) \equiv 1,2,\ldots,\mu_{j_r}(\chi^{(r)}) \mod \frac{2\pi i}{\log q} Z,$$

for $1 \le r \le n$. Then $\mu_{j_1},\ldots,\mu_{j_n}$ are linearly independent. We put $J = \{j_1,\ldots,j_n\}$. We prove that $t = q^\lambda$ is a critical point for J. Suppose otherwise, there exists such $\chi \in X$ that $(\eta,\chi) > 0$ and $q^{\lambda-\chi} \in \overline{X}_J^+$. From Assp 2, $\log_q a'_j + (\mu_j,\lambda-\chi) \in Z^+ \mod \frac{2\pi i}{\log q} Z$ for any $j \in J$. On the other hand, the vanishing of $b_\chi^-(q^{\lambda-\chi})$ implies that for at least one j_r, $\log_q a'_{j_r} + (\mu_{j_r},\lambda-\chi) \equiv 0,-1,-2,\ldots,-\mu_j(\chi)+1$, $\mod \frac{2\pi i}{\log q} Z$. This is a contradiction and Proposition 3 is proved.

We arrange as $Cr = \{\xi^{(1)},\ldots,\xi^{(\kappa)}\}$. Since $\xi^{(r)} \in Cr$ are different from each other, we have

Lemma 3.5. <u>There exist</u> κ <u>Laurent polynomials</u> $\varphi_r \in \mathcal{L}$, $1 \le r \le \kappa$, <u>such that</u> $\varphi_r(\xi^{(s)}) = \delta_{r,s}$, $1 \le s \le \kappa$.

<u>Definition</u> 6. Let $\xi^{(s)} \in Cr_J$ for some $J = \{j_1,\ldots,j_n\}$. After rearranging, we may assume $j_1 < 0, \ldots, j_\ell < 0$ and $j_{\ell+1} > 0,\ldots,j_n > 0$. Then the integration $\langle\varphi\rangle = \int_{X\cdot\xi^{(s)}} \Phi\varphi\varpi = \int_{\mathcal{E}(\xi^{(s)})} \Phi\varphi\varpi$, $\varphi \in V$, is not summable because Φ has poles on $\mathcal{E}(\xi^{(s)})$. In this case we replace Φ by $\Phi' = T_{j_1}\cdots T_{j_\ell}\Phi$ after applying the transformations T_j in (1.6) to eliminate these poles. Then $\int_{\mathcal{E}(\xi^{(s)})} \Phi'\varphi\varpi$ is well defined in view of Assp 3. We call this modification <u>the regularization of integration</u> and denote $reg \int_{\mathcal{E}(\xi^{(s)})} \Phi\varphi\varpi$.

Lemma 3.6. <u>We have</u>

$$(3.19) \qquad \det\Big(\Big(reg \int_{\mathscr{E}(\xi^{(s)})} \Phi\varphi_r \,\overline{\omega} \Big)\Big)_{1\le r,s\le \kappa} \neq 0.$$

Proof. Let us investigate the asymptotic behaviours of $\langle\varphi_r\rangle$, $1 \le r \le \kappa$. If we apply the regularization, we may assume that $j_1 > 0, \dots, j_n > 0$ for $\mathscr{E}(\xi^{(s)})$. Then for $N \to +\infty$

$$(3.20) \qquad \int_{\mathscr{E}(\xi^{(s)})} \Phi\varphi_r\overline{\omega} \sim (1-q)^n \,\delta_{r,s}\, (\xi^{(s)})^{\alpha}\, \Pi_{j=1}^{m} \frac{(a_j'(\xi^{(s)})^{\mu_j})_\infty}{(a_j(\xi^{(s)})^{\mu_j})_\infty}$$

$$\cdot\Big(1 + O\big(\tfrac{1}{N}\big)\Big),$$

where $\Pi_{j=1}^{m} \dfrac{(a_j'(\xi^{(s)})^{\mu_j})_\infty}{(a_j(\xi^{(s)})^{\mu_j})_\infty} \neq 0$ by Assp 3. Hence Lemma 3.6 follows.

As a result, the associated classes with $\varphi_1,\dots,\varphi_\kappa$ in $H^n(\Omega^\cdot,\nabla) \simeq V/\sum_{j=1}^{n}\nabla_j V$ are linearly independent. Therefore we have

Proposition 4. The dimension of $H^n(\Omega^\cdot,\nabla)$ is bounded from the bottom as follows

$$(3.21) \qquad \dim H^n(\Omega^\cdot,\nabla) \ge \kappa.$$

4. Upper estimate of $\dim H^n(\Omega^\cdot,\nabla)$.

We apply ∇^χ for an element $\psi(t) \in \mathscr{L}$ then

$$(4.1) \qquad (\nabla^\chi \psi)(t) = \psi(t) - u^\chi \frac{b_\chi^+(t)}{b_\chi^-(t)} (Q^\chi\psi)(t).$$

If $b_\chi^-(t) | Q^\chi\psi(t)$, i.e., $\psi(t)= (Q^{-\chi}b_\chi^-)(t)\cdot\overline{\psi}(t)$, for some $\overline{\psi}(t)\in \mathcal{L}$, then we have $\nabla^\chi\psi \in \mathcal{L}$. Hence we have

$$(4.2)\qquad (\nabla^\chi\psi)(t) = (D_\chi\overline{\psi})(t) = (Q^{-\chi}b_\chi^-)(t)\cdot\overline{\psi}(t) - u^\chi\, b_\chi^+(t)\cdot(Q^\chi\overline{\psi})(t).$$

Here we denote $D_\chi = D_\chi(u,a,a';t,Q) = (Q^{-\chi}b_\chi^-)(t) - u^\chi b_\chi^+(t)\, Q^\chi$, $\chi \in X$, which is a q-difference operator acting on $\mathcal{L}$.

Lemma 4.1. (i) Let χ and χ' be any two elements of X such that the signs of $\mu_j(\chi)$ and $\mu_j(\chi')$ are the same or 0. Then the following holds.

$$(4.3)\qquad D_{\chi+\chi'} = D_\chi \circ (u^{\chi'} b_{\chi'}(t) Q^{\chi'}) + D_{\chi'} \circ (Q^{-\chi-\chi'} b_\chi^-)(t).$$

(ii) For an arbitrary $\chi \in X$,

$$(4.4)\qquad D_{-\chi} = - D_\chi \circ (u^{-\chi}Q^{-\chi}).$$

Here $\circ$ means the composition of operators.

Proof. (4.3) and (4.4) follow from a direct computation by use of the relations $b^\pm_{\chi+\chi'}(t) = b^\pm_\chi(t)\cdot(Q^\chi b^\pm_\chi)(t)$ and $b^\pm_{-\chi}(t)= (Q^{\mp\chi}b_\chi)(t)$. These relations are obtained from the explicit expressions (1.4) and (1.5).

Corollary. Under the same condition as in Lemma 4.1, we have

(4.5) $$D_{\chi+\chi'}\mathcal{L} \subset D_\chi\mathcal{L} + D_{\chi'}\mathcal{L} \quad \text{and} \quad D_{-\chi}\mathcal{L} = D_\chi\mathcal{L}.$$

Lemma 4.2. Let ℓ_j, $\ell'_j \geq 0$, $1 \leq j \leq m$, $\chi \in X$ and take

(4.6) $$\psi = \frac{\overline{\psi}}{\prod_{\substack{1\leq j\leq m \\ \text{s.t. } \mu_j(\chi)>0}} (a'_j t^{\mu_j})_{\ell'_j-\mu_j(\chi)} (a_j q^{-\ell_j} t^{\mu_j})_{\ell_j}}$$

$$\times \frac{1}{\prod_{\substack{1\leq j\leq m \\ \text{s.t. } \mu_j(\chi)<0}} (a'_j t^{\mu_j})_{\ell'_j} (a_j q^{-\ell_j-\mu_j(\chi)} t^{\mu_j})_{\ell_j+\mu_j(\chi)}},$$

for $\overline{\psi} \in \mathcal{L}$. Then

(4.7) $$\nabla^\chi\psi \cdot \{ \Pi_{j=1}^m (a'_j t^{\mu_j})_{\ell'_j} \cdot (a_j q^{-\ell_j} t^{\mu_j})_{\ell_j} \} = (\Pi_{j=1}^m Q_{a'_j}^{\ell'_j} Q_{a_j}^{-\ell_j}) D_\chi \overline{\psi} \quad .$$

<u>Proof</u>. Since the proof comes from the comparison of factors, we omit it.

We denote by $\mathfrak{a}_q(u)$, $u \in \check{X} \otimes_q \mathbf{C}^*$, the subspace of $\mathcal{L}$ generated by the elements of the form (4.2)

(4.8) $$\mathfrak{a}_q(u) = \sum_{\chi\in X} D_\chi \mathcal{L} \subset \mathcal{L} \ .$$

For each sequence $(L,L')=(\ell_1,\ldots,\ell_n,\ell'_1,\ldots,\ell'_n)$ of non-negative integers, we also define a subspace $\mathfrak{a}_q(u;L,L')$ as follows

$$(4.9)\quad \alpha_q(u;L,L') = \sum_{\chi\in X} \Big(\Pi_{j=1}^m Q_{a_j'}^{\ell_j'} Q_{a_j}^{-\ell_j} D_\chi\Big)\mathcal{L} + \sum_{\substack{1\le j\le m \\ \text{s.t. } \ell_j>0}} (1-a_j q^{-\ell_j} t^{\mu_j})\mathcal{L}$$

$$+ \sum_{\substack{1\le j\le m \\ \text{s.t. } \ell_j'>0}} (1-a_j' q^{\ell_j'-1} t^{\mu_j})\mathcal{L}.$$

Remark that $\alpha_q(u;0,0) = \alpha_q(u)$. Due to Lemma 4.1, $\alpha_q(u)$ and $\alpha_q(u;L,L')$ can also be written as

$$(4.10)\qquad \alpha_q(u) = \sum_{\substack{\chi\in X \\ (\eta,\chi)>0}} D_\chi \mathcal{L},$$

$$(4.11)\quad \alpha_q(u;L,L') = \sum_{\substack{\chi\in X \\ (\eta,\chi)>0}} \Big(\Pi_{j=1}^m Q_{a_j'}^{\ell_j'} Q_{a_j}^{-\ell_j} D_\chi\Big)\mathcal{L} + \sum_{\substack{1\le j\le m \\ \text{s.t. } \ell_j>0}} (1-a_j q^{-\ell_j} t^{\mu_j})\mathcal{L}$$

$$+ \sum_{\substack{1\le j\le m \\ \text{s.t. } \ell_j'>0}} (1-a_j' q^{\ell_j'-1} t^{\mu_j})\mathcal{L}.$$

We define the following two ideals in $\mathcal{L}$ by taking $u^\chi \to 0$ which is equivalent to $N \to +\infty$ because $(\eta,\chi) > 0$.

$$(4.12)\qquad \alpha_q(0) = \sum_{\substack{\chi\in X \\ (\eta,\chi)>0}} (Q^{-\chi} b_\chi^-)(t)\ \mathcal{L},$$

$$(4.13)\quad \alpha_q(0;L,L') = \sum_{\substack{\chi\in X \\ (\eta,\chi)>0}} \{(\Pi_{j=1}^m Q_{a_j'}^{\ell_j'} Q_{a_j}^{-\ell_j})(Q^{-\chi} b_\chi^-)(t)\}\ \mathcal{L}$$

$$+ \sum_{\substack{1\le j\le m \\ \text{s.t. } \ell'_j>0}} (1- a'_j{}^{\ell'_j-1} t^{\mu_j}) \mathscr{L} + \sum_{\substack{1\le j\le m \\ \text{s.t. } \ell_j>0}} (1- a_j q^{-\ell_j} t^{\mu_j}) \mathscr{L}.$$

The union of hyperplanes $\cup_{j=1}^{m} H_j$, $H_j = \{\lambda \in X_{\mathbf{R}} \mid (\mu_j, \lambda) = 0\}$ in $X_{\mathbf{R}} = X \otimes \mathbf{R}$ defines a fan F in the sense of torus embeddings. F consists of rational polyhedral convex cones σ given by connected components in $X_{\mathbf{R}} - \cup_{j=1}^{m} H_j$. By the theory of torus embeddings, F defines a complete toric variety $T_{emb}(F)$ (see [O]). Moreover it is known (see [K2]) that there exists a fan $\hat{F}$ which is a simplicial subdivision of F such that the corresponding toric variety $T_{emb}(\hat{F})$ is a desingularization of $T_{emb}(F)$. Each rational polyhedral cone σ in $\hat{F}$ is spanned by n primitive corner vectors of σ which form also a basis of X. We denote by Y the set of all primitive corner vectors spanning rational polyhedral cones in $\hat{F}$ and Y^+ the set of $\chi \in Y$ such that $(\eta, \chi) > 0$. Both Y and Y^+ are finite sets.

<u>We assume from now on that</u> $(\eta, \chi) \neq 0$ <u>for any</u> $\chi \in Y$. Due to Corollary to Lemma 4.1, $\alpha_q(u)$, $\alpha_q(u;L,L')$, $\alpha_q(0)$ and $\alpha_q(0;L,L')$ can be described as

$$(4.14) \qquad \alpha_q(u) = \sum_{\chi\in Y^+} D_\chi \mathscr{L},$$

$$(4.15) \qquad \alpha_q(u;L,L') = \sum_{\chi\in Y^+} (\Pi_{j=1}^{m} Q_{a'_j}^{\ell'_j} Q_{a_j}^{-\ell_j} D_\chi) \mathscr{L},$$

$$+ \sum_{\substack{1\le j\le m \\ \text{s.t. } \ell_j>0}} (1-a_j q^{-\ell_j} t^{\mu_j}) \mathscr{L} + \sum_{\substack{1\le j\le m \\ \text{s.t. } \ell'_j>0}} (1- a'_j q^{\ell'_j-1} t^{\mu_j}) \mathscr{L},$$

$$a_q(0) = \sum_{\chi\in Y^+} (Q^{-\chi} b_\chi^-)(t)\mathcal{L} , \tag{4.16}$$

$$a_q(0;L,L') = \sum_{\chi\in Y^+} \{(\Pi_{j=1}^m Q_{a_j'}^{\ell_j'} Q_{a_j}^{-\ell_j})(Q^{-\chi} b_\chi^-)(t)\}\mathcal{L} \tag{4.17}$$

$$+ \sum_{\substack{1\le j\le m \\ \text{s.t. } \ell_j>0}} (1-a_j q^{-\ell_j} t^{\mu_j})\mathcal{L} + \sum_{\substack{1\le j\le m \\ \text{s.t. } \ell_j'>0}} (1- a_j' q^{\ell_j'-1} t^{\mu_j})\mathcal{L}.$$

Definition 7. For a Laurent polynomial $f = \sum_{\omega\in \check{X}} c_\omega t^\omega$, $c_\omega \in \mathbf{C}$, the set $\{\omega \in \check{X} \mid c_\omega \neq 0\}$ is said to be the support of f and denoted by Supp(f). The Newton polyhedron $\Delta(f)$ is the convex closure of Supp(f) in $\check{X}_{\mathbf{R}}$. For any two convex sets Δ_1 and Δ_2, Minkowski sum $\Delta_1 + \Delta_2$ is defined. For a convex set $\mathfrak{K}$ in $\check{X}_{\mathbf{R}}$ we denote by $\mathbf{C}\langle\mathfrak{K}\rangle$ the linear subspace of functions $f\in \mathcal{L}$ such that $\mathrm{Supp}(f) \subset \mathfrak{K}$.

The following Lemma 4.3 is immediately proven.

Lemma 4.3. *We have* (i) $\mathrm{Supp}(b_\chi^+) = \mathrm{Supp}(b_\chi^-)$ *and all the Newton polyhedra of* $b_\chi^\pm$, $\tilde{Q}_j^{\pm 1} b_\chi^\pm$, $Q_{a_j}^{\pm 1} b_\chi^\pm$ *and* $Q_{a_j'}^{\pm 1} b_\chi^\pm$ *are the same*.

(ii) $\{0\} \in \Delta(b_\chi^\pm)$.

Lemma 4.4. *Fix* $\chi \in X$. *Then the function* (ω,χ) *on* $\Delta(b_\chi^\pm)$ *takes a maximum* $\sum_{\substack{1\le j\le m \\ \text{s.t. } \mu_j(\chi)>0}} \mu_j(\chi)^2$ *at the unique point* $\omega_{max}^\chi = \sum_{\substack{1\le j\le m \\ \text{s.t. } \mu_j(\chi)>0}} \mu_j(\chi)\mu_j \in \check{X}$.

Proof. The proof follows from (1.6) and (1.7). The value (ω,χ) is not greater than $(\omega_{max}^\chi,\chi) - \min_{\substack{1\le j\le m \\ \text{s.t. } \mu_j(\chi)\neq 0}} |\mu_j(\chi)|$ at any other point in

$\Delta(b^{\pm}_{\chi}) \cap \check{X}$.

Lemma 4.5. <u>Let</u> $\sigma = \sum_{s=1}^{n} R^{+} \tilde{\chi}_{s}$, <u>where</u> $\tilde{\chi}_{1},\ldots,\tilde{\chi}_{n}$ <u>form a basis of</u> X, <u>be an arbitrary cone belonging to the fan</u> $\hat{F}$. <u>Let</u> $\sigma' = \sum_{j=1}^{r} R^{+} \tilde{\chi}_{s_j}$ <u>be a subcone subordinate to</u> σ. <u>Then there exists a corner vector</u> $\chi = \tilde{\chi}_{s_j}$ <u>of</u> σ' <u>such that</u> $(\omega,\chi') \leq (\omega^{\chi}_{max},\chi')$ <u>for any</u> $\omega \in \check{X} \cap \Delta(b^{\pm}_{\chi}) - \{\omega^{\chi}_{max}\}$, <u>where</u> χ' <u>is an arbitrary element of</u> σ'.

<u>Proof</u>. First remark that $\mu_j(\chi') \geq 0$ or ≤ 0 according as $\mu_j(\chi) > 0$ or $\mu_j(\chi) < 0$. $\omega - \omega^{\chi}_{max}$ is written as a linear combination of $-\mu_j$ for $\mu_j(\chi) > 0$ and μ_k for $\mu_k(\chi) < 0$ with non-negative coefficients. The proof is completed.

Let $\mathfrak{K}$ be a convex set in $\check{X}_R$ containing $\{0\}$ defined by the inequalities

$$(4.18) \qquad \mathfrak{K} = \{\omega \in \check{X}_R \mid (\omega,\chi) \leq C_{\chi}, \quad \chi \in Y\}.$$

where C_{χ} are sufficiently large constants so that $\Delta(b^{\pm}_{\chi}) + \omega_{\chi} \subset\subset \mathfrak{K}$ for some $\omega_{\chi} \in \check{X}_R$. H_{χ} denotes the hyperplane $\{ \omega \in X_R \mid (\omega,\chi) = C_{\chi}\}$.

As an immediate consequence of Lemma 4.5, we have

Lemma 4.6. <u>Under the same circumstsance</u>, <u>let</u> $H_{\tilde{\chi}_{s_1},\ldots,\tilde{\chi}_{s_r}} = \{ H_{\tilde{\chi}_{s_1}} \cap \ldots\ldots \cap H_{\tilde{\chi}_{s_r}} \} \cap \partial\mathfrak{K}$ <u>be a face of codimension</u> r <u>of</u> $\mathfrak{K}$. <u>Here</u> $\partial\mathfrak{K}$ <u>is the boundary of</u> $\mathfrak{K}$. <u>Then there exists a Newton polyhedron</u> $\omega_{\chi} + \Delta(b^{\pm}_{\chi}) \subset \mathfrak{K}$, $\chi \in Y$, $\omega_{\chi} \in \mathfrak{K}$, <u>such that it has a unique common point with</u> $H_{\tilde{\chi}_{s_1},\ldots,\tilde{\chi}_{s_r}}$.

If we use Lemma 4.6 successively, we can prove the following by induction on the codimension of faces.

Proposition 4. <u>Suppose</u> $\psi = \sum c_\omega t^\omega \in \mathbf{C}\langle\mathfrak{R}\rangle$. <u>Then there exist Laurent polynomials</u> $\psi_1,\ldots,\psi_\ell \in \mathcal{L}$ <u>and</u> $b^-_{\tau_1}(t),\ldots,b^-_{\tau_\ell}(t)$, $\tau_j \in Y$, <u>such that</u>

$$(4.19) \qquad \Delta(\psi_j) + \Delta(b^-_{\tau_j}) = \Delta(\psi_j) + \Delta(Q^{-\tau_j} b^-_{\tau_j}) \subset \mathfrak{R},$$

$$(4.20) \qquad \psi - \sum_{j=1}^{\ell} \psi_j (Q^{-\tau_j} b^-_{\tau_j}) \equiv 0 \mod \mathbf{C}\langle\mathring{\mathfrak{R}}\rangle,$$

<u>where</u> $\mathring{\mathfrak{R}}$ <u>denotes the interior of</u> $\mathfrak{R}$.

Moreover in case of $\mathfrak{a}_q(u)$, under the additional condition

Assp 4. $\displaystyle\sum_{\substack{1\le j\le m \\ \text{s.t. } \mu_j(\chi)>0}} \mu_j(\chi)(\log_q a'_j - \log_q a_j) \not\equiv (\alpha,\chi) \mod \frac{2\pi i}{\log q}\mathbf{Z}$,

<u>for</u> $\chi \in Y$,

the same argument as above shows

Proposition 5. <u>We have</u>

$$(4.21) \qquad \varphi - \sum_{\chi\in Y^+} D_\chi \psi_\chi \equiv 0 \mod \mathbf{C}\langle\mathring{\mathfrak{R}}\rangle$$

<u>for some</u> $\psi_\chi \in \mathcal{L}$ <u>such that</u> $\Delta(b^-_\chi) + \Delta(\psi_\chi) \subset \mathfrak{R}$.

We can also apply the above argument to $\rho\mathfrak{R} = \{\rho\omega \mid \omega \in \mathfrak{R}\}$, $\rho \ge 1$, in place of $\mathfrak{R}$. Seeing that $\bigcup_{\rho\ge 1}\rho\mathfrak{R} = X_{\mathbf{R}}$, we have the following by induction on decreasing ρ.

Lemma 4.7. <u>We have</u> $\mathcal{L} = \mathbf{C}\langle\mathfrak{R}\rangle + \mathfrak{a}_q(0)$.

On the other hand

Lemma 4.8. <u>The ideal</u> $\mathfrak{a}_q(0)$ <u>is</u> 0-<u>dimensional and the dimension of</u>

$\mathcal{L}/\mathfrak{a}_q(0)$ is equal to κ. $\mathcal{L}/\mathfrak{a}_q(0)$ admits the represetative system $\{\varphi_1, \ldots, \varphi_\kappa\}$ as in Lemma 3.5.

Proof. The zeros of $\mathfrak{a}_q(0)$ in $\overline{X}$ coincide with the set Cr of all critical points, whose multiplicity has been given by (3.10).

We denote by $\mathfrak{h}$ the subspace spanned by $\{\varphi_1, \ldots, \varphi_\kappa\}$ in $\mathcal{L}$. Then since $\mathcal{L} + \mathfrak{a}_q(0) \supset \mathbf{C}\langle\mathfrak{K}\rangle$, an arbitrary monomial $t^\omega \in \mathbf{C}\langle\mathfrak{K}\rangle$ can be expressed by

$$(4.23) \qquad t^\omega \equiv \sum_{\chi\in Y^+} (Q^{-\chi}b_\chi^-)(t)\psi_\chi^{(\omega)}(t) \mod \mathfrak{h}, \quad \text{for } \psi_\chi^{(\omega)} \in \mathcal{L}.$$

Since the number of $\psi_\chi^{(\omega)}$ is finite, we may take a sufficiently large convex set $\mathfrak{K}_0 = \rho\mathfrak{K}$, $\rho \gg 1$, such that $\Delta(b_\chi^-) + \Delta(\psi_\chi^{(\omega)}) \subset \mathfrak{K}_0$ and $\Delta(\varphi) \subset \mathfrak{K}_0$ for all $\varphi \in \mathfrak{h}$. On the other hand by using Proposition 4 successively, we get

$$(4.24) \qquad t^\omega \equiv \sum_{\chi\in Y^+} (Q^{-\chi}b_\chi^-)(t)\psi_\chi^{(\omega)}(t) \mod \mathfrak{h} \quad \text{for } \omega \in \mathfrak{K}_0 - \mathfrak{K},$$

such that $\Delta(b_\chi^-) + \Delta(\psi_\chi^{(\omega)}) \subset \mathfrak{K}_0$. We denote by S_χ the convex set in $\check{X}_R$ such that $S_\chi + \Delta(b_\chi^-) \subset \mathfrak{K}_0$. Then

Lemma 4.9. The mapping

$$(4.25) \qquad \#_0 : \mathfrak{h} \oplus \bigoplus_{\chi\in Y^+} \mathbf{C}\langle S_\chi\rangle \longrightarrow \mathbf{C}\langle\mathfrak{K}_0\rangle,$$

$$\qquad\qquad \Psi \qquad\qquad\qquad\qquad \Psi$$

$$(\varphi, \{\psi_\chi\}_{\chi\in Y^+}) \longrightarrow \varphi + \sum_{\chi\in Y^+} (Q^{-\chi}b_\chi^-)\psi_\chi$$

is surjective.

We can also define a mapping

$$(4.26)\qquad \#_u : \mathfrak{h} \oplus \bigoplus_{\chi\in Y^+} C\langle S_\chi\rangle \longrightarrow C\langle \mathfrak{R}_0\rangle .$$

$$(\varphi , (\psi_\chi)_{\chi\in Y^+}) \longmapsto \varphi + \sum_{\chi\in Y^+} D_\chi \psi_\chi$$

If we take $\alpha = \eta N + \alpha'$ in (1.3) for sufficiently large N, $|u^\chi|$, $\chi \in Y^+$, becomes very small.

Lemma 4.10. Suppose that $|u^\chi|$, $\chi \in Y^+$, are very small. Then $\#_u$ is surjective.

Proof. Lemma 4.9 leads to this lemma because both $\#_0$ and $\#_u$ are linear mappings between finite dimensional vector spaces and $\#_u$ is close to $\#_0$.

Proposition 6. Suppose $|u^\chi|$, $\chi \in Y^+$, are very small. Then

$$(4.27)\qquad \dim \mathcal{L}/\mathfrak{a}_q(u) \le \dim \mathfrak{h} = \dim \mathcal{L}/\mathfrak{a}_q(0) = \kappa.$$

Proof. It is sufficient to show

$$(4.28)\qquad \mathfrak{h} + \mathfrak{a}_q(u) = \mathcal{L}.$$

If (4.28) is false, we can choose such $\tilde{\varphi}$ that $\tilde{\varphi} \in \mathcal{L}$ but $\tilde{\varphi} \notin \mathfrak{h} + \mathfrak{a}_q(u)$. Then since $\Delta(\tilde{\varphi}) \subset \mathfrak{R}_0$ for sufficient large $\mathfrak{R}_0$, we have a contradiction to Lemma 4.10. On the other hand

Lemma 4.11. $\mathfrak{a}_q(0; L,L') = \mathcal{L}$ if $|L| + |L'| > 0$ under Assp 1 ~ 3.

Proof. Indeed, the set of the zeros of $\mathfrak{a}_q(0;L,L')$ are empty, in view of Assp 2 and 3. The lemma follows from Hilbert Nullstellen Satz.

As in Lemma 4.10 we have the following

Lemma 4.12. *Under* Assp 1 ~ 4, $\mathfrak{a}_q(u;L,L') = \mathfrak{a}_q(0;L,L') = \mathcal{L}$ *provided* $|u^\chi|$ *for* $\chi \in Y^+$ *are very small and* $|L| + |L'| > 0$.

We now make the final assumption.

Assp 5. u *is generic in such a way that* $\dim \mathcal{L}/\mathfrak{a}_q(u) \leq \dim \mathcal{L}/\mathfrak{a}_q(0)$ *and* $\mathfrak{a}_q(u;L,L') = \mathcal{L}$ *for* $|L| + |L'| > 0$.

Proposition 6 and Lemma 4.12 show that such u really exists.

Proposition 7. *Under* Assp 1 ~ 5 *the canonical morphism*

$$(4.29) \qquad \mathcal{L}/\mathfrak{a}_q(u) \longrightarrow H^n(\Omega^\cdot,\nabla) \simeq V/\sum_{\chi\in X} \nabla^\chi V,$$

is surjective.

Proof. It can be proved by a standard argument. Indeed an arbitrary $\varphi \in V_\ell$, $\ell > 0$, is cohomologous to $V_{\ell-1}$. The reason is that from (4.7) and Assp 5 there exist functions $\{\psi_\chi \in V\}_{\chi\in X}$ of the form (4.6) which satisfy

$$(4.30) \qquad \varphi - \sum_{\chi \in X} \nabla^\chi\psi_\chi \equiv 0 \bmod V_{\ell-1}.$$

By induction on decreasing ℓ, φ is cohomologous to $V_0 = \mathcal{L}$. Since $\mathfrak{a}_q(u) \subset (\sum_{\chi\in X} \nabla^\chi V) \cap \mathcal{L} \subset \sum_{\chi\in X} \nabla^\chi V \simeq \nabla\,\Omega^{n-1}$, we obtain Proposition 7.

Summing up Propositions 4 and 7, we conclude

Theorem 1. $H^n(\Omega^\cdot,\nabla) \simeq \mathcal{L}/\mathfrak{a}_q(u)$ and $\dim H^n(\Omega^\cdot,\nabla) = \kappa$.

Proof. Proposition 7 shows that $\dim H^n(\Omega^\cdot,\nabla) \le \dim \mathcal{L}/\mathfrak{a}_q(u) \le \kappa$. On the other hand from Proposition 4, $\dim H^n(\Omega^\cdot,\nabla) \ge \kappa$. Theorem 1 thus follows.

5. Holonomic q-difference system.

We are going to find a system of q-linear difference equations in $u \in \check{X} \otimes_q \mathbf{C}^*$ satisfied by the Jackson integrals $\langle 1\rangle = \tilde{\Phi}$ defined in section 2. First we remark that $\tilde{Q}_j^{\pm}\langle\varphi\rangle = \langle t_j^{\pm 1}\varphi\rangle$ for $\varphi \in V$ since $\tilde{Q}_j^{\pm 1}$ correspond to the multiplication operators $\varphi \longrightarrow \varphi t_j^{\pm 1}$ in the integrand $\Phi\varphi$. Especially for $\psi(t) \in \mathcal{L}$, $\langle\psi(t)\rangle = \psi(\tilde{Q})\langle 1\rangle = \psi(\tilde{Q})\tilde{\Phi}$. From (2.3) we have

$$(5.1)\qquad \langle\nabla^{\chi}(Q^{-\chi}b_{\chi}^{-})\rangle = \langle Q^{-\chi}b_{\chi}^{-}\rangle - u^{\chi}\ \langle b_{\chi}^{+}\rangle = 0.$$

Namely

$$(5.2)\qquad ((Q^{-\chi}b_{\chi}^{-})(\tilde{Q}) - u^{\chi}\ b_{\chi}^{+}(\tilde{Q}))\ \tilde{\Phi} = 0 \quad \text{for } \chi \in X.$$

These equations are a canonical extension of the well known basic hypergeometric equations which are q-analogues of Mellin-Sato hypergeometric ones (see [S2] and [B1]). We denote $\tilde{D}_{\chi} = \tilde{D}_{\chi}(a,a';u,\tilde{Q}) = (Q^{-\chi}b_{\chi}^{-})(\tilde{Q}) - u^{\chi}b_{\chi}^{+}(\tilde{Q})$. Then for the q-difference operators $\tilde{D}_{\chi}$ we have the following equalities similar to (4.3) and (4.4).

Lemma 5.1. (i) We take arbitrary χ and $\chi' \in X$ such that $\mu_j(\chi)$ and $\mu_j(\chi')$ have the same signs for all j. Then

(5.3) $\quad \tilde{D}_{\chi+\chi'} = u^{\chi'} b^{+}_{\chi'}(\tilde{Q}) \circ \tilde{D}_{\chi} + (Q^{-\chi-\chi'} b^{-}_{\chi}(\tilde{Q})) \circ \tilde{D}_{\chi'}$,

(ii)

(5.4) $\quad \tilde{D}_{-\chi} = - u^{-\chi} \circ \tilde{D}_{\chi}$.

Proof. The proof is straightforward in view of the identity $u^{\chi} \circ \varphi(\tilde{Q}) = (Q^{\chi}\varphi)(\tilde{Q}) \circ u^{\chi}$ for $\varphi \in \mathcal{L}$.

At this place we once consider $\tilde{\Phi}$ in (5.2) as an unknown function of u. Then (5.2) turns to be a system of q-difference equations in u, which is denoted by $(\mathcal{E})$. From Lemma 5.1, $(\mathcal{E})$ is equivalent to any of the following systems.

$(\mathcal{E}_+)$ $\quad \tilde{D}_{\chi} \tilde{\Phi} = 0$, for $\chi \in X$, $(\eta,\chi) > 0$,

$(\mathcal{E}_Y)$ $\quad \tilde{D}_{\chi} \tilde{\Phi} = 0$, for $\chi \in Y$,

$(\mathcal{E}_{Y^+})$ $\quad \tilde{D}_{\chi} \tilde{\Phi} = 0$, for $\chi \in Y^+$, respectively.

Lemma 5.3. *We fix $\eta \in \check{X}$. Suppose that the system $(\mathcal{E})$ has a quasi-meromorphic solution $\tilde{\Phi}(u)$ of the asymptotic form*

(5.5) $\quad \tilde{\Phi}(u) \sim u_1^{\lambda_1} \cdots u_n^{\lambda_n} \quad (1 + O(\frac{1}{N}))$,

for $u = q^{\alpha}$, $\alpha = \eta N + \alpha'$ for $N \longrightarrow +\infty$, with α' being fixed. Then $\lambda = (\lambda_1,\ldots,\lambda_n)$ satisfies the equations

(5.6) $\quad b^{-}_{\chi}(q^{\lambda-\chi}) = 0$, *for $\chi \in X$ such that $(\eta,\chi) > 0$.*

<u>Proof</u>. We compare the main terms in both sides after substituting (5.5) into ($\mathcal{E}_+$) and have (5.6).

The equations (5.6) are the same as (3.11). According to Proposition 2, there are κ different asymptotic solutions of ($\mathcal{E}_+$). These are exactly given by the Jackson integrals $\int_{\mathfrak{E}(\xi^{(r)})} \Phi\,\overline{\omega}$ over the κ stable q-cycles $\mathfrak{E}(\xi^{(r)})$. Hence we have

Theorem 2. <u>The system of</u> q-<u>difference equations</u> ($\mathcal{E}$) <u>is holonomic and has multiplicity</u> κ. <u>In other words, it has</u> κ <u>different quasi-meromorphic solutions which have the asymptotic form</u> (5.5) <u>where</u> λ <u>are determined by</u> (5.6). <u>These are exactly given by the Jackson integrals of</u> Φ <u>over the stable</u> q-<u>cycles</u> $\mathfrak{E}(\xi^{(r)})$, $1 \leq r \leq \kappa$.

<u>Remark</u> 2. There are many other solutions of ($\mathcal{E}$) which are quasi-meromorphic in u. In fact we take an arbitrary $\xi \in \overline{X}$ such that

$$(5.7) \qquad \log_q a_j' + (\mu_j, \log_q \xi) \not\equiv 0 \mod \frac{2\pi i}{\log q} \mathbb{Z} ,$$

for any $j \in \{\pm 1, \ldots, \pm m\}$. Then the Jackson integral

$$(5.8) \qquad \tilde{\Phi}(u;\xi) = \int_{X\cdot\xi} \Phi\,\overline{\tilde{\omega}} ,$$

which depends not only on u but also on ξ satifies ($\mathcal{E}$). $\tilde{\Phi}(u;\xi)$ can be described as a linear combination of $\tilde{\Phi}(u;\xi^{(r)}) = \int_{\mathfrak{E}(\xi^{(r)})} \Phi\,\overline{\tilde{\omega}}$ in the following.

(5.9) $$\tilde{\Phi}(u;\xi) = \sum_{r=1}^{\kappa} p_r(u;\xi)\cdot\Phi(u;\xi^{(r)}) ,$$

where the connection coefficients $\{p_r(u;\xi)\}$ are quasi-meromorphic functions depending periodically on u and ξ. Hence these must be described in terms of <u>elliptic theta functions</u>.

A simplest example is the Ramanujan's formula :

$$\int_{[0,\xi\infty]_q} t^{\alpha} \frac{(t)_\infty}{(q^{\beta}t)_\infty} \frac{d_q t}{t} = (1-q)\sum_{n=-\infty}^{\infty} q^{n\alpha}\xi^{\alpha} \frac{(\xi q^n)_\infty}{(\xi q^{\beta+n})_\infty}$$

$$= \left(\frac{\xi}{q}\right)^{\alpha-2} \frac{\theta(q^{\beta+1})\theta(q^{\alpha+\beta-2}\xi)}{\theta(q^{\alpha+\beta-1})\theta(q^{\beta}\xi)} \int_{[0,1\infty]_q} t^{\alpha} \frac{(t)_\infty}{(q^{\beta}t)_\infty} \frac{d_q t}{t}$$

for the theta function $\theta(x) = (x)_\infty (q/x)_\infty (q)_\infty$ (see [A6]). This formula has been extended to one dimensional Jordan-Pochhammer case by K.Mimachi from the viewpoint of q-holonomic system (see [M2]).

6. <u>Examples</u>.

(i) $\Phi = \prod_{j=1}^{n} t_j^{\alpha_j} \prod_{0\le i<j\le n} \frac{(a'_{i,j}t_j/t_i)_\infty}{(a_{i,j}t_j/t_i)_\infty}$ where we put $t_0 = 1$.

In this case $m = \binom{n+1}{2}$ and $\mu_j(\chi) = \nu_k - \nu_\ell$ for some $k \neq \ell$. Here we put $\nu_0 = 0$. $[\mu_{j_1},\ldots,\mu_{j_n}] = \pm 1$, or 0 and the sum $\sum_{j=1}^{m} \mu_j(\chi_r)\mu_j(\chi_s)$ is equal to n or -1 according as $r = s$ or $r \neq s$. Therefore the number κ becomes $(n+1)^{n-1}$. This case has been investigated in more detail in [A2].

In particular the basic hypergeometric series

$$(6.1) \quad {}_{n+1}\varphi_n(x) = \sum_{m=0}^{\infty} \frac{(a_1)_m \cdots\cdots (a_{m+1})_m}{(b_1)_m \cdots\ (b_n)_m (q)_m} x^m$$

has an integral representation of Barnes type

$$(6.2) \quad {}_{n+1}\varphi_n(x) = \prod_{j=1}^{n} \frac{(a_j)_\infty (b_j/a_j)_\infty}{(1-q)(q)_\infty (b_j)_\infty} \int_{1>t_1>t_2>\cdots>t_n>0} t_1^{\alpha_1-\alpha_2} \cdots t_{n-1}^{\alpha_{n-1}-\alpha_n} t_n^{\alpha_n} \cdot$$

$$\frac{(qt_1)_\infty (qt_2/t_1)_\infty \cdots (qt_n/t_{n-1})_\infty (a_{n+1}t_n x)_\infty}{(b_1t_1/a_1)_\infty \cdots (b_nt_n/a_nt_{n-1})_\infty (t_n x)_\infty} \varpi$$

for $q^{\alpha_j} = a_j$, where this integration is done over the set such that $t_1,\ t_2/t_1, \ldots, t_n/t_{n-1} = 1, q, q^2, \ldots$. In this case from Theorem 1 we have $\dim H^n(\Omega^\cdot, \nabla) = n$, i.e., the multiplicity of $(\mathcal{E})$ is equal to n, which is a well-known fact.

(ii) $\Phi = \prod_{j=1}^{n} t_j^{\alpha_j} \frac{(a'_{0,j}t_j)_\infty}{(a_{0,j}t_j)_\infty} \cdot \prod_{1\le i<j\le n} \frac{(a'_{i,j}t_j/t_i)_\infty (b'_{i,j}t_it_j)_\infty}{(a_{i,j}t_j/t_i)_\infty (b_{i,j}t_it_j)_\infty}$.

In this case $m = n^2$ and $\sum_{j=1}^{m} \mu_j(\chi_r)\mu_j(\chi_s) = (2n-1)\,\delta_{r,s}$. Assp 1 ~ 5 are all satisfied so that $\kappa = \dim H^n(\Omega^\cdot, \nabla) = (2n-1)^n$.

In particular let us deal with the special case n = 2. We fix $\eta = (\eta_1, \eta_2) \in \check{X}$ such that $0 < \eta_1 < \eta_2$. Then the fan $F = \hat{F}$ is defined by four lines $\lambda_1 = 0$, $\lambda_| = 0$, $\lambda_1+\lambda_2 = 0$, $-\lambda_1+\lambda_2 = 0$. Therefore we can take as $Y^+ = \{\chi_1,\ \chi_2,\ \chi_1+\chi_2\ ,\ -\chi_1+\chi_2\}$. The nine critical points are given by $(q/a'_{01}, q/a'_{02})$, $(q/a'_{01}, q^2/(a'_{01}a'_{12}))$, $(1/a_{01}, qa_{01}/b'_{12})$, $(qa_{12}/a'_{02}, q/a'_{02})$, $(a'_{02}/b'_{12}),\ q/a'_{02})$, $\pm(\sqrt{a'_{12}/b'_{12}}\ , q/\sqrt{a'_{12}b'_{12}}\ )$, and $\pm(q^{1/2}\sqrt{a'_{12}/b'_{12}}\ ,\ q^{3/2}/\sqrt{a'_{12}b'_{12}}\ )$. The Newton polyhedra $\Delta(b^{\pm}_{\chi_1})$, $\Delta(b^{\pm}_{\chi_2})$, $\Delta(b^{\pm}_{\chi_1+\chi_2})$, $\Delta(b^{\pm}_{-\chi_1+\chi_2})$ are given as in Figure 1. The convex set $\mathfrak{K}$ in Figure 2 satisfies the property of Lemma 4.11, i.e., $\#_u$ is surjective for $\mathbf{C}\langle\mathfrak{K}\rangle$. It is smaller than $\mathfrak{K}_0$ in Lemma 4.10. $\mathfrak{h}$ is spanned by the monomials $t_1^{\eta_1}t_2^{\eta_2}$ for $0 \le \eta_1, \eta_2 \le 2$.

Figure 1

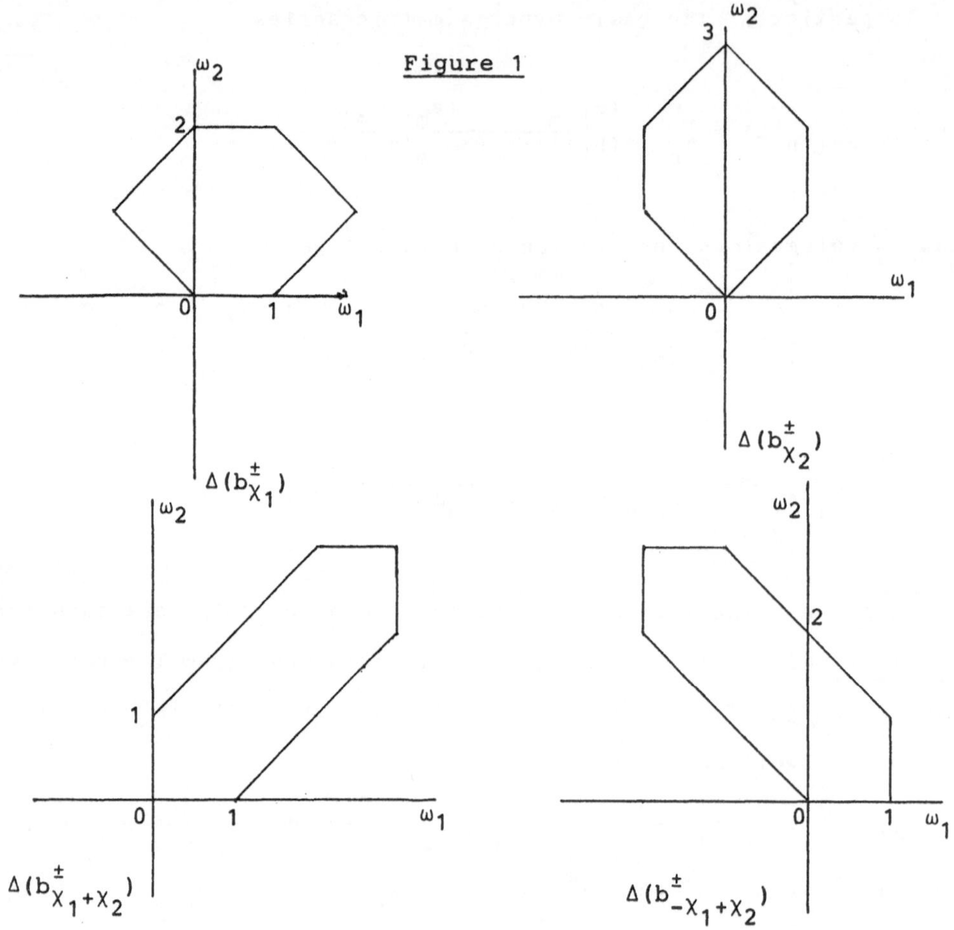

Figure 2

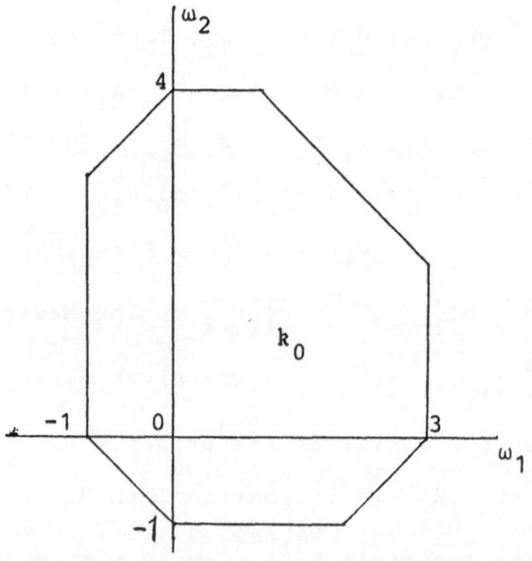

References.

[A1] Andrews, G., Problems and prospects for basic hypergeometric functions, The theory and applications of special functions, (R.Askey,ed.), Academic Press, New York, 1975, 191-224.

[A2] Aomoto, K., Les équations aux différences linéares et les intégrales des fonctions multiformes, J. of Fac. Sci. Univ. of Tokyo, 22(1975), 271-297.

[A3] ————, A note on holonomic q-difference systems, Algebraic analysis I,(M.Kashiwara and T.Kawai, ed.), 1988, 25-28.

[A4] ————, Finiteness of a cohomology associated with certain Jackson integrals, to appear in Tohoku J. of Math.

[A5] ________, q-analogue of de Rham cohomology associated with Jackson integrals, I,II, to appear in Proc. of Japan Academy.

[A6] Askey, R., Ramanujan's extensions of the gamma and béta functions, Amer. Math. Monthly, 87(1980), 346-359.

[B1] Bellardinelli, G., Fonctions hypergéometriques de plusieurs variables et resolutions analytiques des équations algébriques générales, Gauthiers-Villars, 1960.

[B2] Bernshtein, D.N., The number of roots of a system of equations, Funct. Anal. and Its Appli., 9(1975), 183-185.

[D1] Danilov, V.I., The geometry of toric vaieties, Russ. Math. Surveys, 33(1978), 97-154.

[D2] Dantzig, G.B., Linear programming and extensions, Princeton Univ. Press, 1963.

[D3] Deligne, P., Equations différentielles à points réguliers singuliers, Lec. Notes in Math., 163, Springer, 1970.

[G1] Gasper, G. and M. Rahman, Basic hypergeometric series, Cambridge Univ. Press, 1990.

[G2] Gustafson, R.A., A generalization of Selberg's beta integral, Bull. A.M.S., 22(1990), 97-105.

[K1] Katz,N. and T. Oda, On the differentiation of the de Rham cohomology with respect to parameters, J. Math. Kyoto Univ., 8(1968), 199-213.

[K2] Kempf, G. et al, Toroidal embeddings I, Lec. Note in Math., 339, Springer, 1973.

[K3] Khovanskii, A.G., Newton polyhedra and toroidal varieties, Funct. Anal. and Its Appli., 11(1977), 56-57.

[K4] Kita, M. and M. Noumi, On the structure of cohomolgy groups attached to the integral of certain many-valued analytic functions, Japanese J. of Math., 9(1983), 113-157.

[K5] Kouchnirenko, A.G., Polyèdres de Newton et nombres de Milnor, Invent. math., 32(1976), 1-31.

[M1] Milne, M.C., A q-analog of the Gauss summation theorem for hypergeometric series in U(n), Ad. in Math. 72(1988), 59-131.

[M2] Mimachi,K., Connection problem in holonomic q-difference system associated with a Jackson integral of Jordan-Pochhammer type, Nagoya Math. J., 116(1989), 149-161.

[O] Oda,T., Convex bodies and algebraic geometry - An introduction to the theory of toric varieties, Ergebnisse der Math., Springer, 1988.

[S1] Sato, M., Theory of prehomogeneous vector spaces, Algebraic part, The English version of Sato's lecture from Shintani's note, translated by M.Muro, to appear in Nagoya Math. J..

[S2] Slater, L.J., Generalized hypergeometric functions, Cambridge Univ. Press, 1966.

[S3] Smale, S., Algorithms for solving equations, Proc. of Inter. Congress of Math., Berkley, Cal., 1986.

Affine Extensions of Knizhnik-Zamolodchikov Equations and Lusztig's Isomorphisms

by

Ivan CHEREDNIK*
Research Institute for Mathematical Sciences
Kyoto 606, JAPAN
Mathematical Sciences Research Institute
Berkeley, California 94720, USA

Abstract

Some affine analogues of the Knizhnik-Zamolodchikov equations from the two-dimensional conformal field theory are discussed for arbitrary root systems. We demonstrate that they give some versions of Lusztig's isomorphisms from affine Hecke algebras to their degenerate variants.

Section 1. $\mathbf{S}_n$-invariant KZ equations.

Section 2. Trigonometric KZ equations for root systems.

Section 3. The monodromy and Lusztig's isomorphisms.

---

* On leave from A. N. Belozersky Lab., Bldg "A" Moscow State University, Moscow 119899, USSR, research at MSRI supported in part by NSF grant ♯ DMS 8505550.

In this paper we consider differential equations generalizing the affine Knizhnik–Zamolodchikov equation (KZ) from [1] for arbitrary root systems. These equations are also "trigonometric" extensions of the "rational" non-affine KZ equations from [2] and are particular cases of the differential-difference equations from [3]. Their monodromy representations result in Lusztig's type isomorphisms [4] from affine Hecke algebras to the corresponding degenerate algebras and are closely connected with a certain generalization of the Lusztig–Lascoux–Schützenberger operations (see [3]). We will introduce our equations as natural variants of the equation from [1] (in [3] they were obtained on the base of the LLS operations). The main points of these notes are the definitions, the proof of the consistency of arising equations and a calculation of the monodromy representations.

This paper was mainly prepared during my stay at RIMS (Kyoto University) and finished for my stay at MSRI (Berkeley). It is partially based on my talk at the Hayashibara Forum 90 on special functions. I am grateful for the kind invitation and for hospitality. I also express appreciation to G. Heckman, V. Jones, A. Matsuo, T. Miwa, to other my colleagues for useful discussions and to the secretaries in RIMS for the help in preparing the manuscript for publication.

1. $\mathbf{S}_n$-invariant KZ equations.

Let $\mathbf{S}_n$ be the symmetric group generated by the adjacent transpositions $s_i = (i \quad i+1)$ ((ij) permutes $1 \leq i \neq j \leq n$, $n > 1$). The system

$$\frac{\partial \Phi}{\partial z_i} = \kappa \sum_{j \neq i} \frac{(ij)}{z_i - z_j} \Phi, \quad 1 \leq i, j \leq n, \tag{1}$$

is called the Knizhnik–Zamolodchikov equation. Here $\kappa \in \mathbf{C}$, $z = (z_1, \ldots, z_n) \in \mathbf{C}^n$, $\Phi(z)$ takes its values in the group algebra $\mathbf{C}[\mathbf{S}_n] = \oplus_{w \in \mathbf{S}_n} \mathbf{C}w$. We follow [1] and define the affine version of (1) as follows:

$$\frac{\partial \Phi}{\partial z_i} = \kappa \sum_{j \neq i} \frac{(ij)}{z_i - z_j} \Phi + \frac{a_i}{z_i} \Phi, \quad 1 \leq i, j \leq n. \tag{2}$$

This system is consistent iff

$$[a_j, a_i + \kappa(ij)] = 0 = \kappa[(ij), a_i + a_j], \quad 1 \leq i \neq j \leq n, \tag{3}$$

and is $\mathbf{S}_n$-invariant iff

$$w a_i w^{-1} = a_{w(i)} \quad \text{for} \quad w \in \mathbf{S}_n, \quad 1 \leq i \leq n. \tag{4}$$

To be more precise, let us introduce the algebra $\mathfrak{A}$ by adding the elements $a_1, \ldots, a_n$ with the defining relations (3,4) to $\mathbf{C}[\mathbf{S}_n]$. A direct consideration (due to Kohno) gives

Proposition 1. *If the values of Φ are from $\mathfrak{A}$, then (2) is consistent for arbitrary κ and $\mathbf{S}_n$-invariant with respect to the following action of $\mathbf{S}_n \ni w$ on $\mathfrak{A}$-valued functions $f(z)$:*

$$^{w} f(z) = w f(w^{-1}(z)) w^{-1}, \tag{5}$$

where $\mathbf{S}_n$ acts on $\mathbf{C}^n \ni z$ in the usual manner. Conversely, the cross-derivative integrability conditions for (2) and the $\mathbf{S}_n$-invariance are respectively equivalent to (3) and (4). □

It is convenient to rewrite (3,4) as follows.

Proposition 2. *a) The elements*

$$b_i = a_i + \kappa \sum_{n\geq j>i} (ij), \quad 1 \leq i \leq n, \tag{6}$$

satisfy the Murphy–Drinfeld relations

$$s_i b_i - b_{i+1} s_i = \kappa = b_i s_i - s_i b_{i+1}, [s_i, b_j] = 0, \tag{7}$$

$$[b_i, b_k] = 0, \quad 1 \leq i < n, 1 \leq j, k \leq n, j \neq i, i+1. \tag{8}$$

Vise versa, $\{b_i\}$ generate $\mathbb{A}$ over $\mathbb{C}[\mathbf{S}_n]$ and (7,8) are the defining relations for $\mathbb{A}$ as well as (3,4).

b) The same holds true for the elements

$$c_i = \sum_{j\leq i} b_j, \quad 1 \leq i, j \leq n, c_0 = 0, \tag{9}$$

with the relations

$$s_i c_i - \hat{c}_i s_i = \kappa = c_i s_i - s_i \hat{c}_i, [s_i, c_j] = 0, \tag{10}$$

$$[c_i, c_k] = 0, \hat{c}_i = c_{i-1} + c_{i+1} - c_i, \quad 1 \leq j, k \leq n, j \neq i < n. \tag{11}$$

Proof. Formulas (7) result immediately from (3,4). As for (11), we will use the substitution

$$v_k = (z_k - z_{k+1})(z_{k-1} - z_k)^{-1}, \quad 1 \leq k < n, v_1 = z_1 - z_2, v_n = z_n(z_{n-1} - z_n)^{-1}. \tag{12}$$

Then system (2) will turn into

$$\partial\Phi/\partial v_i = v_i^{-1}(c_n - c_{i-1})\Phi + P_i(v)\Phi, \quad 1 \leq i \leq n, \tag{13}$$

where $P_1, \ldots, P_n$ are regular $\mathbb{A}$-valued functions of $v = (v_1, \ldots, v_n)$ in a neighbourhood of $v = 0$. Hence $[c_i, c_k] = 0$ and $[b_i, b_k] = 0$ for all i, k. (Use the compatability of equations (13)).

As a corrolary we obtain that $\{b_i^0\}$ for $a_1 = \cdots = a_n = 0$ satisfy (7,8). We will use these identities (closely connected with the so-called Young bases for $\mathbf{S}_n$) to deduce (3,4) from (7,8). One has

$$s_i a_i - a_{i+1} s_i = s_i(b_i - b_i^0) - (b_{i+1} - b_{i+1}^0)s_i = 0.$$

Hence we arrive at (4), since $[s_i, a_j] = 0$ for $j \neq i, i+1$ (see (7)). One may check the first relation from (3) (the second one results from (4)) only for $i = n-1, j = n$ because of the $\mathbf{S}_n$-invariance:

$$[a_n, a_{n-1} + \kappa s_{n-1}] = [b_n - b_n^0, b_{n-1} - b_{n-1}^0 + \kappa s_{n-1}]$$
$$= [b_n, b_{n-1}] = 0 \quad (b_n^0 = 0, b_{n-1}^0 = \kappa s_{n-1}).$$

The equivalence of (7,8) and (10,11) is evident □

Our first generalization of (1), (2) for other (only classical) root systems is as follows. Let $\widetilde{\mathbf{S}}_n$ be the Weyl group of type B_n, C_n. It is generated by $\mathbf{S}_n$ and the pairwise commutative elements $t_1, \ldots, t_n$ with the relations

$$w t_i w^{-1} = t_{w(i)}, \quad t_i^2 = 1, \quad w \in \mathbf{S}_n, \quad 1 \leq i \leq n.$$

We put

$$(\widetilde{ij}) = t_i t_j (ij), \quad \beta_i = (1 + t_i t_{i+1})/2, \quad 1 \leq i \neq j \leq n,$$

and define $\widetilde{\mathbb{A}}$ as the algebraic span of $\mathbb{C}[\widetilde{\mathbf{S}}_n]$ and $\tilde{b}_1, \ldots, \tilde{b}_n$ satisfying the following conditions:

$$s_i \tilde{b}_i - \tilde{b}_{i+1} s_i = \kappa \beta_i = \tilde{b}_i s_i - s_i \tilde{b}_{i+1}, \quad [s_i, \tilde{b}_j] = 0,$$
$$[\tilde{b}_i, \tilde{b}_k] = 0 = [\tilde{b}_i, t_k], \quad 1 \leq i < n, 1 \leq j, k \leq n, j \neq i, i+1.$$

Proposition 3. *The system*

$$\frac{\partial \Phi}{\partial z_i} \Phi^{-1} = \kappa \left(\sum_{j \neq i} \frac{(ij)}{z_i - z_j} + \frac{(\widetilde{ij})}{z_i + z_j} \right) + \frac{\tilde{a}_i}{z_i}, \quad 1 \leq i, j \leq n, \tag{14}$$

is consistent for any κ if

$$\tilde{b}_i \stackrel{\text{def}}{=} \tilde{a}_i + \sum_{n \geq j > i} \beta_i (ij), \quad 1 \leq i \leq n$$

satisfy the above relations. In this case (14) is $\widetilde{\mathbf{S}}_n$-invariant in the sense of formula (5) with $t_i(z_k) = (1 - 2\delta_{ik}) z_k$, where $1 \leq i, k \leq n$, δ_{ik} is the Kronecker delta. □

This system for $\{\tilde{a}_i = \chi t_i\}$, $\chi \in \mathbb{C}$ is a particular case of the generalized KZ equation from [2]. If $\chi = 1, 1/2, 0$ it can be connected with the root systems of type B_n, C_n, D_n

because the denominators of the *r.h.s.* of (14) are nothing else but the corresponding sets of positive roots.

Let us denote by $\widetilde{Z}, Z$ the subset of $\mathbf{C}^n$ where the *r.h.s.* of (14) is regular and the analogous open subset for (2). The monodromy gives a homomorphism from $\pi_1(\overline{Z})$ into the group of invertible elements $\widetilde{\mathsf{A}}^*$ or A^* for $\widetilde{Z}/\widetilde{\mathsf{S}}_n = \overline{Z} = Z/\mathsf{S}_n$. More precisely, given a point $z^0 \in \mathbf{R}^n$ $(z_1^0 > z_2^0 > \cdots z_n^0 > 0)$ for each $1 \le j < n$ we introduce the element $\overline{\sigma}_j \in \pi_1(\overline{Z}, \overline{z}^0)$ corresponding to the following path $z(\psi) = (z(\psi)_k)$:

$$z(\psi)_j \pm z(\psi)_{j+1} = (z_j^0 \pm z_{j+1}^0)\exp(\pi i \psi), \quad 0 \le \psi \le 1,$$
$$z(\psi)_k = z_k^0 \quad \text{for} \quad k \ne j, j+1, 1 \le k \le n,$$

connecting z^0 with $\tilde{s}_j z^0$ (sign $= +$) or $s_j z^0$ (sign $= -$) respectively in $\widetilde{Z}$ or Z. Here $\overline{z}^0$ is the projection of z^0 on $\overline{Z}$. The image of the path $z(\psi)$ with the constant coordinates $z(\psi)_k = z_k^0$ for $1 \le k < n$ and

$$z(\psi)_n = z_n^0 \exp(\pi i \psi) \quad \text{for} \quad \widetilde{Z} \quad \text{or} \quad z(\psi)_n = z_n^0 \exp(2\pi i \psi) \quad \text{for} \quad Z$$

will be denoted by $\overline{\sigma}_n \in \pi_1(\overline{Z}, \overline{z}^0)$.

Then the monodromy elements T_k $(1 \le k \le n)$ for $\{\overline{\sigma}_k\}$ (see [1] or Sec.3) satisfy the relations

$$T_j T_{j+1} T_j = T_{j+1} T_j T_{j+1}, [T_j, T_k] = 0, \quad 1 \le j < n-1, k - j \ne \pm 1,$$
$$(T_n T_{n-1})^2 = (T_{n-1} T_n)^2, (T_j - q)(T_j + q^{-1}) = 0, \quad 1 \le j < n,$$

where $q = \exp(\pi i \kappa)$. The algebra H generated by $\{T_k\}$ with these relations is called the *affine Hecke algebra of type* GL_n.

We will formulate the main result from [1] without going into detail. The above monodromy elements depend on the choice of Φ. Given generic κ, let Φ be the only solution of (13) with the asymptotic behaviour:

$$\Phi z_n^{c_n} \prod_{i=1}^{n} v_i^{-c_i} = 1 + o(|v|), \quad v \to 0,$$

or the analogous solution for the counterpart of (13) with $\tilde{c}_i = 2\sum_{j\leq i}\tilde{b}_j$ obtained from (14) after the same substitution (12). Then for $q = \exp(\pi i\kappa)$ and a certain scalar function $g(u), u \in \mathbb{C}$,

$$\begin{aligned} &T_n = \exp(2\pi i\kappa\tilde{b}_n) \quad \text{or} \quad T_n = \exp(2\pi i\kappa b_n) \quad \text{for (14) or (2)}, \\ &T_k + (q - q^{-1})(\exp(-2\pi i b'_{k,k+1}) - 1)^{-1} \\ &\qquad = g(b'_{k,k+1})(s_k - \beta_k(\tilde{b}_{k,k+1})^{-1}), \quad 1 \leq k < n \end{aligned} \tag{15}$$

where $b'_{k,k+1} = b'_k - b'_{k+1}$, $b'_k = \tilde{b}_k + t_k/4$ for (14) or $\tilde{b}_k = b'_k = b_k$, $\beta_k = 1$ for (2), $1 \leq k \leq n$.

These formulas produce an isomorphism of $\widetilde{\mathrm{A}}$ (or A) and the affine Hecke algebra H after a suitable completion. Similar isomorphisms (with some other g) were obtained by Lusztig in [4] for arbitrary root systems (see also [3]). Our next aim is to get the corresponding generalization of (2) to find a monodromy interpretation of Lusztig's isomorphisms.

2. Trigonometric KZ Equations for Root Systems.

Let us first transform (2) into some trigonometric equation. We multiply it by z_i and use the following new variables:

$$z_n = u_n, z_{n-1} = \exp(u_n + u_{n-1}), \ldots, z_1 = \exp(u_n + \cdots + u_1). \tag{16}$$

One obtains the system

$$\partial\Phi/\partial u_k = \kappa\Big(\sum_{i\leq k<j}(ij)(z_i z_j^{-1} - 1)^{-1}\Big)\Phi + c_k\Phi, \quad 1 \leq k \leq n, \tag{17}$$

where c_k are from (9), $z_i z_j^{-1} = \exp(u_i + \cdots + u_{j-1})$. We replace c_k by $c'_k = c_k - (k/n)c_n$. This corresponds to the substitution of $\Phi' \exp\Big(\sum_{k=1}^n k u_k c_k/n\Big)$ for Φ because $c_n = \sum_{i=1}^n b_i$ is from the centre of A. One arrives at the system for $\Phi'(u_1, \ldots, u_{n-1})$:

$$\partial\Phi'/\partial u_k = \Big(\kappa \sum_{\alpha_k\in\alpha}(\exp(u_\alpha) - 1)^{-1}s_\alpha + c'_k)\Phi', \quad 1 \leq k \leq n-1, \tag{18}$$

where α runs over the set of positive roots of type A_{n-1} with the simple roots $\alpha_1, \ldots, \alpha_{n-1}$, $u_\alpha = \exp(u_i + \cdots + u_{j-1})$ for $\alpha = \alpha_i + \cdots + \alpha_{j-1}$. We have omitted u_n because $\partial\Phi'/\partial u_n = 0$ and all the u_α do not depend on u_n.

Note that c'_i satisfy relations (10,11) with $c'_n = 0$. The corresponding algebra $\mathbb{A}'$ generated by $\mathbb{C}[\mathbf{S}_n]$ and $\{c'_i, 1 \le i < n\}$ is called the *degenerate Hecke algebra of type* A_{n-1}. We will generalize (18) for the case of an arbitrary root system $R = \{\alpha\} \subset \mathbb{R}^n$ of type $A_n, B_n, \ldots, G_2$.

Given R, let s_α be the orthogonal reflections in the hyperplanes $(\alpha, u) = 0$ with respect to the canonical euclidean form $(\ ,\ )$ on $\mathbb{R}^n \ni u$, $\{\alpha_1, \ldots, \alpha_n\}$ the simple roots, R_+ the set of positive ($\alpha > 0$) roots, W the Weyl group generated by the reflections $s_i = s_{\alpha_i}$ $(1 \le i \le n)$.

We assume $(\ ,\ )$ and the action of W to be extended $\mathbb{C}$-linearly to $u \in \mathbb{C}^n$; $u_\alpha \overset{\text{def}}{=} (u, \alpha)$, $u_i = (u, \alpha_i)$. Then $\{u_i\}$ are coordinates of $\mathbb{C}^n$ and one can define the derivatives $\partial/\partial u_i$. In particular, $\partial u_\alpha/\partial u_i \overset{\text{def}}{=} \nu_i(\alpha)$ is the multiplicity of α_i in $\alpha \in R$.

We fix $\kappa, \kappa' \in \mathbb{C}$ and put $\kappa_\alpha = \kappa$ or κ' respectively for short or long α; $\kappa_i = \kappa_{\alpha_i}$.

Definition 4. *The degenerate affine Hecke algebra* $\mathbb{A}$ *(cf.[4]) is the extension of the group algebra* $\mathbb{C}[W]$ *by the elements* $d_1, \ldots, d_k$ *with the following relations*

$$s_i d_j - d_j^{(i)} s_i = \delta_{ij}\kappa_i, \quad d_j^{(i)} \overset{\text{def}}{=} \sum_{k=1}^{n} \nu_j(s_i(\alpha_k)) d_k, \tag{19}$$

$$[d_i, d_j] = 0, \quad 1 \le i, j \le n. \tag{20}$$

□

Here $\nu_i(s_\alpha(\alpha_k)) = \nu_i(\alpha_k) - 2(\alpha, \alpha_k)(\alpha, \alpha)^{-1}\nu_i(\alpha)$ for $\alpha \in R$, $d_j^{(i)} = d_j$ and $[s_i, d_j] = 0$ if $j \ne i$. The formulas for $d_j^{(i)}$ correspond to the action of s_i on the dual fundamental weights δ_j satisfying the conditions $(\alpha_i, \delta_j) = \delta_{ij}$:

$$s_i(\delta_j) = \sum_{k=1}^{n} (s_i(\delta_j), \alpha_k)\delta_k = \sum_{k=1}^{n} (\delta_j, s_i(\alpha_k))\delta_k = \sum_{k=1}^{n} \nu_j(s_i(\alpha_k))\delta_k.$$

Theorem 5[3]. *The following system for* $\mathbb{A}$*-valued function* $\Phi(u)$ *is consistent*

$$\partial\Phi/\partial u_i = D_i\Phi, \quad D_i \overset{\text{def}}{=} \sum_{\alpha>0} \kappa_\alpha \nu_i(\alpha)(\exp(u_\alpha) - 1)^{-1} s_\alpha + d_i, \quad \alpha \in R, \tag{21}$$

and is W-invariant:

$$ {}^{w}\Delta_i = \sum_{k=1}^{n} \nu_i(w^{-1}(\alpha_k))\Delta_k, \quad {}^{w}(\partial/\partial u_i)({}^{w}f) \stackrel{\text{def}}{=} {}^{w}(\partial f/\partial u_i) \tag{22} $$

where $1 \le i \le n$, ${}^{w}f$ is like in (5), $\Delta_i = \partial/\partial u_i - D_i$.

Proof. First $\partial/\partial u_i(\nu_j(\alpha)(\exp(u_\alpha) - 1)^{-1}) = -\nu_i(\alpha)\nu_j(\alpha)\exp(u_\alpha)(\exp(u_\alpha) - 1)^{-2}$ $= \partial/\partial u_j(\nu_i(\alpha)(\exp(u_\alpha) - 1)^{-1})$. Hence the consistency results from the pure algebraic relations $[D_i, D_j] = 0$. The same holds for (22). It is enough to check the W-invariance for D_i only, because

$$ {}^{s_j}(\partial/\partial u_i)(u_k) = {}^{s_j}(\partial({}^{s_j}u_k)/\partial u_i) = \nu_i(s_j(\alpha_k)). $$

Lemma 6. *a) Let $\{\rho_\alpha, \alpha \in R\}$ be a set of elements in an extension $\tilde{\mathbb{A}}$ of $\mathbb{C}[W]$ with the property*

$$ \rho_{w(\alpha)} = w\rho_\alpha w^{-1}, \quad w \in W, \quad \rho_\alpha + \rho_{-\alpha} = \kappa_\alpha s_\alpha. $$

Then $\tilde{d}_i \stackrel{\text{def}}{=} \sum_{\alpha>0} \nu_i(\alpha)\rho_\alpha$ satisfy (19).

b) Moreover, if $[\rho'_1, \rho'_2] = 0$ for every two-dimensional subsystem of roots $R' \subset R$ (ρ' is the restriction of ρ on R') then $[\tilde{d}_i, \tilde{d}_j] = 0 = [\tilde{D}_i, \tilde{D}_j]$ for arbitrary $1 \le i, j \le n$ and $\tilde{D}_i$ from (21) with $\tilde{d}_i$ instead of d_i.

Proof. One has (see [3], Cor. 3.6):

$$ s_i\tilde{d}_i s_i - (\rho_i + s_i\rho_i s_i) = \sum_{\alpha>0} \nu_i(s_i(\alpha))\rho_\alpha. $$

Similarly, $[s_i, \tilde{d}_j] = 0$ for $j \ne i$. Here we have used that $s_i(R_+\backslash\alpha_i) \subset R_+$. As for b), see [3], Prop. 3.2 and [1,2]. □

Let us apply a) to prove the W-invariance of D_i for $\tilde{\mathbb{A}} \ni \mathbb{C}[W]$, $\rho_\alpha = -\kappa_\alpha s_\alpha/2$. Then $\overline{d}_i = d_i + \rho_i$ satisfy the relations $s_j\overline{d}_i s_j^{-1} = \overline{d}_i^{(j)}$ for all j. One has:

$$ D_i = \overline{d}_i + \sum_{\alpha\in R} \kappa_\alpha \nu_i(\alpha) cth(u_\alpha)s_\alpha/4, \quad 1 \le i \le n, $$

where $cth(v) \stackrel{\mathrm{def}}{=} (\exp v + 1)/(\exp(v) - 1) = -cth(-v)$. The required property resutls from the formulas ${}^{w}u_{\alpha} = u_{w(\alpha)}$, $ws_{\alpha}w^{-1} = s_{w(\alpha)}$, $w \in W$, $\alpha \in R$.

It was proven in [3] by means of the reduction to the cases A_2, B_2, G_2 that for some $\tilde{\mathbb{A}} \supset \mathbb{C}[W]$ there exists a set $\{\rho_{\alpha}\}$ satisfying both conditions a) and b) of the lemma. Then $[\tilde{D}_i, \tilde{D}_j] = 0$ for the corresponding $\{\tilde{d}_i\}$. But this commutativity does not depend on a concrete choice of $\mathbb{A}$ and $\{\tilde{d}_i\}$ with conditions (19, 20) because of the following

Lemma 7. *For $\{d_i, D_i, 1 \le i \le n\}$ from Theorem 5 the commutators $[D_i, D_j]$ take values in $\mathbb{C}[W]$.*

Proof. It results from (19) that

$$s_{\alpha}d_i - \sum_{k=1}^{n} \nu_i(s_{\alpha}(\alpha_k))d_k s_{\alpha} \in \mathbb{C}[W], \quad \alpha \in R, \quad 1 \le i \le n.$$

Hence

$$\begin{aligned}
&(\nu_i(\alpha)s_{\alpha})d_j - (\nu_j(\alpha)s_{\alpha})d_i \bmod \mathbb{C}[W] \\
=&\sum_{k=1}^{n}(\nu_i(\alpha)\nu_j(s_{\alpha}(\alpha_k)) - \nu_j(\alpha)v_i(s_{\alpha}(\alpha_k))d_k s_{\alpha} \\
=&\nu_i(\alpha)\sum_{k=1}^{n}(\nu_j(\alpha_k) - 2(\alpha_k, \alpha)(\alpha, \alpha)^{-1}\nu_j(\alpha))d_k s_{\alpha} \\
&- v_j(\alpha)\sum_{k=1}^{n}(\nu_i(\alpha_k) - 2(\alpha_k, \alpha)(\alpha, \alpha)^{-1}\nu_i(\alpha))d_k s_{\alpha} \\
=&(\nu_i(\alpha)d_j - \nu_j(\alpha)d_i)s_{\alpha} = d_j(\nu_i(\alpha)s_{\alpha}) - d_i(\nu_j(\alpha)s_{\alpha}).
\end{aligned}$$

This gives the required statement since $[d_i, d_j] = 0$. □□

As a concluding remark to this section, we will formulate the generalization from [3] of the above theorem. Given arbitrary $\mu, \mu' \in \mathbb{C}$, we claim that the following system is consistent $(1 \le i \le n)$:

$$\partial\Phi/\partial u_i = \sum_{a>0} \kappa_{\alpha}\nu_i(\alpha)(\exp(u_{\alpha}) - 1)^{-1}(1 + \mu_{\alpha}\sigma_{\alpha})s_{\alpha} + d_i.$$

Here $\mu_\alpha = \mu, \mu'$ respectively for short or long $\alpha \in R$, $\{\sigma_\alpha\}$ are new elements with the same relations as those for s_α and the additional ones:

$$[\sigma_\alpha, s_\beta] = 0 = [\sigma_\alpha, d_i], \quad \sigma_\alpha u_\beta \sigma_\alpha = u_{s_\alpha(\beta)}, \quad \alpha, \beta \in R, \quad 1 \le i \le n.$$

This system generalizes Heckman's trigonometric analogues of the so-called Dunkl operators (see [3] for details). The proof is parallel to that of Theorem 5.

3. The monodromy and Lusztig's isomorphisms.

Now we will describe how to get Lusztig's type isomorphisms with the aid of system (21). Put

$$v_\alpha = \exp(-u_\alpha), \quad v_i = \exp(-v_i), \quad v = (v_1, \cdots, v_n).$$

Then for $1 \le i \le n,\ 1 \le k \le n,\ \alpha \in R$ (cf. (13))

$$\frac{\partial \Phi}{\partial v_i} + (\frac{d_i}{v_i} + P_i(v))\Phi = 0, \quad P_i = \sum_{\alpha > 0} \frac{\kappa_\alpha \nu_i(\alpha) \prod v_k^{\nu_k(\alpha)} s_\alpha}{v_i(1 - \prod v_k^{\nu_k(\alpha)})}. \tag{23}$$

Here P_i are regular $\mathbb{A}$-valued functions in a neighbourhood of $v = 0$. Hence, there exists for generic κ, κ' the unique solution $\Phi(v)$ of (23) with the following normalizing condition:

$$\Phi(v) \prod_{i=1}^{n} v_i^{d_i} = 1 + o(|v|), \quad v \to 0. \tag{24}$$

Later we will use only this Φ.

System (23) is well-defined in

$$V = \left\{ v \in \mathbf{C}^n, \quad \prod_{i=1}^{n} v_i \prod_{\alpha > 0} (v_\alpha - 1) \neq 0 \right\}.$$

Therefore Φ can be considered as a multi-valued functions on V. Let us fix a point $u^0 \in \mathbf{R}_+^n$ which is close to $u = \infty$ ($u_i^0 > M$ for some big M) and has its image v^0 in V. We denote the image of v^0 in $\overline{V} \stackrel{\text{def}}{=} V/W$ by $\overline{v}^0$ and the fundamental group $\pi_1(\overline{V}, \overline{v}^0)$ by π_1. The latter is generated by (the classes of) the images $\{\overline{\sigma}_j, \overline{\tau}_j, 1 \le j \le n\}$ in π_1 of the paths

$$\sigma_j(\psi) = u^0 + (u_j^0 (\alpha_j, \alpha_j)^{-1} (\cos(\pi\psi) - 1) + i(\alpha_j, \alpha_j)^{-1/2} \sin(\pi\psi))\alpha_j,$$

$$\tau_j(\psi) = (u_1^0, \ldots, u_{j-1}^0, u_j^0 + 2\pi i\psi, u_{j+1}^0, \ldots, u_n^0),$$

where $0 \leq \psi \leq 1,\ i = \sqrt{-1}$.

Given $\overline{\gamma} \in \pi_1$ one can introduce the *monodromy "matrix"*

$$T_{\overline{\gamma}} = \Phi(w_{\gamma}^{-1}(v))^{-1} w_{\gamma}^{-1} \Phi_{\gamma}(v),$$

where the pullback $\gamma \in V$ of $\overline{\gamma}$ connects v^0 with $w_\gamma(v^0), w_\gamma \in W$, and Φ_γ is the analytical continuation of Φ (see (24)) from a neighbourhood of v^0 to some neighbourhood of $w_j(v^0)$ along γ. Then $T_{\overline{\gamma}} \in \mathsf{A}^*, T_{\overline{\gamma}_1 \circ \overline{\gamma}_2} = T_{\overline{\gamma}_1} T_{\overline{\gamma}_2}$ for $\overline{\gamma}_1, \overline{\gamma}_2 \in \pi_1$, if we assume $\overline{\gamma}_1$ to be the first path in the product $\overline{\gamma}_1 \circ \overline{\gamma}_2$. We will denote $T_{\overline{\sigma}_j}, T_{\overline{\tau}_j}$ by T_j, Y_j.

Proposition 8. a) $Y_j = \exp(2\pi i d_j), 1 \leq j \leq n$.

b) $(T_j - q_j)(T_j + q_j^{-1}) = 0, \quad q_j \stackrel{\text{def}}{=} \exp(\pi i \kappa_j)$. □

The quotient-algebra H of $\mathbb{C}[\pi_1]$ by relations b) is called the *affine Hecke algebra* (see e.g. [4, 5] for the pure algebraic and more general definition). It is generated by $T_1, \ldots, T_n, Y_1, \ldots, Y_n$ with the following relations

$$T_i \hat{Y}_i T_i = Y_i, \quad [T_i, Y_j] = 0 = [Y_i, Y_j], \quad 1 \leq i \neq j \leq n,$$

where $\hat{Y}_i = \prod_{k=1}^{w} Y_k^{\nu_i(s_i(\alpha_k))}$ is a group version of $d_i^{(i)}$ from (19), and the usual non-affine Hecke relations:

$$(T_i - q_i)(T_i - q_i^{-1}) = 0, \quad T_i T_j T_i T_j \cdots = T_j T_i T_j T_i \cdots$$

with the number of factors on both sides of the latter equal to $m_{ij} = \mathrm{ord}(s_i s_j)$. We note that the definition from [4] or [3] is somewhat different.

The map $\tau : \overline{\gamma} \to T_{\overline{\gamma}}$ gives the homomorphism $\tau : \mathsf{H} \to \mathsf{A}$, which is an isomorphism for generic κ, κ'. To prove this statement it is sufficient to check that generic irreducible representations of A remain irreducible as H-modlues with respect to τ. It results from the explit formulas for $\{T_j\}$ which will be discussed now.

Lemma 9. *Let $T = T(S, A)$ be T_1 for the root system A_1, i.e. $T = T_{\overline{\sigma}_1}$ for the equation*

$$\partial\Phi_0/\partial v + (v^{-1}A/2 + \kappa(1-v)^{-1}S)\Phi_0 = 0, \tag{25}$$

where $S^2 = 1, SA + AS = 2\kappa$ ($S = s_1, A = 2d_1$). Then (cf. (15))

$$T + (q - q^{-1})(\exp(-2\pi i A) - 1)^{-1} = g(A)(S - \kappa A^{-1}), \tag{26}$$

where $q = \exp(\pi i \kappa)$, g is a certain scalar function of one variable.

Proof. The r.h.s F of (26) satisfies the relation $FAF^{-1} = -A$ for arbitrary g. Hence, $FY^{\pm 1}F^{-1} = Y^{\mp 1}$, where $Y = Y_1 = \exp(\pi i A)$ (see Prop. 8). The l.h.s of (26) has the same property since $TY^{-1}T = Y$ in π_1. Therefore we have (26) for suitable g. Indeed, the centralizer $\{X \in \mathfrak{A}, XA = AX\}$ of A in $\mathbb{C}[S, A] = \mathfrak{A}$ coincides with the subalgebra generated by A. □

The algebra $\mathfrak{A}$ for A_1 has only scalar or two-dimensional irreducible representations. Hence the above equation can be integrated in terms of the hypergeometric functions. In particular, g can be expressed by means of the Γ-functions of A (see below).

Each element s_i changes only one coordinate (namely, v_i) of the point $v = 0$. Hence, to calculate T_i for Φ from (24) it is enough to solve only the i-th equation of system (23) after the substitution $v_j = 0$ for all $j \neq i$. It follows from the construction of Φ.

Indeed, we are to calculate the coefficients $\{\Phi_m\}$ in the decomposition $\Phi(v)\prod_{i=1}^n v_j^{d_j} = \sum_m \Phi_m \prod_{j\neq i} v_j^{m_j}$, where $m = (m_1, \ldots, m_{i-1}, m_{i+1}, \ldots, m_n) \in \mathbb{Z}_+^{n-1}$, Φ_m are functions of v_i. The coefficient $\Phi_0(v_i)$ determines $\Phi(v)$ uniquely. Moreover it has the asymptotic expansion $\Phi_0(v_i)v_i^{d_i} = 1 + o(v_i)$ and satisfies the differential equation (25) for $S = s_i, A = 2d_i$. Knowing $\Phi_0(v_i)$ one can calculate T_i. It coincides with $T(s_i, a_i)$ from Lemma 9 with

$$a_i = 2\sum_{k=1}^{n}(\alpha_i, \alpha_k)(\alpha_i, \alpha_i)^{-1}d_k \tag{27}$$

Recall that d_j are connected with the dual fundamental weights δ_j (see Sec. 2). In the same sence α_i corresponds to $2\alpha_i(\alpha_i, \alpha_i)^{-1}$. The difference $2d_i - a_i$ belongs to the centralizer of $\{s_i, d_i\}$ in $\mathfrak{A}$ and $a_i s_i + s_i a_i = 2\kappa_i$. This explains the appearence of a_i instead of $2d_i$.

Theorem 10. *For $1 \le j \le n$ and a_j from (27)*

$$T_j + (q_j - q_j^{-1})(\exp(-2\pi i a_j) - 1)^{-1} = g(a_j)(s_j - \kappa_j a_j^{-1}),$$

where $g(a_j) = \Gamma^2(1+a_j)\Gamma^{-1}(1+\kappa_j+a_j)\Gamma^{-1}(1-\kappa_j+a_j)$, Γ is the gamma function.

An outline of the proof. The only remaining part is the calculation of g from Lemma 9. One may assume that

$$S = \begin{pmatrix} 1 & 0 \\ 0 & -1 \end{pmatrix}, \quad A = \kappa\left(S - \begin{pmatrix} 0 & \lambda \\ \mu & 0 \end{pmatrix}\right), \quad \lambda, \mu \in \mathbf{C}$$

and consider the vector version of (25) with

$$\varphi = (v^{\kappa/2}(v-1)^{\kappa} f_1, v^{\kappa/2}(v-1)^{-\kappa} f_2)^{tr}.$$

Then $\partial f_1/\partial v = \kappa\lambda(2v)^{-1}v^{\kappa}(v-1)^{-2\kappa} f_2, \partial f_2/\partial v = \kappa\mu(2v)^{-1}v^{-\kappa}(v-1)^{2\kappa} f_1$. Hence we obtain the hypergeometric equation:

$$\partial^2 f_1/\partial v^2 + (2\kappa(v-1)^{-1} + (1-\kappa)v^{-1})\partial f_1/\partial v - \kappa^2\lambda\mu(2v)^{-2} f_1 = 0.$$

We will give the final formula for Φ_0. Let

$$\zeta = \kappa(1+\lambda\mu)^{1/2}, \quad a = \kappa + \zeta, \quad b = \kappa, \quad c = 1 + \zeta,$$
$$F(v) = F(a, b, c; v), \quad F_+(v) = F(a+1, b+1, c+1; v)$$

for the classic hypergeometric function F (see [6]). We will denote $f(-\zeta)$ by f^* for any function f depending on ζ and maybe on some other parameters (e.g. $a^* = \kappa - \zeta$ and so on). Then

$$\Phi_0(v) = \begin{pmatrix} \varphi_1 & \varphi_1^* \\ \varphi_2 & \varphi_2^* \end{pmatrix} G^{-1}, \quad G = \begin{pmatrix} 1 & 1 \\ a(\kappa\lambda)^{-1} & a^*(\kappa\lambda)^{-1} \end{pmatrix},$$
$$\varphi_1 = v^{\zeta/2}(v-1)^{\kappa} F(v), \quad \varphi_2 = v^{\zeta/2}(v-1)^{\kappa}(\kappa\lambda)^{-1}(aF + 2vabF_+C^{-1}).$$

We need G because of the asymptotic expansion

$$(\varphi_1, \varphi_2)^{tr} = (v-1)^{\kappa} v^{\zeta/2}((1, a(\kappa\lambda)^{-1})^{tr} + o(v)), \quad v \to 0.$$

By means of the well-known formula

$$\frac{\Gamma(b-a)\Gamma(c)}{\Gamma(b)\Gamma(c-a)}\frac{F(a,1-c+a,1-b+a;v^{-1})}{(-v)^a F(a,b,c;v)}+\frac{\Gamma(a-b)\Gamma(c)}{\Gamma(a)\Gamma(c-b)}\frac{F(b,1-c+b,1-a+b;v^{-1})}{(-v)^b F(a,b,c;v)}=1$$

we have for $w=v^{-1}\to 0$

$$\begin{pmatrix}\varphi_1\\ \varphi_2\end{pmatrix}=(w-1)^{\kappa}w^{\zeta/2}\left(SG\left(\exp(-\pi i\zeta)\frac{\Gamma(b-a)\Gamma(c)}{\Gamma(b)\Gamma(c-a)},\frac{\Gamma(a-b)\Gamma(c)}{\Gamma(a)\Gamma(c-b)}\right)^{tr}+o(w)\right).$$

Hence, $T=G\begin{pmatrix}t_1 & t_2^*\\ t_2 & t_1^*\end{pmatrix}G^{-1}$, where

$$t_1=\exp(-\pi i\zeta)\Gamma(-\zeta)\Gamma(1+\zeta)\Gamma^{-1}(\kappa)\Gamma^{-1}(1-\kappa),$$

$$t_2=\Gamma(\zeta)\Gamma(1+\zeta)\Gamma^{-1}(\kappa+\zeta)\Gamma^{-1}(1+\zeta-\kappa).$$

To finish the proof we need the formula

$$AG=G\,\mathrm{diag}(-\zeta,\zeta),$$

$$G^{-1}(S-\kappa A^{-1})G=\zeta^{-1}\begin{pmatrix}0 & \zeta-\kappa\\ \zeta+\kappa & 0\end{pmatrix}.$$

Finally, $g=t_2\zeta(\zeta+\kappa)^{-1}=\Gamma(1+\zeta)\Gamma^{-1}(1+\zeta+\kappa)\Gamma^{-1}(1+\zeta-\kappa)$. We note that this g coincides, in fact, with g from (15) (see [1]). □

References

[1] Cherednik, I.V. : Monodromy representations for generalized Knizhnik-Zamolodchikov equations and Hecke algebras. Preprint ITP-89-74E, Kiev (1990), to appear in Publ. of RIMS.

[2] —— : Generalized braid groups and local r-matrix systems. Doklady Akad. Nauk SSSR, 307:1, 27-34 (1989).

[3] —— : A unification of Knizhnik-Zamolodchikov and Dunkle operators via affine Hecke algebras. Preprint RIMS–724, Kyoto (October, 1990).

[4] Lusztig G. : Affine Hecke algebras and their graded version. J. of AMS, 2:3, 599-685 (1989).

[5] Matsumoto, H. : Analyses harmonique dans les systems de Tits bornologiques de type affine. Lect. Notes Math. 590, Springer-Verlag, Berlin, Heidelberg, New-York (1979).

[6] Bateman, H. : Higher transcendental functions, vol.1, McGraw-Hill (1953).

HYPERGEOMETRIC FUNCTIONS

LEON EHRENPREIS*

I INTRODUCTION

Hypergeometric functions pervade many branches of mathematics. Perhaps the reason for this is that hypergeometric functions represent a confluence of three fundamental viewpoints

(a) Partial differential Operators and Lie groups

(b) Power series and ordinary differential operators.

(c) Algebraic integrals.

In this paper we shall be concerned mainly with (a) and (b). Section II centers around the ideas of (a) and how to go from (a) to (b). In Section III we show how to reverse this process and derive (a) from (b). This involves an analysis of new methods of factorization which we call Hadamard-Hermite factorization. We explain how the q analog of Hadamard-Hermite factorization leads to a proof of the Rogers -Ramanujan identities of partition theory.

II PARTIAL DIFFERENTIAL OPERATORS AND LIE GROUPS.

Much of this section is taken from my paper [2]. We start with a differential operator $P(D)$ or a system of differential operators $\overrightarrow{P}(D) = (P_1(D), \cdots, P_m(D))$. Classically, $P(D)$

* 1004 East 18st., Brookly NY 11230, USA

is the Lapucian or Δ or

$$\Delta_{pq} = \sum_{i=1}^{p} \frac{\partial^2}{\partial x_i{}^2} - \sum_{i=p+1}^{n} \frac{\partial^2}{\partial x_i{}^2}.$$

For the operators Δ_{pq}, there is a naturally associated Lie group, namely $G = O(p,q)$, which commutes with Δ_{pq}. To decompose the kernel of Δ_{pq} we could start with a maximal abelian subgroup H of G and then decompose the kernel under H according to the eigenfunctions (character) of H.

It turns out that this is not sufficient. The reason is that although G is the group of all transformation that commute with $P(D)$ there is a larger group G^c that preserves the kernel of $P(D)$. G^c is called the conformal group of $P(D)$. Precisely, a transformation a of R^n is called $P(D)$ conformal if

$$u \circ P(D) \circ u^{-1} = \alpha P(D) \beta. \tag{1}$$

α and β are operators of multiplication which are called conformal weights. If $\beta \equiv 1$ then u maps the kernel of $P(D)$ into itself. If $\beta \not\equiv 1$ we can use various formalisms, e.g. forming a "conformal completion" of R^n to reduce the situation to $\beta \equiv 1$. We shall therefore generally ignore β as it does not cause any difficulties.

For example, the conformal group $O^c(p,q) = O(p+1, q+1)$.

Since G^c preserves the kernel of $P(D)$ we can decompose the kernel under maximal abelian subgroups of G^c.

Let us examine the case $n = 3, p = 1$. Then $G^c = O(2,3)$. G^c has rank 2, meaning that there are maximal abelian subgroups H of dimension 2. If we use hyperbolic spherical coordinates r, ζ, u in R^3 with

$$t = r \cosh \zeta$$

(2) $$x = r \sinh \zeta \cos \phi$$

$$y = r \sinh \zeta \sin \phi$$

then one choice for H is the rotation group of ϕ and scalar multiplication in r. A solution of $\Delta_{1,2}$ which is an eigenfunction of H with eigenvalues (m, s) in (ϕ, r) of the form

(3) $$f(r, \zeta, \phi) = r^{is} e^{im\phi} L^m_{is}(cosh\zeta)$$

where L^m_{is} is a Legendre function (linear combination of P^m_{is} and Q^m_{is}).

If we start with $\Delta_{2,2}$ the group $G^c = O(3,3)$ which has rank 3. We think of R^4 as M_2 the space of real 2×2 matrices. Then we could choose for H the group of left rotations (i.e. multiplication of the matrix on the left by a rotation) times right rotations times scalar multiplication. It turns out that the analog of L^m_{is} of (3) is the hypergeometric function ${}_2F_1$. Note that ${}_2F_1$ depends on 3 parameters which are exactly the eigenvalues of G^c.

In the above examples the maximal abelian subgroup H was a Cartan subgroup of G^c. Other choices are possible; we call them <u>confluent</u> <u>Cartan</u> subgroups. In the example of $\Delta_{1,2}$ we could have chosen H as the product of scalar multiplication and the milpotent subgroup of $O(1,2)$. The analog of L of (2) is now a Bessel function.

Similarly, confluent Cartans for the conformal group of $\Delta_{2,2}$ lead to confluent hypergeometric fuctions.

Each Cartan group H (confluent or not) comes equiped with a Weyl group W which is the group of automorphisms of H in G^c modulo H. For conformal Cartans W is smaller then it is for ordinary Cartans. W acts on the special functions L. If we start with one L

then its W orbit is a set of similar special functions. For the example $\Delta_{1,2}$ the group W has 2 elements. The non trivial element w interchanges s and m in (3) and also introduces some more conformal factors. By studying the effect of w on the space of L in (3) we arrive at Whipple's formula

(4) $$Q_s^m(\cosh\zeta) = e^{im\pi}(\frac{\pi}{2})^{\frac{1}{2}}\Gamma(m+s+1)(\sinh\zeta)^{-\frac{1}{2}}P^{s-\frac{1}{2}}_{-m-\frac{1}{2}}(coth\zeta).$$

For the example $\Delta_{2,2}$ the Weyl group W is S_4, the symmetric group on 4 letters. The order of W is 24. If we trace the W orbit of a suitable ${}_2F_1$ we obtain Kummer's 24 solutions of the hypergeometric equation.

In the examples given above H was of dimension $n-1$. Thus the interesting functions which are the analogs of L of (3) depend on one variable. How are they to be prescribed?

To answer this we rexamine the nature of conformal maps. Since our conformal maps belong to Lie groups we can think of the infinitessimal analogs. It is easily seen that an infinitessimal conformal map Ψ should satisfy

(5) $$[P(D), \Psi] = aP(D) + P(D)b$$

Where a and b are operators of multiplication.

As explained above we can generally reduce to the case $b = 0$ so we shall assume it. Then (5) says that $P(D)$ is sort of an eigenoperator of Ψ with "eigenvalue "a. Of course a is a function rather than a constant but that is not important for most of our considerations.

It follows that, in the above examples, we have $n-1$ operators Ψ_j which commute and for which $P(D)$ is an eigenoperator. What about the last coordinate?

To understand this means going beyond conformal. Such an operator Ψ_o is 2 geometric meaning that $P(D) = P^1(D) + P^2(D)$ where $P^1(D)$ and $P^2(D)$ are eigenoperators. We can define 3 geometric, etc in a similar fashion. 2 geometric operators are also called hypergeometric.

For example, for the wave operator

(6) $$\Box = \Delta_{1,2} = \frac{\partial^2}{\partial t^2} - \frac{\partial^2}{\partial x^2} - \frac{\partial^2}{\partial y^2}$$

the operator $t\partial/\partial t$ is hypergeometric because

(7) $$[\frac{\partial^2}{\partial t^2} + \frac{\partial 2}{\partial y^2}, t\frac{\partial}{\partial t}] = 0$$
$$[\frac{\partial^2}{\partial t^2}, t\frac{\partial}{\partial t}] = 2\frac{\partial^2}{\partial t^2}.$$

In fact it is $t\partial/\partial t$ which is the third operator used to define $L_{is}^m(\cosh\zeta)$ in (3). [Note that, by (2) we have $\cosh\zeta = t/r$.]

Hypergeometric operators lead to hypergeometric series. These are power series in which the ratio of successive coefficients is a quotient of products of Γ functions. In other terms a hypergeometric function is

(8) $${}_pF_q[\begin{matrix} a_1 & \cdots & a_p \\ b_1 & \cdots & b_q \end{matrix}; x] = \sum \frac{(a_1)_m \cdots (a_p)_m}{(b_1)_m \cdots (b_q)_m} \frac{x^m}{m!}$$

Here $(a)_m$ is Appell's symbol

(9) $$(a)_m = \frac{\Gamma(a+m)}{\Gamma(a)} = a(a+1)\cdots(a+m-1).$$

To see why hypergeometric operators lead to hypergeometric series, consider the simplest case where $n = 1, P(D) = d/dx - \lambda$, and $\Psi = xd/dx$. Expand a solution f of $P(D)f = 0$ in a series of eigenfunctions $\{x^m\}$ of xd/dx. Thus

(10) $$f(x) \equiv \sum c_m x^m.$$

Now use the fact that $P(D)f = 0$. Note that

(11) $$\lambda x^m = const\ x^m$$

(12) $$\frac{d}{dx}x^m = const\ x^{m-1}.$$

The first equation (11) is trivial. The second equation (12) comes from the fact that

(13) $$[\frac{d}{dx}, x\frac{d}{dx}] = \frac{d}{dx}$$

so that d/dx is a boost operator for xd/dx which means it takes eigenfunctions of xd/dx into eigenfunctions, diminishing the eigenvalue by 1. Choosing the definitions carefully the constant in (12) is m and this accounts for the fact that the exponential function is hypergeometric. (In fact $expx = {}_oF_o[x]$.)

We see from this example that it is only after a suitable normalization of the eigenfunctions of Ψ_o that the ratio of the coefficients $\{c_m\}$ is a quotient of products of Γ functions. We must use $\{x^n\}$ as the eigenfunctions and then the multiplicative nature of the $\{x^n\}$ and the fact that d/dx is first order forces (Leibnitz) the constant in (12) to be $c_o m$ which leads to the desired structure of the ratio of coefficients.

We see from this that so long as Ψ_o is a homogeneous first order operator and the eigenvalues of the two parts of $P(D)$ are constants (rather than functions) then the natural multiplicative normalization leads to solutions which are integrals or sums of powers of u_o, an eigenfunction of Ψ_o, with coefficients which are quotients of products of Γ functions.

In what sense is this like the group case?

To answer this we might ask: What do we gain from the group? So far we know of two nice things that happen when we have a group:

(a) Integral representation.

(b) Weyl group.

Integral representation arize in various ways in semi-simple groups. One way is the Harish-Chandra representation of spherical functions. Another arises if $\overrightarrow{P}(D)$ has constant coefficients. For then we can apply the Fundemental Principle of [1] to get representations for the whole kernel of $\overrightarrow{P}(D)$. We then decompose these Fourier representations under H.

A detailed study of this will appear in a book now being prepared. An important point to note is that such integral representations really depend on parabolic subgroups of G or G^c and not on the whole group.

It seems that the recent work of the school of Gelfand on hypergeometric functions is in this same spirit.

The whole group is put together, by Bruhat decomposition, using Weyl groups and parabolic subgroups. We have noted that the Weyl group leads from one hypergeometric function to the whole set of hypergeometric functions related to G^c and H. In so doing it also leads to some hypergeometric relations (Kummer relations).

Thus, perhaps, knowing integral representation and the whole set of hypergeometric functions and their Kummer relations is, as far as hypergeometric functions are concerned, as good as having a group.

Let me mention one other type of relation amongst hypergeometric functions: Gauss' contiguity relations. Two functions ${}_2F_1[{}^{a\ b}_{\ c}; x]$ and ${}_2F_1[{}^{a'b'}_{\ c'}; x]$ are called <u>contiguous</u> if the respective parameters a, b, c and a', b', c' differ by 0 or 1.

In [2] we showed that there relations have operator -theoretic interpretations. An operator Ψ is called <u>biconformal</u> for $P(D)$ if $[\Psi, P(D)]$ is conformal. It is readily verified that for $P(D) = \Delta_{pq}$ such an operator maps the kernel of Δ_{pq} into the kernel of Δ_{pq}^2.

There are, however, many biconformal maps and not so many solution of Δ^2_{pq} with given eigenvalue under H. This leads to Gauss contiguity.

III. FACTORIZATION AND THE ROGERS-RAMANUJAN IDENTITIES

Examine the definition (8) of ${}_pF_q$ in case $p = 0$. The coefficient of x^m is $[(b_1)_m \cdots (b_q)_m m!]^{-1}$. The Appel symbol $(b)_m$ is a shift of $m!$. Thus ${}_oF_q(x)$ looks like some sort of $(q+1)^{st}$ power of the exponential function.

The simplest case is for $q = 1$ and $b_1 = 1$. Then ${}_oF_1(x)$ is essentially the Bessel function

(14) $$J_o(x) = \sum \frac{1}{2^{2m} m!} \frac{x^{2m}}{m!}.$$

(There is a change from x^m to x^{2m} and there is a factor 2^{2m} in the denominator; neither of these changes is significant.) If we think in terms of factorization then J_o is the Hadamard product of $exp(x/2)$ with itself, except for the change from x^m to x^{2m}. We can realize this Hadamard product as the constant term (in z) in the usual product

(15) $$\mathbf{J}(z,x) = e^{x/2z} e^{xz/2}.$$

Similarly if we replace ${}_oF_q$ by a Bessel like construct

(16) $$J_{b_1 \cdots b_q}(x) = \sum \frac{x^m}{m!} \frac{x^{x+b_1}}{(m+b_1)!} \cdots \frac{x^{m+b_q}}{(m+b_q)!}$$

then $J_{b_1 \cdots b_q}$ is the constant term in z in the expansion of

(17) $$\mathbf{J}(x; z_1, \cdots, z_q) = exp[\frac{x}{z_1 \cdots z_q} + x(z_1 + \cdots + z_q)].$$

We can regard $\mathbf{J}$ as defining a hierarchy constructed from ${}_oF_q$. This means that we have put ${}_oF_q$ in a whole family of functions, namely the family whose generating function is $\mathbf{J}$.

What is the advantage of this generating function? In the first place it satisfies the differential equations

(18) $$\frac{\partial}{\partial x}\mathbf{J} = [(z_1 \cdots z_q)^{-1} + z_1 + \cdots + z_q]\mathbf{J}$$

(19) $$z_j \frac{\partial}{\partial z_j}\mathbf{J} = x[z_j - (z_1 \cdots z_q)^{-1}]\mathbf{J}.$$

Equation (19) can be regarded, for many purposes, as an analog of Gauss' contiguity.

In the second place $\mathbf{J}$ has natural expansions. For example, for the Bessel case of (15), if we write $z = e^{i\theta}$ then

(20) $$\mathbf{J}(x,\theta) = e^{x\cos\theta} = \sum x^m \cos^m \theta.$$

This contrasts to the original expansion

(21) $$\mathbf{J}(x,\theta) = \sum J_m(x)\cos m\theta.$$

Using the Tchebychev expansion of $\cos m\theta$ in terms of $\{\cos^j \theta\}$ yields the classical identity

(22) $$(x/2)^m = \sum_{j=o}^{\infty}(m+2j)\Gamma(m+j)J_{m+2j}(x)/j!.$$

As we shall soon see, this is a classical analog of the Rogers-Ramanujan identities.

From another point of view, once we have constructed all the J_m we are well on the way to constructing groups for which the J_m play a significant role. Such is the viewpoint of Infeld and Hull; their construction of the J_m from J_o uses the differential equation J_o satisfies and so is in a somewhat different spirit from the present article. (The Infeld-Hull idea is discussed in [2].)

Our factorization idea is based on the relation

$$\bar{X}^{-l}x^m = x^{m-l} \tag{23}$$

where $\bar{X}$ is the operation of multiplication by x. Instead of using $\bar{X}$ we can use $\partial/\partial x$. This changes things in an interesting way.

For the analog of **J** we now conider

$$\begin{aligned}\mathbf{H}(x,z) &= e^{x\partial/\partial z}e^{xz} \\ &= \sum_j\sum_m \frac{x^m}{m!}\frac{x^{m+j}}{j!}z^j \\ &= e^{x(x+z)}\end{aligned} \tag{24}$$

because $exp(x\partial/\partial z)$ represents translation of z by x. We write

$$\mathbf{H}(x,z) = \sum H_m(z)x^m. \tag{25}$$

Except for a simple factor and a minus sign the H_m are the Hermite polynomials.

It is reasonable to regard (25) as the analog of (20) with $H_m(z)$ playing the role of $\cos^m\theta$.

We refer to this factorization of $\exp x^2$ which is the coefficient of z^o in (24) as the Hadamard-Hermite factorization. We can derive analogs of (18),(19) and (22) for the present situation.

We now pass the q analog. The Rogers-Ramanujan identities are

$$\sum \frac{q^{N^2}}{(1-q)\cdots(1-q^N)} = \Pi(\frac{1}{1-q^{5N\pm1}}) \tag{26}$$

$$\sum \frac{q^{N(N+1)}}{(1-q)\cdots(1-q^N)} = \Pi(\frac{1}{1-q^{5N\pm2}}). \tag{27}$$

In partition theoretic terms the first identity asserts that the number of partitions of an integer into summands of minimal difference ≥ 2 is equal to the number of partitions whose summands are $\equiv \pm 1$ *mod*5. The second identity says that the number of partitions with minimal difference $\geq z$ and minimal summand ≥ 2 equals the number of partitions with summands $\equiv \pm 2$ *mod*5.

To understand the relation of this to our previous considerations it is important to note that $(1-q)\cdots(1-q^N)$ is the q analog of $N!$.

We shall analyze (27) as (26) works in essentially the same manner (but slightly more complicated notation). With (27) we introduce the function

$$F_o(x) = \sum \frac{q^{n(n+1)}}{(1-q)\cdots(1-q^n)} x^n. \tag{28}$$

If we seek to factor F_o we note that it looks very much like a q analog of $\exp x^2$. This suggests we seek a Hadamard-Hermite factorization.

Call η the operator $x \to qx$. The natural q analog of differentiation is

$$\delta = \frac{1-\eta}{x} \tag{29}$$

so

$$\delta x^n = (1-q^n)x^{n-1} \tag{30}$$

exactly as it should be.

We are therefore led to the function

$$\begin{aligned} \tilde{\mathrm{F}}(x,z) &= \left\{\sum \frac{q^{\frac{1}{2}n(n+1)} x^n}{(1-q)\dots(1-q^n)} \delta_z^n\right\} \left\{\sum \frac{q^{\frac{1}{2}n(n+1)} x^n}{(1-q)\cdots(1-q^n)} z^n\right\} \\ &= \sum F_n(x) z^n. \end{aligned} \tag{31}$$

Clearly this $F_o(x)$ agrees with (28).

There is one difficulty that occurs here which did not appear before: There is no workable analog of the contiguity equation. If we analyze things deeply we see that the contiguity equation is closely related to the expansions (20) and (25).

To remedy this problem we modify $\tilde{\mathbf{F}}$ by replacing z^n in the second equation of (31) by $A_n(\theta)$. Thus we replace $\tilde{\mathbf{F}}$ by

$$\mathbf{F}(x,\theta) = \sum_n F_n(x) A_n(\theta). \tag{32}$$

We allow $A_n(\theta)$ to depend on q.

Now we have great freedom in choosing $\{A_n(\theta)\}$. We must choose them so that

(a) We can expand $\mathbf{F}$ in a power series in x.

(b) The coefficients of this power series can be explicity expressed in terms of the $\{A_n(\theta)\}$. These two steps can be carried out for a sequence $\{A_n(\theta)\}$ which can be thought of as the q analog of $\{\cos^n \theta\}$. The expansion coefficients in (a slightly modified version of) (a) are $\{\cos m\theta\}$. Thus we need a q analog of the Tchebychev polynomials. Using them we can derive the Rogers-Ramanujan identities. It is clear in what way they are the q analogs of (22).

The details of this construction will appear in [3].

BIBLIOGRAPHY

1. L. Ehrenpreis, *Fourier Analysis in Several Complex Variables*, Wiley-Interscience, New York (1970).
2. *"Hypergeometric Functions", in,* Algebraic Analysis **Vol. I** (1988), Academic Press, New York.
3. *"Function theory for Rogers-Ramanjan-like partition identities"*, to appear in Springer volume dedicated to the memory of Emil Grosswald..

Differential operators on the moduli space of G–bundles on algebraic curve and Lie algebra cohomologies

B. L. Feigin*

Introduction

In this paper we discuss the following problem. Let $\mathfrak{G}$ be a semi-simple complex Lie algebra, $\hat{\mathfrak{G}}$ - the corresponding affine Kac-Moody algebra, K is the central element of $\hat{\mathfrak{G}}$, $U_k(\hat{\mathfrak{G}})$ the universal enveloping algebra of $\hat{\mathfrak{G}}$ where we suppose, that K is equal to the number $k \in \mathbf{C}$; $U_k(\hat{\mathfrak{G}})$ contains infinite combination of the generators, which are acting in the representations of $\hat{\mathfrak{G}}$ with highest weight. It was noticed some time ago that there is a remarkable value of k, which is equal to $-g$, where g is the dual Coxeter number of $\mathfrak{G}$. For such k the algebra $U_{-g}(\hat{\mathfrak{G}})$ has a large center. Algebra $U_k(\hat{\mathfrak{G}})$ is a quantization of the Lie algebra of currents on the circle $\mathfrak{G}^S$, k is the paramenter of quantization. In some sense the current algebra $\mathfrak{G}^S$ is a sum of algebras $\oplus_\varphi \mathfrak{G}(\varphi)$, where $\mathfrak{G}(\varphi)$ is $\mathfrak{G}$ attached to a point $\varphi \in S$. Each $\mathfrak{G}(\varphi)$ has a family of Casimir operators, which are destroyed after the quantization. For example, the famous Sugawara construction provides the Virasoro algebra from the quadratic central elements of $\mathfrak{G}$. But the center appears again for the special value of k. It was proved by Hayashi [5], Malikov [4], and Goodman and Wallach [9].

Hayashi noted that the center $Z \subset U_{-g}(\hat{\mathfrak{G}})$ is isomorphic to the algebra of functions on some infinite dimensional hamiltonian space.

The bracket of two elements $f,\ g \in Z$ is defined by the formula $f \overset{\epsilon}{*} g - g \overset{\epsilon}{*} f = \epsilon \cdot [f \cdot g] + \ldots$ Here ϵ is a little parameter and $u_1 \overset{\epsilon}{*} u_2$ is the product of u_1, u_2 in $U_{-g+\epsilon}(\hat{\mathfrak{G}})$ (we identify $U_{-g+\epsilon}(\hat{\mathfrak{G}})$ with $U_{-g}(\hat{\mathfrak{G}})$ as the vector spaces). By this construction we obtain the classical object — hamiltonian space from the quantum situation. Ten years ago Drinfeld

* Institute for Solid State Phys., Shernogolovka Noginskyi District, Moscow 142423, USSR

and Sokolov [3] used the construction of hamiltonian reduction for constructing the new hamiltonian spaces from the dual spaces to Kac-Moody algebras. It is interesting that Hayashi hamiltonian space is naturally isomorphic to Drinfeld-Sokolov space for the dual group.

The moduli space of holomorphic G-bundles on the curve is a double coset space of the current group G^S. Drinfeld formulated an analytic version of Langlands conjectures. He invented the notion of Langlands correspondence, which connects D-modules on the moduli space of G-bundles on the curve $\mathcal{E}$ and representations of the fundamental group of $\mathcal{E}$ into the dual group G^0. He also suggested to use the properties of $U_{-k}(\hat{\mathfrak{G}})$, $k = -g$ for the construction of Langlands correspondence.

In this paper we caluclate the cohomologies of current algebra with coefficients in $U_{-g}(\hat{\mathfrak{G}})$ (with adjoint action). In the section 3 we discuss the applications of these results to the sheaf of differential operators on the moduli space. This artical reflects the results of discussions with V. Drinfeld, S. Beilinson and E.Frenkel. I am grateful to all of them.

1. Cohomologies of the Lie algebras $\mathfrak{G}(\mathbf{C}[x,\xi])$ and $\mathfrak{G}(\mathbf{C}[x]/(x^n))$

Let $\mathfrak{G}$ be a semi-simple finite-dimensional complex Lie algebra. Denote by $\mathfrak{G}(A)$ the Lie algebra $\mathfrak{G} \otimes A$, where A is a commutative algebra, the bracket is given by the formula :

$$[g_1 \otimes a_1, g_2 \otimes a_2] = [g_1, g_2] \otimes a_1 a_2, g_{1,2} \in \mathfrak{G}, a_{1,2} \in A.$$

If A is a ring of functions on the mainfold M, then $\mathfrak{G}(A)$ is an algebra of maps (currents) $M \to \mathfrak{G}$ with the pointwise bracket.

Let $Z(\mathfrak{G})$ be the ring of invariant (with respect to the adjoint action) polynomials on $\mathfrak{G}$.

First we shall attach to any homogeneous element $u \in Z(\mathfrak{G})$ of degree n the map $\phi(u) : \bigwedge(\mathfrak{G}(A)) \to \Omega^n(A)$. Here $\Omega^n(A)$ is a space of n — forms on the algebraic manifold Spec(A). The spaces $\Omega^i(A)$ constitute the de Rham complex of A : $A \cong \Omega^0(A) \to \Omega^1(A) \to \ldots \to \Omega^i(A) \to \ldots$. The formula for $\psi(u)$ is :

$$\psi(u)((g_1 \otimes a_1) \wedge (g_2 \otimes a_2) \wedge \ldots \wedge (g_n \otimes a_n))$$
$$= \mathrm{Alt}\, \tilde{u}(g_1, \ldots, g_n) da_1 \wedge \ldots \wedge da_n$$

Here Alt is the alternation and $\tilde{u}(\ )$ is a symmetric polylinear form assosiated with u. It may be shown that $\psi(u)(r) = 0$ if r is an image of the differencial $\bigwedge^{n+1}(\mathfrak{G}(A)) \to \bigwedge^{n}(\mathfrak{G}(A))$ in the standard homological complex of the Lie algebra $\mathfrak{G}(A)$. So we have a map from $H_n(\mathfrak{G}(A))$ into $\Omega^n(A)$. In fact the image of $H_n(\mathfrak{G}(A))$ belongs to the space $d\Omega^{n-1}(A)$ of exact n forms.

Remark. The maps of this type (and the generalizations) are investigated in the theory of cyclic homologies of rings (see [7]).

Let us denote by $\Delta_1, \dots, \Delta_s$ the set of generators of the algebra $Z(\mathfrak{G})$, s is the rank of $\mathfrak{G}$, $\{l_1, \dots, l_s\}$, $l_i = \deg\Delta_i$, $\{l_i - 1\}$ is the set of exponents of $\mathfrak{G}$.

Now we consider the case when $A = \mathbf{C}[x, \xi]$ is the algebra of polynomials in one even and one odd variables ($deg x = 0$ and $deg \xi = 1$). Let us denote by $W_{1,1}$ the Lie algebra of polynomial vector fields on $(1,1)$-dimensional vector superspace. Note that $W_{1,1}$ is acting in $\mathbf{C}[x, \xi]$ and therefore in Lie algebra $\mathfrak{G}(\mathbf{C}[x, \xi])$; $W_{1,1}$ isomorphic to the Lie algebra of outer derivations of $\mathfrak{G}(\mathbf{C}[x, \xi])$.

The Lie algebra $\mathfrak{G}(\mathbf{C}[x, \xi])$ contains the subalgebra $\mathfrak{G} \cong \mathfrak{G} \otimes 1$. It will be convenient for us to consider the relative (co)homologies with respect to $\mathfrak{G} \subset \mathfrak{G}(\mathbf{C}[x, \xi])$. Recall that relative homologies are the homologies of the complex

$$[\bigwedge{}^*(\mathfrak{G}(\mathbf{C}[x, \xi])/\mathfrak{G} \otimes 1)]^{Inv} \cong C_*(\mathfrak{G}(\mathbf{C}[x, \xi], \mathfrak{G}))$$

where Inv is the space of invariants with respect to adjoint action of $\mathfrak{G}$.

Theorem. *The space of relative cohomologies $H^*(\mathfrak{G}(\mathbf{C}[x, \xi], \mathfrak{G})$ is a free skew commutative algebra. The space of generators is decomposed in a sum $\oplus K_j$, $deg\, K_j = l_j$ and K_j is dual to the space $d\Omega^{l_j - 1}(\mathbf{C}[x, \xi])$, $1 \le j \le s$. (Each K_j has the natural structure of superspace.)*

This theorem is well known in the case when $\mathfrak{G}$ is the Lie algebra of the infinite matrices with the finite set of non-zero entries (see, for example [7]).

Let us reformulate the theorem. Denote by V the graded $W_{1,1}$ – module $V = \oplus V_i$, $V_i \cong d\Omega^{l_i - 1}(\mathbf{C}[x, \xi])$. Let $\bigwedge^*(V)$ be a direct $sum \oplus$ of $\bigwedge^i(V)$, where $\bigwedge^i(V)$ is an exterior power of V as a superspace. We shall consider $\bigwedge^*(V)$ as a complex with trivial differential. The set of maps $\psi(\Delta_i)$ provide by the obvious way morphisms of complexes :

$$\bigwedge{}^*(\mathfrak{G}(\mathbf{C}[x,\xi])) \to \bigwedge{}^*(V); \psi : C_*(\mathfrak{G}(\mathbf{C}[x,\xi]),\mathfrak{G}) \to \bigwedge{}^*(V)$$

The theorem asserts that ψ is a homotopic equivalence of complexes. Note that these morphisms of complexes are also the morphisms of coalgebras ($\bigwedge^*(V)$ is dual to the free skew commutative algebra with the space of generators V^*).

Proof of the theorem. Fix the odd element $Q = f(x)\partial/\partial\xi$, $Q \in W_{1,1}$. It is easy to see, that $[Q,Q] = Q^2 = 0$, therefore Q defines the differential in the Lie algebra $\mathfrak{G}(\mathbf{C}[x,\xi])$. The corresponding differential graded Lie algebra we shall denote by $\mathfrak{G}(\mathbf{C}[X,\xi],Q)$ (recall that $\deg\xi = 1$, $\deg x = 0$). That homologies of Q are non-trivial only in dimension zero and isomorphic to some finite dimensional Lie algebra $A(Q)$; $A(Q) \cong \mathfrak{G}(B)$, where B is the algebra of homologies of the complex $\mathbf{C}[x,\xi]\xrightarrow{Q}\mathbf{C}[x,\xi]$. The standard complex of differential Lie superalgebra $\mathfrak{G}(\mathbf{C}[x,\xi],Q)$ is a bycomplex with two differential $\tilde{Q}$ and d, where $\tilde{Q}$ is induced by Q and d is a differential in the standard complex of $\mathfrak{G}(\mathbf{C}[x,\xi])$. Then, we may assosiate with the bycomplex $\{\bigwedge^*(\mathfrak{G}(\mathbf{C}[x,\xi]),\tilde{Q},d\}$ the spectral sequence E^*_{**}. The first term of E^*_{**} is isomorphic to $H_*(\mathfrak{G}(\mathbf{C}[x,\xi]))$ and the limit term is isomorphic to $H_*(A(Q))$.

Lemma. *The second term of the spectral sequence E^*_{**} is isomorphic to the limit term, so $E^2_{**} \cong H_*(A(Q))$.*

The differential E^2_{**} is given by the following construction. Let w be a chain from the space $\bigwedge^n(\mathfrak{G}(\mathbf{C}[x,\xi])$ such that $\tilde{Q}w = a$ and $a = dw_1$, $w_1 \in \bigwedge^{n+1}(\mathfrak{G}(\mathbf{C}[x,\xi])$. The chain $\tilde{Q}w_1$ represents the differential of the class of the chain w in E^2_{**}. Therefore we have to find w_1 (w_1 is a solution of the equation $dw_1 = a$) such that $\tilde{Q}w_1 = 0$. It will give us a proof of the lemma.

The space $\bigwedge^n(\mathfrak{G}(\mathbf{C}[x,\xi]))$ consists of the skew symmetric polynomial functions in n variables $Y_1,\ldots,Y_n; Y_i \in \mathbf{C}^{1,1}$ with values in the space $\mathfrak{G}\otimes\ldots\otimes\mathfrak{G}$ (n times). The chain w_1 is a function in variables $Y_1,\ldots,Y_{n+1}$, the differential dw_1 depends only on the value of w_1 on the set of $(Y_1,\ldots,Y_{n+1})$ such that $Y_i = Y_j$ ($i \neq j$ is a pair of indices). Our statement is a consequence of the following geometrical fact. Consider in $\mathbf{C}^{n+1|n+1}$ a submainfold

$M = UM_{ij}$, $M_{ij} = \{Y_1, \dots Y_{n+1}, Y_\alpha \in \mathbf{C}^{1,1}, Y_i = Y_j, i \neq j\}$, here Y_α is a pair (x_α, ξ_α). The vector field $\tilde{Q} = \sum f(x_\alpha)\partial/\partial\xi_\alpha$ is acting on M. Let F be a restriction on M of some function G on $\mathbf{C}^{n+1|n+1}$ and suppose that $\tilde{Q}F = 0$. Then it can be found a function $\tilde{G}$ on $\mathbf{C}^{n+1|n+1}$ such that $\tilde{Q}\tilde{G} = 0$ and $\tilde{G}|_M = F$. In fact, let $\mathcal{O}(M) \subset \mathbf{C}[Y_1, \dots, Y_{n+1}]$ be the space of functions which are zero on M. Note that vector field $\tilde{Q}$ is acting in $\mathcal{O}(M)$. It is easy to see that the map $\mathcal{O}(M) \to \mathbf{C}[Y_1, \dots, Y_{n+1}]$ induces the imbedding of the homologies of the differential $\tilde{Q}$. This fact is equivalent to our geometrical statement.

Now we can prove the theorem. Consider the map

$$\psi : C_*(\mathfrak{G}(\mathbf{C}[x,\xi]), \mathfrak{G}) \to \bigwedge{}^*(V).$$

It can be shown that ψ is surjective. Namely, the symmetric powers $S^*(\mathfrak{G} \otimes \mathbf{C}[x]) \cong \bigwedge^*(\mathfrak{G} \otimes \mathbf{C}[x]\xi) \subset \bigwedge^*(\mathfrak{G}(\mathbf{C}[x,\xi]))$; it is evident that $\psi(S^*(\mathfrak{G} \otimes \mathbf{C}[x]))$ is equal to the symmetric algebra of the even part of the space V. Now it is sufficient to use the fact that the map ψ is a morphism of $W_{1,1}$-modules.

Consider the simplest vector field $Q = x^2/\partial\xi$. In this case an algebra $A(Q)$ is isomorphic to $\mathfrak{G}(\mathbf{C}[x]/x^2\mathbf{C}[x]) \cong \mathfrak{G} \oplus \mathfrak{G}x$: The relative homologies $H_*(\mathfrak{G}(\mathbf{C}[x]/x^2\mathbf{C}[x], \mathfrak{G})$ are the same as the space of $\mathfrak{G}$ -invariants in $\bigwedge^*(\mathfrak{G}x)$. It is well known that $[\bigwedge^*(\mathfrak{G})]^{Inv} \cong H_*(\mathfrak{G})$, so $H_*(\mathfrak{G} \oplus \mathfrak{G}x, \mathfrak{G}) \cong H_*(\mathfrak{G})$. Note that vector field Q induces the differential Q_1 in the space $\bigwedge^*(V)$ and it is easy to see that the homologies of Q_1 are isomorphic to $H_*(\mathfrak{G})$. Let K be the kernel of ψ; K is a bycomplex with two defferentials : d and $\tilde{Q}$. The homologies of the sum $d + \tilde{Q}$ are zero. Let H be the homologies of d, $\tilde{Q}$ induces operator $\tilde{Q}_1 : H \to H$. We can conclude from the lemma that homologies of $\tilde{Q}_1$ in H are zero. But it is possible only in the case when $H = O$. Note that the component of H of degree n is a B_n-module, where B_n is the ring of symmetric functions in the variables $Y_1, \dots, Y_n, Y_i \in \mathbf{C}^{1,1}$; $H^{(n)} \subset H$ is a B_n- subquotient of the B_n-module $\bigwedge^n(\mathfrak{G}(\mathbf{C}[x,\xi])/\mathfrak{G})$. The space $H^{(n)}$ is graded $W_{1,1}$-module $H^{(n)} = H_a^{(n)} \oplus H_{a+1}^{(n)} \oplus H_{a+2}^{(n)} \oplus \dots$, $\tilde{Q}_1$ is acting from $H_j^{(n)} \to H_{j-1}^{(n)}$. The component $H_a^{(n)}$ is C_n -module, where $C_n \subset B_n$ is an algebra of symmetric polynomials in n variables $x_1, \dots, x_n$. The image of $\tilde{Q}_1 : H_{a+1}^{(n)} \to H_a^{(n)}$ is liying in the space $C_n^\circ H_a^{(n)}$, where C_n° is an ideal in C_n consisting of the symmetric functions which is equal to zero in the origin. Therefore $H_a^{(n)}/C_n^\circ H_a^{(n)} = 0$ and hence $H_a^{(n)} = 0$. We proved that $H = 0$, so ψ is homotopic equivalence. The theorem is proved.

Algebra $\mathfrak{G}_n = \mathfrak{G}(\mathbf{C}[x]/x^n\mathbf{C}[x])$ is a Lie algebra of homologies of the differential induced by the vector field $Q = x^n\partial/\partial\xi$. So we can calculate the space $H_*(\mathfrak{G}_n)$. It is easier to formulate the result about cohomologies. Algebra $\mathfrak{G}_n = \mathfrak{G} \oplus \mathfrak{G}x \oplus \mathfrak{G}x^2 \oplus \ldots \oplus \mathfrak{G}x^{(n-1)}$ is graded by the natural way, namely $\deg(\mathfrak{G}x^j) = j$.

Corollary. $H^*(\mathfrak{G}(\mathbf{C}[x]/x^n\mathbf{C}[x])) \cong H^*(\mathfrak{G}) \otimes H^*(\mathfrak{G}_n, \mathfrak{G})$. *The algebra* $H^*(\mathfrak{G}_n, \mathfrak{G})$ *is free skew commutative with generators* $\xi_i(m), 1 \le i \le s, 0 \le m \le n-1, \xi_i(m) \in H^{2l_i-1}(\mathfrak{G}_n, \mathfrak{G})$, $\operatorname{gr}\xi_i(m) = 2l_i - 1 + m$.

2. The center of the universal enveloping algebra of the affine Kac-Moody algebra

Let $\hat{\mathfrak{G}}$ be the central extension of the Lie algebra $\mathfrak{G}^S = \mathfrak{G}(A)$, where $A = \mathbf{C}[z^{-1}, [[z]]]$; A is an algebra of function on the punctured infinitesemal neighborhood of the origin in the line. Let K be the central element of $\hat{\mathfrak{G}}$. We suppose that 2-cocycle is normalized by the standard way (see [1]). Denote by $U_k(\hat{\mathfrak{G}}), k \in \mathbf{C}$ the quotient $U(\hat{\mathfrak{G}})/(K - k\cdot 1)U(\hat{\mathfrak{G}})$, where $(K - k\cdot 1)U(\hat{\mathfrak{G}})$ is a 2-sided ideal generated by $K - k\cdot 1$. In fact we need some modification $U_k^{loc}(\hat{\mathfrak{G}})$ of the associative algebra $U_k(\hat{\mathfrak{G}})$.

Lie algebra $\hat{\mathfrak{G}}$ has the family of subalgebras $\hat{\mathfrak{G}}_n = \mathfrak{G}(A_n)$, where A_n is equal to $z^n \cdot \mathbf{C}[[z]]$, $A_n \subset A$. It is easy to see, that $U(\hat{\mathfrak{G}}_n)$ is a subalgebra in $U_k(\hat{\mathfrak{G}})$. Denote by $U^+(\hat{\mathfrak{G}}_n)$ the kernel of augmentation $U(\hat{\mathfrak{G}}_n) \to \mathbf{C}$ and let $\tilde{U}_k(\hat{\mathfrak{G}})$ be a projective limit of the spaces ($l \to \infty$) $U_k(\hat{\mathfrak{G}})/U_k(\hat{\mathfrak{G}})U^+(\hat{\mathfrak{G}}_l)$. It is evident that $\tilde{U}_k(\hat{\mathfrak{G}})$ has the natural structure of the assosiative algebra.

Let π be a finite dimensional representation of $\mathfrak{G}$ and $\hat{\pi} = \pi \otimes \mathbf{C}[z^{-1}, [[z]]]$ be the corresponding representation of $\mathfrak{G}^S$. We shall call the representation W of $\mathfrak{G}^S$ "geometrical" if W admits a finite filtration $\{W_j\}, W_0 \subset W_1 \subset \ldots \subset W_N$, such that W_0 and any quotient W_{j+1}/W_j is trivial one-dimentional representation or isomorphic to some $\hat{\pi}_j$. Note that $\tilde{U}_k(\hat{\mathfrak{G}})$ is a $\mathfrak{G}^S$ module with respect to adjoint action. Let $U_k^{loc}(\hat{\mathfrak{G}})$ be the maximal "geometrical" $\mathfrak{G}^S$ -submodule in $\tilde{U}_k(\hat{\mathfrak{G}})$. The elements from $U_k^{loc}(\hat{\mathfrak{G}})$ we shall call "local".

It may be shown that the bracket of two "local" elements is also "local", so $U_k(\hat{\mathfrak{G}})$ has a structure of Lie algebra, note that $U_k^{loc}(\hat{\mathfrak{G}}) \cap U_k(\hat{\mathfrak{G}}) = \mathbf{C} \cdot 1$.

Remark. The limit $U_k^{loc}(\hat{\mathfrak{G}})$ $k \to \infty$ has the beautiful geometric interpretation. Naturally, consider the dual space $\hat{\mathfrak{G}}^*$ with the canonical hamiltonian structure. It is well known that $\hat{\mathfrak{G}}^*$ can be identified with the space of differential operators $\alpha d + \tilde{A}$ which are acting from $\Omega^0(\mathfrak{g}) \to \Omega^1(\mathfrak{g})$. Here $\Omega^i(\mathfrak{g})$ is a space of i-forms on the circle with coefficients in $\mathfrak{g}$, $\alpha \in \mathbf{C}$, d is de Rham differential and $\tilde{A}$ is operator $\omega \to [A, \omega]$, A is an element from $\Omega^1(\mathfrak{g})$. The submanifold $N \subset \hat{\mathfrak{G}}^*$ which consists of the operators $\partial/\partial z + V$ is the hamiltonian submanifold. The Lie algebra $U_\infty^{loc}(\hat{\mathfrak{G}})$ is isomorphic to the algebra of differential hamiltonians. The differential hamiltonian is a function $N \to \mathbf{C}$ which may be represented as a composition $V \to \Omega^1 \to \mathbf{C}$. Here the first arrow is a non-linear differential operator with values in the space of 1-forms on the Spec$\mathbf{C}[z^{-1}, [[z]]]$ and the second arrow is a residue of the 1-form in the origin. The Lie algebra $U_\infty^{loc}(\hat{\mathfrak{G}})$ is naturally appeared in the theory of integrable equations of KdV -type.

Another description of the Lie algebra $U_k^{loc}(\hat{\mathfrak{G}})$ is the following. Lie algebra $U_k(\hat{\mathfrak{G}})$ has the obvious filtration $U_k^{(r)}(\hat{\mathfrak{G}}), r = 0, 1, 2,$ such that the adjoint algebra is isomorphic to the symmetric algebra of the space $\mathfrak{G}^S$. So, $U_k(\hat{\mathfrak{G}})$ is a deformation of the algebra $S^*(\mathfrak{G}^S)$. Let $LU_k(\hat{\mathfrak{G}})$ be the Lie algebra $LU_k(\hat{\mathfrak{G}}) \cong U_k(\hat{\mathfrak{G}})$ and bracket is equal to the commutator. The adjoint Lie algebra (with respect to the same filtration) is isomorphic to the Lie algebra of functions on the hamiltonian space $(\mathfrak{G}^S)^*$. The similar statement is true for $U_k^{loc}(\hat{\mathfrak{G}}) = U$. This algebra also has a filtration $U^{(r)}$ such that adjoint algebra $\oplus U^{(r+1)}/U^{(r)} = U^{ad}$ is isomorphic to the Lie algebra of differential hamiltonians on $(\mathfrak{G}^S)^* \cong \mathfrak{G}^S \otimes \Omega^1 \cong \mathfrak{G}^S$. Consider the symmetric power $S^n(\mathfrak{G}^S)$. Let B_n be the ring of symmetric function in n variables $(x_1, \ldots, x_n), x_i \in Spec\mathbf{C}[z^{-1}, [[z]]]$ and B_n^0 be the ideal of functions f such that $f(x, \ldots, x) = 0; B_n$ is acting in $S^n(\mathfrak{G}^S)$. The set of subspaces $(B_n^o)^j(S^n(\mathfrak{G}^S)), j = 1, 2, \ldots$ defines the adic topology in $S^n(\mathfrak{G}^S)$ and let $\overline{S^n}(\mathfrak{G}^S)$ be the completition of $S^n(\mathfrak{G}^S)$.

Lemma. $U^{(r+1)}/U^{(r)} \cong (\overline{S}^r(\mathfrak{G}^S))^*$, so $U^{ad} = \oplus(\overline{S^r}(\mathfrak{G}^S))^*$.

Here $(\overline{S}^r(\mathfrak{G}^S))^*$ is a space of continious functionals on $\overline{S}^r(\mathfrak{G}^S)$. If we identify $S^r(\mathfrak{G}^S)$ with symmetric $\mathfrak{G} \otimes \ldots \otimes \mathfrak{G}$ (r times)-valued functions in r variable $(x_1, \ldots, x_r), x_i \in$ *Spec*$\mathbf{C}[z^{-1}, [[z]]]$ then $U^{(r+1)}/U^{(r)}$ be the space of generalized symmetric function in $(x_1, \ldots, x_r)$ with support on the diagonal $x_1 = x_2 = \ldots = x_r$. It follows from the lemma that $H^*(\mathfrak{G}^S, \mathfrak{G}; U^{ad}) \cong \oplus H^*(\mathfrak{G}^S, \mathfrak{G}; (\overline{S^r}(\mathfrak{G}^S))^*)$. Note that $H^*(\mathfrak{G}(A)) \cong \oplus H^*(\mathfrak{G}^S, (S^r(\mathfrak{G}^S))^*)$, where $A = \mathbf{C}[z^{-1}, [[z]], \xi]$, where ξ is an odd variable.

Proposition. *The natural map*

$$H^*(\mathfrak{G}^S, \mathfrak{G}; U^{ad}) \to H^*(\mathfrak{G}(\mathbf{C}[z^{-1}, [[z]],]\xi]), \mathfrak{G})$$

is an isomorphism. (Here A and $\mathfrak{G}(A)$ are topological algebras in adic topology.)

Using the methods of section 1 it is easy to prove the following statement.

Proposition.

(1) The natural map $H^*(\mathfrak{G}^S, \mathfrak{G}; \mathbf{C}) \to H^*(\mathfrak{G}^S, \mathfrak{G}; U^{ad})$ is an embedding.

(2) $H^*(\mathfrak{G}^S, \mathfrak{G}; U^{ad})$ is free skew commutative $H^*(\mathfrak{G}^S, \mathfrak{G}; \mathbf{C})$ algebra.

(3) The space of generators of $H^*(\mathfrak{G}^S, \mathfrak{G}; U^{ad})$ is isomorphic to $\oplus d\Omega^{(l_j - 1)}$. Here $l_1, \ldots, l_s$ are the set of exponents of $\mathfrak{G}, \Omega^i$ is the space of i-forms on Spec $\mathbf{C}[z^{-1}, [[z]], \xi], d\Omega^{l_j - 1} \subset (H^{l_j}(\mathfrak{G}^S, \mathfrak{G}; U^{ad})/B^+)^{l_j}$*, where B^+ is the kernel of augmentation $H^*(\mathfrak{G}^S, \mathfrak{G}; \mathbf{C}) \to \mathbf{C}$, and the quotient is up the action of B^+.*

Remark. The results of section 1 can be used when we treat the following problem. Let $S^{1,1}$ be a product of the circle and $(0,1)$-dimensional superspace; $S^{1,1}$ is a C^∞-supermanifold. Denote by $\mathfrak{G}(S^{1,1})$ the Lie algebra of C^∞-currents $S^{1,1} \to \mathfrak{G}$ with pointwise bracket; $\mathfrak{G}(S^{1,1})$ is a topological $\mathbf{Z}_2$-graded algebra Lie with C^∞-topology. We can use the methods from Gelfand-Fuchs theory of continious cohomologies of Lie algebras of vector fields on smooth manifolds and compute the continious cohomologies $H^*_c(\mathfrak{G}(S^{1,1}))$. The theorem from section 1 gives us the cohomologies of the subcomplex of the standard complex of $\mathfrak{G}(S^{1,1})$ which consists of the space of cochains with support in one point $p \in S^{1,1}$. Then the standard topological technique gives us the space $H^*_c(\mathfrak{G}(S^{1,1}))$. In fact we get that $H_c^*(\mathfrak{G}(s^{1,1}))$ is a free $H_c^*(\mathfrak{G}(S))$-module. Here $\mathfrak{G}(S)$ is a Lie algebra of C^∞-currents on the circle and $H_c^*(\mathfrak{G}(S))$ is isomorphic to the cohomologies of the topological space of maps $S \to G$, where G is a group corresponding to $\mathfrak{G}$, Algebra $H_c^*(\mathfrak{G}(S^{1,1}))$ is a free skew commutative $H_c^*(\mathfrak{G}(S))$ algebra and the space of generators is a sum of $d\Omega^{l_j-1}$ where Ω^j is a space of j-forms on $S^{1,1}$. Our last proposition is a counterpart of this fact in the case if we replace $S^{1,1}$ by the scheme Spec $\mathbf{C}[z^{-1},[[z]],\xi]$.

Proposition. *There is a spectral sequence $\{E\}$ which converges to the space $H^*(\mathfrak{G}^S, \mathfrak{G}; U_k^{loc}(\hat{\mathfrak{G}}))$. First term of $\{E\}$ is isomorphic to the space $H^*(\mathfrak{G}(\mathbf{C}[z^{-1},[[z]],\xi]), \mathfrak{G})$. For generic k (irrational)*

$$H^i(\mathfrak{G}^S, \mathfrak{G}; U_k^{loc}(\hat{\mathfrak{G}})) = 0 \ \ if\ i > 0$$

and isomorphic to $\mathbf{C}$ if $i = 0$.

Let g be a so called dual Coxter number of Lie algebra $\mathfrak{G}$ (see [1]).

Proposition. *If $k = -g$ then the first term of $\{E\}$ coincides with the limit term. So,*

$$H^*(\mathfrak{G}(\mathbf{C}[z^{-1},[[z]],\xi]), \mathfrak{G}) \text{ is isomorphic to } H^*(\mathfrak{G}^S, \mathfrak{G}; U_{-g}^{loc}(\hat{\mathfrak{G}}))$$

(as a graded vector spaces, we suppose that $deg\,\xi = -1$ then the degree of the subspace $W = S^*(\mathfrak{G} \otimes C[z^{-1},[[z]],\xi])$ from the standard homological complex of Lie algebra $\mathfrak{G}(\mathbf{C}[z^{-1},[[z]],\xi], \mathfrak{G})$ is zero. We identify W with the adjoint space of $U_{-g}^{loc}(\hat{\mathfrak{G}})$ with respect to filtration).

Proof of this proposition is based on the theorem of Malikov, Hayashi, and Goodman and Wallach [4],[5],[9] about the center of Lie algebra $U_{-g}^{loc}(\hat{\mathfrak{G}})$. Actually, they proved, the center Z of $U_{-g}^{loc}(\hat{\mathfrak{g}})$ (by definition, $Z \cong H^0(\mathfrak{g}^s, \mathfrak{g}; U_{-g}^{loc}(\hat{\mathfrak{g}})))$ is isomorphic to the subspace from $H^*(\mathfrak{G}(\mathbf{C}[z^{-1},[[z]],\xi]), \mathfrak{G})$ consisting of the elements of degree zero. We can conclude from this fact that the differential in the first term of the spectral sequence $\{E\}$ is zero if $k = -g$.

Remark. Let V_0^0 be an iduced representation of $\hat{\mathfrak{G}}$ from the one dimensional representation of subalgebra $K \cdot \mathbf{C} \oplus \mathfrak{G}(\mathbf{C}[[z]]) \subset \hat{\mathfrak{G}}$, where K is acting by multiplication on-g. It may be shown that any $\mathfrak{G}(\mathbf{C}[[z]])$ -invariant vector from V_0 provides the family of central elements from $U_k^{loc}(\hat{\mathfrak{G}})$. The existence of the large family of invariant vectors in V_0 is a consequence of Kac-Kazhdan theorem [1] or follows from the construction of Wakimoto representations [6].

3. Applications

We calculated the space $D \cong H^1(\mathfrak{G}^S, U_{-g}^{loc}(\hat{\mathfrak{G}}))$; each element from D is extended to the derivation $U_{-g}^{loc}(\hat{\mathfrak{G}}) \to U_{-g}^{loc}(\hat{\mathfrak{G}})$. For example, we know that arbituary semisimple Lie algebra has an exponent $l_1 = 2$, which corresponds to the ordinary Cazimir operator. In $H^1(\mathfrak{G}^s, U_{-g}^{loc}(\hat{\mathfrak{G}}))$ we have the corresponding subspace $d\Omega^1$ (see the previous section). The elements from $d\Omega^1$ define the action of the Lie algebra of vector field on the scheme $Spec\mathbf{C}[z^{-1}, [[z]]]$ in $U_{-g}^{loc}(\hat{\mathfrak{G}})$. Note, that this Lie algebra is acting in the space $U_{-g}^{loc}(\hat{\mathfrak{G}})$ for arbitrary k, but if $k \neq -g$ it acts by inner derivations, but if $k = -g$ then these derivations are outer.

The space $H^0(\mathfrak{G}^S, \tilde{U}_{-g}(\hat{\mathfrak{G}}))$ -the center of $\tilde{U}_{-g}(\hat{\mathfrak{G}})$, is a commutative algebra. Let $O(\mathfrak{G})$ be a spectrum of $H^0(\mathfrak{G}^S, \tilde{U}_{-g}(\hat{\mathfrak{G}}))$, $Z = H^0(\mathfrak{G}^S, U_{-g}^{loc}(\hat{\mathfrak{G}}))$ is a space of functions on $O(\mathfrak{G})$. If $\mathfrak{G} = sl_2(\mathbf{C})$ then $O(sl_2(\mathbf{C}))$ is isomorphic to the space of projective connections on $X = Spec\mathbf{C}[z^{-1}, [[z]]]$ Recall that the projective connection is the same as a differential operator $\partial^2 + q$ on X which is acting from $F_{-1/2}$ into $F_{3/2}$ Here F_λ is a space of tensor fields on X of weight $\lambda \in \mathbf{C}$. In this case Z is isomorphic to the space of functions on the space of projective connections on X which have a form $q \to \operatorname{Res} P(q)$, where P is a nonlinear differential operators with values in the space of 1-forms on X. In general case the situation is similar, $O(\mathfrak{G})$ is isomorphic to the set of some geometrical structures of X and Z is a space of differential polynomials on $O(\mathfrak{G})$. If $\mathfrak{G} \cong sl_n(\mathbf{C})$, then $O(sl_n(\mathbf{C}))$ is isomorphic to the space of differenrtial operators of order $n, \partial^n + q_{n-2}\partial^{n-2} + q_{n-3}\partial^{n-3} \ldots + q_0$ which are acting from $F_{(-n+1)/2} \to F_{(n+1)/2}$. We shall call the space of differential polynomials on $O(\mathfrak{G})$ the classical W-algebra corresponding to $\mathfrak{G}$.

It is evident that that D is action in $O(\mathfrak{G})$ by vector fields. If we change K a little $-g \to -g+\epsilon$, then the differential appears in our spectral sequence, $d(\epsilon) : H^0(\mathfrak{G}^S, U^{loc}_{-g}(\hat{\mathfrak{G}})) \to H^1(\mathfrak{G}^S, U^{loc}_{-g}(\hat{\mathfrak{G}}))$. The operator $d(\epsilon)$ assign to a function $f \in Z$ the vector field on $O(\mathfrak{G})$. It means that we have a hamiltonian structure on the manifold $O(\mathfrak{G})$. (In fact, this construction of the hamiltonians structure was invented by Hayashi). The image of D on the space of vector field on $O(\mathfrak{G})$ consists of field which are acting along the leaves of the foliation. The leaf of the foliation is the set of the operators $\partial^2 + q$ with the adjoint holonomy (monodromy) transformations. The same is true for sl_n case.

Remarks. (1) There is a natural map from $H^i(\mathfrak{G}^s, U^{loc}_{-g}(\hat{\mathfrak{G}}))$ into the Hochschild cohomologies $H^i(\tilde{U}_{-g}(\hat{\mathfrak{G}}))$ (recall that Hochschild cohomologies of associative algebra A are $Ext^*(A,A)$, here A is $A \otimes A^0$-module, A^0 -dual algebra). The space $H^*(\tilde{U}_{-g}(\hat{\mathfrak{G}}))$ is equipped by the structure of Lie algebra, $H^i \times H^j \to H^{i+j-1}$. In our case $H^*(\tilde{U}_{-g}(\hat{\mathfrak{G}}))$ is isomorphic to the Lie algebra of polyvector field on the manifold $O(\mathfrak{G})$. The space of polyvector field is the space of sections of the exterior algebra of the tangent bundle. The deformation $\tilde{U}_{-g}(\hat{\mathfrak{G}}) \to \tilde{U}_{-g+\epsilon}(\hat{\mathfrak{G}})$ defines an element w from the space $H^2(\tilde{U}_{-g}(\hat{\mathfrak{G}}), \tilde{U}_{-g}(\hat{\mathfrak{G}}))$ which corresponds to infinitesinally small value of ϵ. The bracket $x \to [w, x]$ defines a differential d in $H^*(\mathfrak{G}^s, U_{-g}(\hat{\mathfrak{G}}))$, d is acting also in $H^*(\mathfrak{G}^s, U^{loc}_{-g}(\hat{\mathfrak{G}}))$.

(2) The situation is similar to some facts from the quantum group theory. Consider the quantum group $U_q(\mathfrak{a})$, where $q \in \mathbf{C}$ and $\mathfrak{a}$ is a semisimple finite dimensional Lie algebra. If q is not equal to the root of the unity, then the center of $U_q(\mathfrak{a})$ is isomorphic to the center of $U(\mathfrak{a}) = U_1(\mathfrak{a})$. But if $q = \sqrt[l]{1}$ then the center Z_l of $U_q(\mathfrak{a})$ is much larger. In this case $U_q(\mathfrak{a})$ has a non-trivial algebra L of outer derivations. The little deformation $U_q(\mathfrak{a}) \to U_{q+\epsilon}(\mathfrak{a})$ defines hamiltonian structure on the spectrum of Z_l. The Lie algbra L is acting along the leaves of the hamiltonian structure. Kac and de Contini [8] investigated the representations of $U_q(\mathfrak{a}), q = \sqrt[l]{1}$, in particular they decomposed the spectrum of the center and suggested that the category of representations of $U_q(\mathfrak{a})$ which correspond to the values of the central character from one leaf are isomorphic. So, their theory is a counterpart of $U_{-g}(\mathfrak{G})$ case.

In the forthcoming paper we shall prove that the hamiltonian space $O(\mathfrak{G})$ is isomorphic to the hamiltonian space obtained by the Drinfeld–Sokolov reduction of the dual Lie algebra.

Now we discuss the connections between the cohomologies

$$H^*(\mathfrak{G}^s, U_{-g}(\hat{\mathfrak{G}}))$$

and the sheaf of differential operators on the moduli space G-bundles on the algebraic curve. Let $\hat{G}$ be a group with Lie algebra $\hat{\mathfrak{G}}$, $\hat{G}$ is a central extension $1 \to \mathbf{C}^* \to \hat{G} \to G^s \to 1$, where G^s corresponds to the Lie algebra $\mathfrak{G}^s$. The classes of central extensions are defined by $k \in \mathbf{Z}$, the value of central charge ; denote the corresponding group by $\hat{G}_k$. Let $\mathcal{E}$ be a compact algebraic curve and fix a point $p \in \mathcal{E}$. In G^s there are two subgroups G^s_+ and G^s_-, G^s_+ is isomorphic to a group of maps $\{\mathcal{E}\backslash p\} \to G$ and the Lie algebra of G^S_- is $\mathfrak{G} \otimes \mathbf{C}[[z]]$, G^S_- consists of the maps of formal neighbourhood of p into G. It is well known that the moduli space $\mathcal{M}(\mathcal{E}, G)$ of G- bundles on $\mathcal{E}$ may be represented as a double coset space $G^S_-\backslash G^S/G^S_+$. The restriction of 2–cocycle of G^S on G^S_- and G^S_+ is trivial, so we can chose two subgroups G_- and G_+ in G_k which are mapping isomorphically onto G^S_- and G^S_+. The fiber of the map $\varphi_k : G_-\backslash\hat{G}_k/G_+ \to G^S_+\backslash G^S/G^S_- = \mathcal{M}(\mathcal{E}, G)$ is isomorphic to $\mathbf{C}^*$. So, φ_k defines the line bundle ξ_k on $\mathcal{M}(\mathcal{E}, G)$. It is evident, that $\xi_k = \xi_1 \otimes \ldots \otimes \xi_1$ (k times).

The central extension $\hat{G} \to G^S$ defines the bundle $\tilde{\xi}_k$ on the group G^S itself. There is a natural map form $U^{loc}_k(\hat{\mathfrak{G}})$ into the algebra of differential operators $\tilde{\xi}_k \to \tilde{\xi}_k$.

Proposition. *There is a map*

$$\varphi : H^i(\mathfrak{G}^S, \mathfrak{G}; U^{loc}_{-g}(\hat{\mathfrak{G}})) \to H^i(\widetilde{Dif}(\xi_{-g}, \xi_{-g}))$$

where $\widetilde{Dif}(\xi_{-g}, \xi_{-g})$ is a sheaf of differential operators in $\mathcal{M}(\mathcal{E}, G)$ acting from ξ_{-g} into ξ_{-g}. For $i = 0$ this map is a surjections. (Cohomologies in the right handside are the Zariski topology of a sheaf). This statement means that there is a map of supermanifolds:

$$Spec\, H^*(\widetilde{Dif}(\xi_{-g}, \xi_{-g})) \to Spec\, H^*(\mathfrak{G}^S, \mathfrak{G}; \tilde{U}_{-g}(\hat{\mathfrak{G}})).$$

We noted that the right-hand side is a tangent bundle to the leaves of foliation on $O(\mathfrak{G})$ (with the odd fibers) times $Spec\, H^*(\mathfrak{G}^S, \mathfrak{G})$. *The image of the left-hand side is a tangent bundle to some manifold* $O(\mathfrak{G}, \mathcal{E})$. *If* $\mathfrak{G} = sl_2$, *then* $O(\mathfrak{G}, \mathcal{E})$ *is a space of projective connections on* $\mathcal{E}$, *which is isomorphic to* $\mathbf{C}^{3h-3}$, *where h is genus of* $\mathcal{E}(\mathbf{C}^{3h-3} \cong O(\mathfrak{G})\mathcal{E})$ *as affine spaces).*

The surjectivety of $\varphi(i = 0)$ is based on the results of Hitchin [10].

Let us give the sketch of construction of the map φ. The group G^S has a natural structure of complex infinitedimensional manifold and let A be a ring of holomorphic function and P_k be a space of holomorphic sections of the line bundle $\tilde{\xi}_k$. Lie algebras L_1 and L_2 of the groups G^S_+ and G^S_- are acting in A and in P_k; G^S_+ is acting by the left action and G^S_- by the right action. Let $C(A)$ be the standard cohomological complex of the sum $L_1 \oplus L_2$ with coefficients in A and $C(P)$ be a complex with coefficients in P_k. Let π be a projection $\pi : G^S \to \mathcal{M}(\mathcal{E}, G)$ and U be a open set in $\mathcal{M}(\mathcal{E}, G)$. Denote by $A(U), P_k(U), C(A, U), C(P_k, U)$ the results of the previous constructions on $\pi^{-1}(U)$.

Note that the correspondence $U \to C(A, U)$ defines the sheaf of skew commutative differential graded algebras on $\mathcal{M}(\mathcal{E}, G)$ and $C(P_k, U)$ is a sheaf of $C(A, U)$ modules. Denote by $\tilde{D}(U)$ the sheaf which assign to U the space of differential operators acting from $C(P_k, U)$ into $C(P_k, U)$. There are evident maps $C(\mathfrak{G}^S, U^{loc}_{-g}(\hat{\mathfrak{G}})) \to C(L_1, U^{loc}_{-g}(\hat{\mathfrak{G}})) \to \tilde{D}(U)$ and $C(\mathfrak{G}^S, U^{loc}_{-g}(\hat{\mathfrak{G}})) \to C(L_2, U^{loc}_{-g}(\hat{\mathfrak{G}})) \to \tilde{D}(U)$. Here $C(a, M)$ is a standard cohomological complex of the Lie algebra a with coefficients in M. It is easy to see, that the image of $C(\mathfrak{G}^S, U^{loc}_{-g}(\hat{\mathfrak{G}}))$ in $\tilde{D}(U)$ admits the natural projection into $Dif(\pi^*\xi_{-g}, \pi^*\xi_{-g})$.

Remark. Let T be a tangent bundle $M(\mathcal{E}, G)$ and S^*T be a symmetric algebra of T. The sheaf $\widetilde{Dif}(\xi_k, \xi_k)$ has a natural filtration up to degree of the differential operator and the adjoint sheef is isomorphic to S^*T. So, there is a spectral sequence with the first term $E_1 \cong H^*(\mathcal{M}(\mathcal{E}, G), S^*T)$ and the limit $H^*(\widetilde{Dif}(\xi_k, \xi_k))$. It is the natural conjecture, that $H^i(\widetilde{Dif}(\xi_k, \xi_k)) = 0$ if $i > 0$ and $k \neq -g$, if $k = -g$, then the spectral sequence degenerates in E_1. Our statement is a "local" counterpart of this fact.

The description of the manifold $O(\mathfrak{G}, \mathcal{E})$ is the following . Let $\mathfrak{G}'$ be a Lie algebra of the dual group to G (this is a group with dual root system). Let b' be Borel subalgebra in $\mathfrak{G}'$. The $\mathfrak{G}'$- bundle ν on $\mathcal{E}$ with holomorphic connection is calling the Hodge bundle if (a) there is a structure of b'- bundle on ν, it means that in $End(\nu)$ we fix b'-sub bundle R; connection defines an element $u \in \Omega^1 \otimes (End(\nu)/R)$. (b) The secomd property — an element u is 1–form on $\mathcal{E}$ with values in the space of non — degenerate elements of $End(\nu)/R$. Note, that the fiber of $End(\nu)/R$ is isomorphic to $\mathfrak{G}'/b'$, the non-degenerate element from $\mathfrak{G}'/b'$ is liying in the dense orbit of b' in $\mathfrak{G}'/b'$. The constructions of this paper gives us a possibility to correspond D-module on $\mathcal{M}(\mathcal{E}, G)$ to a D-module on $O(\mathfrak{G}, \mathcal{E})$. We hope to discuss this correspondence in the next paper.

References

[1] V. Kac, Infinite dimensional Lie algebras, Cambridge 1990.

[2] D. B. Fuchs, Cohomology of infinite dimensional Lie algebras, Plenum Press 1988.

[3] V. Drinfeld, V. Sokolov, Lie algebras and equations of K^dV types, J. of Sov. Math. 30, 1975 (1985).

[4] F. Malikov, Singular vectors in Verma modules over affine algebras, Funct. Anal. Prilozen. 23, 76 – 77 (1989).

[5] T. Hayashi, Sugawara construction and Kac–Kazhdan conjecture, Inven. Math. 99, 13 – 52 (1988).

[6] B. Feigin, E. Frenkel, Affine Kac–Moody algebras and bosonisation, on Phisics and Mathematics in strings, V. Knizhnik Memorial Volume, eds. L. Brink, D.F , A. Polyakov, 271–316, Singapore : World Scientific, 1990.

[7] B. Feigin, B. Tsyyan, Additive K–theory, Lecture notes in mathematics. 1289, 67–209, 1987.

[8] V. Kac, C. de Condini, Quantum group at roots of unity, preprint 1990.

[9] R. Goodman, N. Wallach, Trans. of AMS, 1989.

[10] N. J. Hitchin, Flat Connections and Geometric Quantization Commun, Math. Phys. 131, 347–380 (1990).

[11] A. Beilinson, D. Kazhdan, Projective flat connections, Preprint.

HYPERGEOMETRIC FUNCTIONS, TORIC VARIETIES AND NEWTON POLYHEDRA

I.M.Gelfand[1], M.M.Kapranov[2], A.V.Zelevinsky[2]

Introduction

In this talk we give a survey of our recent results on multidimensional hypergeometric functions [GZK 1,2,7]. Before developing the general theory we briefly discuss main features of the classical Gauss function $F(x)={}_2F_1(a,b;c;x)$. By definition, $F(x)$ is the solution of the hypergeometric equation

$$x(1-x)\frac{d^2F}{dx^2}+[c-(a+b+1)x]\frac{dF}{dx}-abF=0 \qquad (1)$$

regular at $x=0$ and normalized by $F(0)=1$. Here a,b and c are complex parameters.

It is well-known [BE] that $F(x)$ can be expanded as the power series

$$F(x)=\sum_{n\geq 0}\frac{(a)_n(b)_n}{(c)_n}\frac{x^n}{n!}, \qquad (2)$$

where $(a)_n=\Gamma(a+n)/\Gamma(a)=a(a+1)\ldots(a+n-1)$. The analytic continuation of this series can be given by the Euler integral

$$F(x)=\frac{\Gamma(c)}{\Gamma(b)\Gamma(c-b)}\int_0^1\frac{t^{b-1}(1-t)^{c-b-1}}{(1-tx)^a}dt, \qquad (3)$$

converging for $Re(c)>Re(b)>0$.

Several classical attempts were made to develop multidimensional generalizations of the Gauss function (cf. e.g. [AK]). Their starting point is consideration of power series in several variables similar to (2). The most general series of this type were introduced by J.Horn [H]. Important results on these series were obtained by M.Sato in his study of prehomogeneous vector spaces [S].

[1] Laboratory of Biorganic Chemistry, Moscow State University, Corpus A, Moscow I 7-34, USSR

[2] Scientific Council for Cybernetics, 40, Vavilova St., 117333 Moscow, USSR

We develop an unified approach to special functions represented by Horn series. This includes the construction and study of a holonomic system of linear differential equations satisfied by these series as well as their integral representations generalizing (3). Our approach is closely connected with geometry of toric varieties and combinatorics of convex polytopes. We associate a system of generalized hypergeometric functions to every finite subset A of an integer lattice $\mathbf{Z}^{n-1}$, and call them A-*hypergeometric functions*. In the corresponding Horn series the summation is over some part of the lattice of affine relations among elements of A.

Note that hypergeometric functions on the Grassmannian $Gr_p(\mathbf{C}^{p+q})$ studied in [A], [G], [VGZ] are a particular kind of A-hypergeometric functions. The corresponding set A is the set of vertices of the product $\Delta^{p-1}\times\Delta^{q-1}$ of two simplices.

In §1 for any finite $A\subset\mathbf{Z}^{n-1}$ we introduce a system of linear differential equations satisfied by A-hypergeometric functions. We call it the A-*hypergeometric system*. This system first appeared in [GGZ], where it was proven to be holonomic. We calculate its characteristic cycle in terms of certain Newton polytopes. In particular, this gives the number of linearly independent solutions at a generic point.

In §2 we prove that each hypergeometric series in the sense of Horn satisfies a certain A-hypergeometric system. Moreover, we construct a class of bases in the space of A-hypergeometric functions consisting of series of this type. It is amazing that these bases correspond to certain triangulations of the convex hull of A.

In §3 we study integral representations of A-hypergeometric functions. The corresponding integrals are of the form

$$\int_\sigma \prod P_i(x_1, \ldots, x_k)^{\alpha_i} x_1^{\beta_1}\ldots x_k^{\beta_k}\, dx_1 \ldots dx_k \qquad (4)$$

for some Laurent polynomials P_i; the integrals are considered as functions of the coefficients of P_i. Here σ is some k-cycle; the precise meaning of the integral will be explained in §3. It is natural to call the integrals of type (4) *generalized Euler integrals*. They include as special cases the classical Euler integral (3) as well as more general integrals given by Appell

[AK]. For the hypergeometric functions on Grassmannians this construction gives us the integrals of products of powers of linear functions studied in [A], [VGZ].

We shall prove that any integral of type (4) satisfies a certain A-hypergeometric system. Conversely, we show that for any A the integrals of type (4) form the complete system of solutions of the A-hypergeometric equations.

§1. A-hypergeometric system

<u>1.1. Notation and assumptions.</u> Let A be a finite subset of an integral lattice Z^{n-1}. Denote by $L=L(A)\subset Z^A$ the lattice of affine relations among elements of A, i.e., the set of integer vectors (a_ω), $\omega\in A$, such that $\sum_{\omega\in A} a_\omega \omega=0$, $\sum_{\omega\in A} a_\omega=0$. All the constructions below will depend only on the affine geometry of A, i.e., on the lattice $L(A)\subset Z^A$. Therefore we can and will assume that A lies in fact in Z^n and satisfies the following two conditions:

(a) A generates Z^n as an Abelian group.

(b) There is a group homomorphism $h:Z^n\longrightarrow Z$ such that $h(\omega)=1$ for any $\omega\in A$.

Let $V=C^A$ be the space of vectors (v_ω), $\omega\in A$, where $v_\omega\in C$. . For any $a\in L$ we define the differential operator $\square_a$ on V by

$$\square_a=\prod_{\omega:a_\omega>0}(\partial/\partial v_\omega)^{a_\omega}-\prod_{\omega:a_\omega<0}(\partial/\partial v_\omega)^{-a_\omega}.$$

Note that (b) implies $\sum_{\omega\in A} a_\omega=0$ for any $a\in L$, so $\square_a$ is homogeneous.

Define also differential operators

$$Z_i=\sum_{\omega\in A}\omega_i v_\omega(\partial/\partial v_\omega),\quad i=1,\dots,n$$

on V, where ω_i is the i-th coordinate of $\omega\in A\subset Z^n$.

<u>1.2. Definition.</u> *Let $\gamma=(\gamma_1,\dots,\gamma_n)$ be a complex vector. The A-hypergeometric system with parameters γ is the following system of linear differential equations on a function $\Phi(v)$, $v\in V$:*

$$\square_a\Phi=0\ (a\in L),\qquad Z_i\Phi=\gamma_i\Phi\qquad (i=1,\dots,n).\tag{5}$$

Holomorphic solutions of (5) will be called *A-hypergeometric functions*.

This system was introduced and studied in [GGZ], [GZK 1,2]. The second group of equations in (5) means that $\Phi(v)$ is homogeneous under the action of the complex torus $(C^*)^n$ on V given by $(\lambda v)_\omega=\lambda^\omega v_\omega$, $\omega\in A$, where $\lambda=(\lambda_1,\ldots,\lambda_n)$, $\lambda^\omega=\prod \lambda_i^{\omega_i}$. Therefore, Φ can be written as $v^\theta F(x)$, where v^θ is a monomial and $x\in V/(C^*)^n$ is the image of v. Thus A-hypergeometric functions can be reduced to functions in $|A|$-n variables.

1.3. Theorem. ([GGZ]) *The system* (5) *is holonomic, and so the number of its linearly independent solutions at a generic point is finite.*

We shall denote by $Hyp(\gamma)$ the sheaf on V whose sections are A-hypergeometric functions. By general theory of holonomic systems, $Hyp(\gamma)$ is a constructible sheaf. By $Hyp(\gamma)_v$ we denote the space of local holomorphic solutions near v, i.e., the stalk of $Hyp(\gamma)$ at v.

1.4. The number of solutions. Let $Q\subset\mathbf{R}^n$ be the convex hull of A. This is a convex polytope of dimension n-1 lying in the hyperplane h(u)=1. We introduce the volume form Vol_{n-1} on this hyperplane by setting the volume of an elementary simplex on the lattice $\{u\in\mathbf{Z}^n: h(u)=1\}$ to be equal to 1.

1.5. Theorem. ([GZK 1,2]). *The number of linearly independent holomorphic solutions of the hypergeometric system* (5) *at a generic point of* V *is equal to* $Vol_{n-1}(Q)$.

1.6. The characteristic variety. To make precise the notion of "generic points" we shall introduce the D-module corresponding to the system (5), and its characteristic variety. Let D_V be the sheaf of rings of differential operators on V with holomorphic voefficients. Define a left ideal $J_\gamma\subset D_V$ by the formula $J_\gamma=\sum D_V\Box_a+\sum D_V(Z_i-\gamma_i)$. Its elements are "differential consequences" of the equations (5), i.e., they vanish on all A-hypergeometric functions. As usual, we associate to the system (5) the D_V-module $M_\gamma=D_V/J_\gamma$, see [K], [Bj]. By definition, the *characteristic variety* (or *singular support*) $SS(M_\gamma)$ is the subvariety in the cotangent bundle T^*V defined by principal symbols of all differential operators from J_γ. The characteristic variety of a holonomic D-module governs the singularities of the corresponding system: one of the irreducible components of $SS(M_\gamma)$ is V (the zero section of the cotangent bundle), and the set of singularities of (5) is the union of projections of all other components. We define the *generic stratum* $V_{gen} \subset V$ as the complement of this union in V.

Let V^* be the dual vector space to V, and (ξ_ω), $\omega \in A$ be the coordinates dual to (v_ω). For each $a \in L$ consider the polynomial $\hat{\Box}_a(\xi) = \prod_{\omega: a_\omega > 0} \xi_\omega^{a_\omega} - \prod_{\omega: a_\omega < 0} \xi_\omega^{-a_\omega}$ which is the symbol of the differential operator $\Box_a$. Let S be the subvariety in V^* defined by equations $\hat{\Box}_a(\xi) = 0$ for all $a \in L$. As shown in [GZK 2], S is a toric variety (not necessarily normal). More precisely, consider the action of the torus $(C^*)^n$ on V^* given by $(\lambda\xi)_\omega = \lambda^{-\omega}\xi_\omega$, $\omega \in A$. For any face $\Gamma \subset Q$, including Q and $\emptyset$ (where Q is the convex hull of A) consider the subvariety $S(\Gamma) \subset S$ defined by equations $\xi_\omega = 0$ for $\omega \notin \Gamma$.

<u>1.7</u> <u>Proposition</u>. ([GZK 1,2]). a) *S is the closure of the orbit of the point* (1,...,1).

b) *There is a* 1-1 *correspondence* $\Gamma \longrightarrow S_0(\Gamma)$ *between faces of* Q *and orbits of* $(C^*)^n$ *on* S *such that the closure of* $S_0(\Gamma)$ *is* $S(\Gamma)$. *In particular*, $S_0(Q)$ *is the orbit of* (1,...,1).

We shall identify each of the cotangent bundles T^*V and T^*V^* with $V \times V^*$ and hence with each other. Let $\varphi: T^*V^* \longrightarrow T^*V$ be the corresponding isomorphism. For any subvariety $Z \subset V^*$ denote $T_Z^*(V^*)$ its conormal bundle (if Z is singular then by $T_Z^*(V^*)$ we mean the closure of the conormal bundle to the set of smooth points of Z). This subvariety is Lagrangian and conic, i.e., invariant under the action of C^* on the fibers of the cotangent bundle (in our case each of these fibers is identified with V). It is well known [Gi] that each conic Lagrangian subvariety in T^*V^* has the form $T_Z^*(V^*)$.

<u>1.8.</u> <u>Theorem.</u> *The irreducible components of* $SS(M_\gamma)$ *are precisely the* $\varphi(T^*_{S(\Gamma)}V^*)$ *for all faces* $\Gamma \subset Q$ *(including* $G = \emptyset, Q$*).*

Since each $S(\Gamma)$ is invariant under the action of C^* on V^*, the variety $\varphi(T^*_{S(\Gamma)}V^*)$ has the form $T^*_{V(\Gamma)}V$ for some $V(\Gamma) \subset V$. In fact, $V(\Gamma)$ is the projectively dual variety to $S(\Gamma)$. In particular, $V(\emptyset) = V$.

<u>1.9.</u> <u>Corollary.</u> *We have* $V_{gen} = V - \bigcup_\Gamma V(\Gamma)$, *where* Γ *runs over all non-empty faces of* Q.

It is easy to see that $V(\Gamma)$ is an irreducible variety defined over $\mathbf{Z}$. When $\operatorname{codim}(\Gamma)=1$ we call the irreducible polynomial on V vanishing on $V(\Gamma)$ the $A\cap\Gamma$-discriminant. These discriminants are studied in [GZK 3,4,5,6].

1.10. Example. Let $\Delta^{p-1}\subset\mathbf{R}^p$ be the standard simplex, i.e., the convex hull of the standard basis vectors. Let $A\subset\mathbf{Z}^{p+q}$ be the set of vertices of the product $\Delta^{p-1}\times\Delta^{q-1}=Q$. (Here the number n from n.1.1. is equal to $p+q-1$). Then V is naturally identified with the space of all $p\times q$ complex matrices. As usual we regard V as an affine chart on the Grassmannian $Gr_p(\mathbf{C}^{p+q})$. The corresponding A-hypergeometric system is the system from [GGe] written in the local coordinates, see also [GGZ]. Faces $\Gamma\subset Q$ correspond to pairs of subsets $I\subset\{1,\dots,p\}$, $J\subset\{1,\dots,q\}$. The corresponding subvariety $V(\Gamma)$ consists of matrices whose (I,J)-submatrix is of non-maximal rank. Therefore, the generic stratum V_{gen} is the set of matrices with all minors non-zero. In terms of the Grassmannian this means that all Plucker coordinates are non-zero. The volume of Q in our normalisation is equal to $\binom{p+q-2}{p-1}$. In the case $p=q=2$ each A-hypergeometric function $\Phi(v)$ can be factorized as $\Phi(v)=v^{\theta}F(x)$, where $x=v_{11}v_{22}/v_{12}v_{21}$ (cf. remark after definition 1.2). It is easy to verify that the equations (5) on Φ reduce to the Gauss equation (1) for F. This observation was the starting point of the whole theory.

1.11. The characteristic cycle. Let $\underline{SS}(M_\gamma)$ be the characteristic cycle of the D-module M_γ (cf. [K], [Gi]). This is the formal linear combination of irreducible components $T^*_{V(\Gamma)}V$ of $SS(M_\gamma)$ with some integral positive multiplicities m_Γ. These multiplicities are important numerical invariants of M_γ. In particular, $m_\emptyset$ (the multiplicity of the zero section) equals the number of linearly independent solutions of (5) at a generic point. In general, m_Γ is defined as the multiplicity in the sense of commutative algebra of the O_{T^*V}-module $gr(M_\gamma)$ along the subvariety $T^*_{V(\Gamma)}V$, where $gr(M_\gamma)$ is the associated graded module to M_γ with respect to some good filtration, see [Gi], [K]. We shall describe these multiplicities in terms of volumes of certain polytopes similar to Q.

Let Γ be a face of Q. Denote by $\mathbf{R}\Gamma\subset\mathbf{R}^n$ the real vector space generated by Γ, and by $\operatorname{Lin}(\Gamma\cap A)$ the subgroup in $\mathbf{R}^n$ generated by

$\Gamma\cap A$. Put $Z\Gamma=Z^n\cap R\Gamma$ and $i(\Gamma,A)=[Z\Gamma:Lin_Z(\Gamma\cap A)]$. Consider the lattice $Z^n/Z\Gamma$ in the vector space $R^n/R\Gamma$. This lattice induces the volume form Vol on $R^n/R\Gamma$ as in n.1.4. Consider two polytopes in $R^n/R\Gamma$: $P(\Gamma)$, which is the convex hull of images of all $\omega\in A$, and $Q(\Gamma)$, the convex hull of images of all $\omega\in A-\Gamma$.

1.12. Theorem. *For any face $\Gamma\subset Q$ the multiplicity m_Γ is equal to* $i(\Gamma,A)(\mathrm{Vol}P(\Gamma)-\mathrm{Vol}Q(\Gamma))$.

Remarks. a) In [GZK 1,2] the factor $i(\Gamma,A)$ was lost; the correct formulation appeared in [GZK 6].

b) For $\Gamma=\emptyset$ we have $R^n/R\Gamma=R^n$, $Q(\Gamma)=Q$, the convex hull of A, and $P(\Gamma)$ is the pyramid with vertex 0 and base Q. Our normalisation implies $\mathrm{Vol}(P(\Gamma))=\mathrm{Vol}_{n-1}(Q)$, $\mathrm{Vol}(Q(\Gamma))=0$, and we arrive at theorem 1.5.

§2. Multidimensional hypergeometric series

We shall consider Laurent series in r variables of the form $\sum c(m_1,\dots,m_r)x_1^{m_1}\dots x_r^{m_r}$ or simply $\sum c(m)x^m$, where m runs over Z^r.

2.1. Definition (J.Horn [H]). *A formal Laurent series $F(x)=\sum c(m)x^m$ is hypergeometric if there exist non-zero rational functions $b_i(m)$ such that $c(m+e_i)=b_i(m)c(m)$, $i=1,\dots,r$, where $e_i\in Z^r$ are the standard basis vectors.*

The functions b_i must satisfy the conditions

$$b_i(m+e_j)b_j(m)=b_j(m+e_i)b_i(m). \qquad (6)$$

This means that the b_i form a 1-cocycle of the group Z^r with coefficients in the multiplicative group $C(m)^*$ of rational functions in r variables. Here Z^r acts on $C(m)^*$ by the formula $(m'b)(m)=b(m+m')$. Such a cocycle was called by M.Sato [S] a *b-function*.

If two cocycles are cohomologous then the corresponding series are related by a (pseudo-)differential operator. Namely, suppose $b'_i(m)=a(m+e_i)a(m)^{-1}b_i(m)$ for some rational function $a(m)$. Write $a(m)=p(m)/q(m)$, where p,q are polynomials. Then the series $F(x)$ and $F'(x)=\sum c'(m)x^m$, where $c'(m+e_i)=b'_i(m)c'(m)$ are connected by the equality

$$\mathrm{const}.p(x_1\partial/\partial x_1,\dots,x_r\partial/\partial x_r)F(x)=q(x_1\partial/\partial x_1,\dots,x_r\partial/\partial x_r)F'(x).$$

This means that if $F(x)$ converges somewhere then so does F' and the analytic behaviour of F' is determined by that of F.

2.2 Description of cocycles. Let $D=(d_{jp})$ be an integral $N\times r$ matrix for some N, and let $\mu=(\mu_1,\dots,\mu_r)\in(\mathbb{C}^*)^r$, $\theta=(\theta_1,\dots,\theta_N)\in\mathbb{C}^N$. Put

$$b_i(m)=b_i(D,\mu,\theta;m)=\mu_i\prod_{j=1}^{N}\left(\theta_j+1+\sum_{p=1}^{r}d_{jp}m_p\right)_{d_{ji}}^{-1}$$

(where $(x)_d=\Gamma(x+d)/\Gamma(x)$, cf. Introduction).

The $b_i(m)$ form a cocycle. The cocycle condition (6) follows from the equality $b_i(m)=c(m+e_i)/c(m)$, where

$$c(m)=c(D,\mu,\theta;m)=\mu^m/\prod_{j=1}^{N}\Gamma\left(\theta_j+1+\sum_{p=1}^{r}d_{jp}m_p\right). \tag{7}$$

Denote by $L=L(D)\subset\mathbb{Z}^N$ the lattice generated by the columns $d_p=(d_{1p},\dots,d_{Np})$ of D.

2.3. Proposition. *If $\theta,\theta'\in\mathbb{C}^N$ are congruent modulo the lattice $L(D)$ then the cocycles $\{b_i(D,\mu,\theta;m)\}$ and $\{b_i(D,\mu,\theta';m)\}$ are cohomologous.*

Proof. Suppose that $\theta'=\theta+\sum_{p=1}^{r}k_pd_p$, $k_p\in\mathbb{Z}$. Then by definition $b_i(D,\mu,\theta';m)=b_i(D,\mu,\theta;m+k)$, where $k=(k_1,\dots,k_r)$. It suffices to consider the case when $k=e_p$, the standard basis vector of $\mathbb{Z}^r$. By the cocycle condition (6) for $b_i(m)=b_i(D,\mu,\theta;m)$ we have $b_i(m+e_p)/b_i(m)=b_p(m+e_i)/b_p(m)$, and the RHS is the coboundary of the 0-cochain b_p.

2.4. Theorem. *Each cocycle in $H^1(\mathbb{Z}^r,\mathbb{C}(m)^*)$ is cohomologous to $\{b_i(D,\mu,\theta;m)\}$ for some D,μ,θ as above.*

It seems that the first proof of this theorem was given by O.Ore [O]. It was also rediscovered by M.Sato [S].

2.5.Assumptions on D,μ,θ. For D,μ,θ as above let $F(D,\mu,\theta;x)=\sum_{m\in\mathbb{Z}^r}c(D,\mu,\theta;m)x^m$, where the coefficients $c(D,\mu,\theta;m)$ are given by (7). In our study of the behaviour of $F(D,\mu,\theta;x)$ we shall restrict ourselves to the case when:

(1) $\mu=1=(1,\dots,1)$.

(2) $N>r$ and $\sum_{j=1}^{N} d_{jp}=0$ for all p.

(3) $rk(D)=r$, and the lattice L(D) is primitive, i.e., the g.c.d. of all $r\times r$-minors of D is equal to 1.

(4) There is an r-element subset $J\subset\{1,\ldots,N\}$ such that $\det(d_{jp})_{j\in J}$ is non-zero and $\theta_j\in\mathbf{Z}$ for $j\in J$.

This can be justified as follows.

For (1): by definition, $F(D,\mu,\theta;x)=F(D,1,\theta;\mu x)$.

For (2): this will guarantee the existence of a holonomic system satisfied by $F(D,\mu,\theta;x)$ with regular singularities everywhere including the infinite points. The series not satisfying (2) should be considered as "confluent". They are either divergent everywhere (as the series $\sum n!x^n$) or have too large domain of convergence (as the series $\sum (n!)^{-1}x^n$).

For (3): if $rk(D)<r$ then after a monomial change of variables $x_i=y^{\omega_i}$, $\det(\omega_1,\ldots,\omega_r)=\pm1$, we shall have $F=(\sum_{k=-\infty}^{+\infty} y_1^k)G(y_2,\ldots,y_r)$; so it does not converge. If L is not primitive then by a similar change of variables with $\det(\omega_1,\ldots,\omega_r)=$g.c.d.(all $r\times r$-minors of D) we can obtain a series in y for which the corresponding lattice will be primitive.

For (4): this condition assures us that coefficients of our series vanish outside some affine simplicial cone in $\mathbf{R}^r$. We shall see that this provides a non-trivial domain of convergence for F.

2.6. A-hypergeometric system for Horn series. We shall show that each series of the form $F(D,\theta;x)=F(D,1,\theta;x)$ satisfying conditions (1)-(4) of n.2.5, can be extended to a solution of certain A-hypergeometric system. Moreover, series of this type will allow us to construct a complete system of A-hypergeometric functions.

Consider the lattice $\mathbf{Z}^N/L(D)$ of rank $n=N-r$. We shall denote this lattice simply $\mathbf{Z}^n$. Denote the projection $\mathbf{Z}^N\longrightarrow\mathbf{Z}^n$ as well as the corresponding projection of complexifications $\mathbf{C}^N\longrightarrow\mathbf{C}^n$ by pr. Let $\omega_j=pr(e_j)\in\mathbf{Z}^n$ be the image of the j-th standard basis vector in $\mathbf{Z}^N$, $j=1,\ldots,N$. Set $A=\{\omega_1,\ldots,\omega_N\}$. Conditions (2) and (3) from n.2.5 imply that A satisfies the conditions a) and b) from n.1.1. Furthermore, L(D) coincides with the lattice L(A) introduced in n.1.1; we shall denote it simply by L.

Let $V=C^A$ be the vector space with coordinates $v_\omega, \omega\in A$, or simply $v_1,\dots,v_N$. Denote $x_p=\prod_{j=1}^{N} v_j^{d_{jp}}$, $p=1,\dots,r$. Thus $F(D,\theta;x)$ becomes a function (or, rather, a formal series) on V.

<u>2.7. Proposition.</u> *The series* $v^\theta F(D,\theta;x)$ *formally satisfies the* A-*hypergeometric system* (5) *with vector of parameters* $pr(\theta)$.

<u>Proof.</u> We can write the series $v^\theta F(D,\theta;x)$ in the form

$$\Phi_\theta(v)=\sum_{a\in L} v^{\theta+a}/\prod_{j=1}^{N}\Gamma(\theta_j+a_j+1) \tag{8}$$

Now the statement becomes Proposition 1 of [GZK 2], §1.

<u>2.8. Convergence of Horn series.</u> We say that a subset $I\subset\{1,\dots N\}$ is *a base* if the ω_j, $j\in I$, form a basis of R^n. (This defines a matroid structure on $\{1,\dots,N\}$). It is easy to see that I is a base if and only if the r×r-minor $\det(d_{jp})$, $j\in\{1,\dots,N\}-I$, is non-zero. If $\theta_j\in Z$ for $j\in\{1,\dots,N\}-I$ we shall say that $\Phi_\theta(v)$ and $F(D,\theta;x)$ are *associated with* I.

Introduce the "logarithmic" space R^N with coordinates $u_1,\dots,u_N$. For each base $I\subset\{1,\dots,N\}$ and a point $u\in R^N$ define a linear form $\varphi_{I,u}$ on R^n which takes the value u_j at ω_j for $j\in I$. Let $C(I)\subset R^N$ be the cone consisting of such u that $\varphi_{I,u}(\omega_j)\le u_j$ for $j\in\{1,\dots,N\}-I$. Clearly, C(I) is isomorphic to $(R_+)^{N-n}\times R^n$.

<u>2.9. Proposition.</u> *If* $I\subset\{1,\dots,N\}$ *is a base then any series* $\Phi_\theta(v)$ *associated with* I *has non-empty domain of convergence. Moreover, there exists a vector* $k\in C(I)$ *such that* $\Phi_\theta(v)$ *converges whenever* $(-\log|v_1|,\dots,-\log|v_N|)\in C(I)+k$.

<u>Proof</u>. See [GZK 2], Proposition 1.2.

<u>Remark.</u> Precise description of the domain of convergence for Horn series amounts to the study of A-discriminants which was undertaken in [GZK 3-6].

<u>2.10 Regular triangulations of the Newton polytope and bases in the space of solutions.</u> Now we fix $\gamma\in C^n$ and construct a complete system of solutions to the A-hypergeometric system with parameters γ. Let $\Pi(\gamma)\subset C^N$ be the inverse image $pr^{-1}(\gamma)$, see n.2.6.

This is an affine subspace of dimension $r=N-n$ parallel to the complexification of the lattice L.

Let $I\subset\{1,\dots,N\}$ be a base. Denote by $\Delta(I)\subset\mathbf{R}^n$ the simplex with vertices ω_i, $i\in I$. It has dimension n-1 and lies in the affine span of A, i.e., in the hyperplane $\{h(u)=1\}$, see §1.

<u>2.11.</u> <u>Proposition.</u> *The number of distinct series* $\Phi_\theta(v)$, $\theta\in\Pi(\gamma)$, *which are associated with* I *is equal to* $\mathrm{Vol}_{n-1}\Delta(I)$ *where* Vol_{n-1} *is the volume form introduced in* §1.

<u>Proof.</u> Let $\Pi_{\mathbf{Z}}(\gamma,I)=\{\theta\in\Pi(\gamma):\ \theta_j\in\mathbf{Z}$ for $j\in\{1,\dots,N\}-I\}$. Clearly $\Pi_{\mathbf{Z}}(\gamma,I)$ is an integral lattice in $\Pi(\gamma)$ which is invariant under translations by elements of L. By (8), for each $a\in L$, $\theta\in\Pi_{\mathbf{Z}}(\gamma,I)$ we have $\Phi_{\theta+a}(v)=\Phi_\theta(v)$. Thus the number of distinct series equals $|\Pi_{\mathbf{Z}}(\gamma,I)/L|$ which is easily seen to coincide with $\mathrm{Vol}_{n-1}\Delta(I)$.■

<u>2.12.</u> <u>Definition.</u> *A set of bases* T *is a triangulation of* (Q,A) *if* $\bigcup_{I\in T}\Delta(I)=Q$, *and for any* $I_1,I_2\in T$ *the simplices* $\Delta(I_1)$ *and* $\Delta(I_2)$ *meet at a common face (possibly empty).*

For a triangulation T we set $C(T)=\bigcap_{I\in T}C(I)$ (see n.2.8); this is a cone in the "logarithmic" space. Let $\mathrm{Vert}(T)\subset A$ denote the set of vertices of T, i.e., $\mathrm{Vert}(T)=\bigcup_{I\in T}I$. We can describe C(T) as follows.

For any $u\in\mathbf{R}^N$ define a continuous piecewise linear function $\varphi_{T,u}:Q\longrightarrow\mathbf{R}$ whose restriction to each $\Delta(I)$ coincides with $\varphi_{I,u}$ introduced in n.2.8. The following statement is obvious.

<u>2.13.</u> <u>Proposition.</u> *The cone* C(T) *consists of those* $u\in\mathbf{R}^N$ *for which* $\varphi_{T,u}$ *is convex, and* $\varphi_{T,u}(\omega_j)\le u_j$ *for* $j\in\{1,\dots,N\}-\mathrm{Vert}(T)$.

<u>2.14.</u> <u>Definition.</u> *A triangulation* T *of* (Q,A) *is regular if* C(T) *has nonempty interior.*

In other words, T is said to be regular if it admits a *strictly* convex continuous T-piecewise linear function.

<u>2.15.</u> <u>Proposition.</u> *For each* A *as above there exists a regular triangulation of* (Q,A).

For the proof see [GZK 2], §1, Proposition 4.

<u>2.16.</u> <u>Definition.</u> *A vector of parameters* $\gamma\in\mathbf{C}^n$ *is* T-*nonresonant if the sets* $\Pi_{\mathbf{Z}}(\gamma,I)$ *for* $I\in T$ *are mutually disjoint.*

It suffices, e.g., that the components γ_i of γ and 1 be linearly independent over Q.

Set $\Pi_Z(\gamma,T)=\bigcup_{I\in T}\Pi_Z(\gamma,I)$. By Proposition 2.11 for a T-nonresonant γ we have $\mathrm{Card}(\Pi_Z(\gamma,T))=\mathrm{Vol}_{n-1}Q$.

2.17. Theorem. *Let* T *be a regular triangulation of* (Q,A) *and* γ *be a* T-*nonresonant vector of parameters. Then the series* $\Phi_\theta(v)$, $\theta\in\Pi_Z(\gamma,T)$, *have a common domain of convergence and form a basis of the space of* A-*hypergeometric functions in this domain.*

By proposition 2.9., the common domain of convergence contains the set $U(T,k)=\{v\in V:(-\log|v_1|,\dots,-\log|v_N|)\in C(T)+k\}$ for some $k\in C(T)$. This explains the significance of the regularity condition for T.

2.18. Remark. By theorem 2.17, the set U(T,k) as above does not meet the singularities of A-hypergeometric system, i.e., the union $\bigcup_\Gamma V(\Gamma)$ from corollary 1.9. This explains us why triangulations are so relevant in the study of discriminants, cf. [GZK 3,4,5,6].

Some applications of theorem 2.17 to classical multivariate hypergeometric functions (of Horn, Appell, Lauricella etc.) can be found in [GZK 2], §3 and [GG 2].

§3. Generalized Euler integrals

Proofs of all the results of this section will appear in [GZK 7].

3.1. Notation. Let $A_1,\dots,A_m\subset Z^k$ be arbitrary finite subsets. To each element $\omega=(\omega_1,\dots,\omega_k)\in Z^k$ we associate the Laurent monomial $x^\omega=x_1^{\omega_1}\dots x_k^{\omega_k}$ in k variables $x_1,\dots,x_k$. We shall regard the vector space C^{A_i} as the space of Laurent polynomials of the form $P_i(x)=\sum v_\omega x^\omega$, $\omega\in A_i$.

Let $\alpha=(\alpha_1,\dots,\alpha_m)\in C^m$, $\beta=(\beta_1,\dots,\beta_k)\in C^k$ be complex vectors. We shall study the integral

$$F_\sigma(\alpha,\beta;P)=\int_\sigma \prod P_i(x_1,\dots,x_k)^{\alpha_i}\, x_1^{\beta_1}\dots x_k^{\beta_k}dx_1\dots dx_k, \tag{9}$$

where $P=(P_1,\ldots,P_m)\in\prod C^{A_i}$, as a multivalued function of P. Since the integrand is also multivalued we have to explain the meaning of this integral and the domain of integration.

3.2. Precise definition of the integral. Denote the region $(C^*)^k-\cup\{P_i=0\}$ by $U(P)=U(P_1,\ldots,P_m)$. Consider the one-dimensional local system (i.e., locally constant sheaf of C-vector spaces) $L(\alpha,\beta,P)$ on U(P) defined by monodromy exponents α_i around $\{P_i=0\}$ and β_j around $\{x_j=0\}$. We shall sometimes abbreviate $L(\alpha,\beta,P)$ as L(P) or simply L. A section of L over a simply connected region $U\subset U(P)$ can be viewed as a function $f:U\longrightarrow C$ such that f is a scalar multiple of some branch of $P^\alpha x^\beta=\prod P_i(x_1,\ldots,x_k)^{\alpha_i}x_1^{\beta_1}\ldots x_k^{\beta_k}$. A (singular) p-chain with coefficients in L is a finite formal sum $\sum(\delta,f_\delta)$, where each $\delta:\Delta^p\longrightarrow U(P)$ is a singular p-simplex in U(P), and f_δ is a section of $\delta^*(L)$ over Δ^p. For each k-chain σ we define the integral (9) as $\sum_\delta\int_{\Delta^k}f_\delta dx_1\ldots dx_k$, where the x_j's are viewed as functions $x_j(\delta(t))$ on Δ^k.

Denote by $C_p(U(P),L)$ the space of p-chains defined above. The boundary operator $d:C_p(U(P),L)\longrightarrow C_{p-1}(U(P),L)$ is defined in a standard way (see, e.g., [St]). Let $H_p(U(P),L)$ be the homology of this chain complex.

We shall consider the integrals $F_\sigma(\alpha,\beta;P)$ only when σ is a k-cycle, i.e., $d\sigma=0$. Then $F_\sigma(\alpha,\beta;P)$ depends only on the homology class of σ. We fix α,β and consider F_σ as a multivalued analytic function of the coefficients of all the P_i, i.e., on the space $\prod C^{A_i}$. More precisely, choose some initial $P=(P_1,\ldots,P_m)$ and a k-cycle $\sigma=\sum(\delta,f_\delta)$ in U(P). Then for P′ sufficiently close to P all simplices δ occuring in σ will lie in U(P′). There is a unique k-cycle $\sigma'=\sum(\delta,f'_\delta)$ with coefficients in $L(\alpha,\beta,P')$ such that the f'_δ are obtained from f_δ by analytic continuation.

We define the germ of our multivalued function by

$$F_\sigma(\alpha,\beta;P')=\int_{\sigma'}\prod P'_i(x_1,\ldots,x_k)^{\alpha_i}x_1^{\beta_1}\ldots x_k^{\beta_k}dx_1\ldots dx_k$$

3.3. Remark. The correspondence $\sigma\longrightarrow\sigma'$ defines a mapping $G_{P,P'}:H_k(U(P),L(P))\longrightarrow H_k(U(P'),L(P'))$ for P′ sufficiently close to P. Intuitively, P can be "more singular" than P′. Then $G_{P,P'}$ is not an isomorphism. When P and P′ are generic, $G_{P,P'}$ is the isomorphism of parallel transport with respect to the Gauss-Manin connection.

<u>3.4. Lemma.</u> For any $\varphi_1,\ldots,\varphi_m \in Z^k$ we have the equality

$$F_\sigma(\alpha,\beta;x^{\varphi_1}P_1,\ldots,x^{\varphi_m}P_m) = F_\sigma(\alpha,\beta+\sum \alpha_i\varphi_i;P_1,\ldots,P_m).$$

This follows at once from the definitions.■

According to this lemma we can and will assume that each A_i contains 0. We shall assume also that the union of A_i generates the Abelian group Z^k (otherwise the integral either can be reduced to this case or is equal to 0).

<u>3.5. From $(A_1,\ldots,A_m)$ to one set of monomials (the Cayley trick).</u> We shall construct an A-hypergeometric system having F_σ as its solution. Consider the lattice Z^m with the basis $e_1,\ldots,e_m$ and let $A=\cup(\{e_i\}\times A_i)\subset Z^m\times Z^k=Z^{m+k}$. Clearly this set lies in the hyperplane $h(u)=1$, where $h:Z^{m+k}\longrightarrow Z$ is the sum of first m coordinates. On the level of Laurent polynomials this amounts to associating to a collection $(P_1,\ldots,P_m)$ of polynomials in $x=(x_1,\ldots,x_k)$ a new polynomial $P(y,x)=\sum y_iP_i(x)$, where $y=(y_1,\ldots,y_m)$. This substitution was used by Cayley in elimination theory.

The proof of the next lemma is straigtforward.

<u>3.5.</u> <u>Lemma</u>. *If the union of A_i generates Z^k as an Abelian group, and each A_i contains 0 then A generates Z^{m+k}.*■

We shall use the natural identification $C^A=\prod C^{A_i}$ and denote a sequence $(P_1,\ldots,P_m)$ and the corresponding polynomial P(y,x) by the same letter P. Under this identification C^A_{gen} corresponds to the set of $(P_1,\ldots,P_m)$ such that all hypersurfaces $\{P_i=0\}\subset(C^*)^k$ are smooth, intersect each other transversely, and the same conditions hold "at infinity", i.e., on a suitable compactification.

<u>3.6.</u> <u>Theorem</u>. *For any $P=(P_1,\ldots,P_m)\in C^A$ and any $\sigma\in H_k(U(P),L(\alpha,\beta,P))$ the function $P'\longrightarrow F_\sigma(\alpha,\beta;P')$ in a neighborhood of P satisfies the A-hypergeometric system* (2) *with*

parameters γ, *where* $(\gamma_1,\ldots,\gamma_n)=(\alpha_1,\ldots,\alpha_m,-\beta_1-1,\ldots,-\beta_k-1)$.

3.7. Remark. The integrals we are considering are proper, i.e., taken over compact cycles. Therefore we do not have problems of convergence. It is more common in the theory of special functions to integrate over some non-compact regions naturally connected with the integrand. For example, if all P_i have real coefficients one can consider the integral of type (9) over some connected component of $\mathbf{R}^k\cap U(P)$. If such an integral has good properties of convergence Theorem 3.6 is also true for it since it is proven in a purely formal way.

3.8. The non-resonance condition. Let $A=\cup(\{e_i\}\times A_i)\subset Z^m\times Z^k=Z^{m+k}$ be as above. It is clear that any set $A\subset Z^n$ satisfying the conditions a) and b) from n.1.1 can be transformed to such a form by an automorphism of Z^n, at least for m=1, k=n-1. Denote $K\subset\mathbf{R}^n$ the convex cone generated by $A\subset Z^n$. Clearly, each face of K is a cone over some face of Q, the convex hull of A. For each face $\Gamma\subset K$ of codimension 1 let $Lin(\Gamma)\subset C^n$ be the C-linear span of Γ.

We say that a vector of parameters $\gamma=(\gamma_1,\ldots,\gamma_n)\in C^n$ is *non-resonant* (for A) if for each face $\Gamma\subset K$ of codimension 1 we have $\gamma\notin Z^n+Lin(\Gamma)$.

Now we can formulate the converse statement to Theorem 3.6, namely that generalized Euler integrals form a complete set of solutions for any A-hypergeometric system. Denote by

$$E=E(\alpha,\beta,P):H_k(U(P),L(P))\longrightarrow Hyp(\gamma)_P$$

the mapping $\sigma\longrightarrow F_\sigma(\alpha,\beta;P)$.

3.9. Theorem. *Suppose that* α *and* β *in Theorem* 3.6 *are such that* γ *is non-resonant for* A. *Then for each* $P=(P_1,\ldots,P_m)\in\prod C^{A_i}$ *the mapping* $E(\alpha,\beta,P)$ *is an isomorphism.*

Note that the non-resonance of γ in Theorem 3.9 implies that all $\alpha_i\neq 0$.

Our proof of Theorem 3.9 is based on the following

3.10. Theorem. *If* $P\in C^A_{gen}$ (see 1.9) *and* γ *is non-resonant then the monodromy representation of* $\pi_1(C^A_{gen},P)$ *on* $Hyp(\gamma)_P$ *is irreducible.*

3.11. Remark. Let $z_1,\dots,z_n$ be the coordinates on the open orbit S_0 of the torus in S (see 1.6). The non-resonance condition means that the multivalued function z^γ on S_0 has non-trivial monodromy along each orbit $S(\Gamma)$ of codimension 1. It seems probable that a weaker condition suffices for Theorem 3.9 (but not for Theorem 3.10), viz., that z^γ has a singularity along each $S(\Gamma)$ for $\mathrm{codim}(\Gamma)=1$. (Thus it can either ramify, or have a pole). This means that γ does not lie in $(Z^n\cap K)+\mathrm{Lin}(\Gamma)$. Among these "semi-non-resonant" γ there are some integer points, namely $\gamma\in Z^n$ such that $(-\gamma)$ lies strictly within K. The study of Euler integrals and hypergeometric functions for such γ is very important since they are connected with polylogarithms.

3.12. The sheaf $H(\alpha,\beta)$. Now we give sheaf-theoretic versions of Theorems 3.9 and 3.10. Let us introduce a constructible sheaf $H(\alpha,\beta)$ on $\prod C^{A_i}$ whose stalk at any point P is $H_k(U(P),L(\alpha,\beta,P))$. Let $U=\{(P,x)\in(\prod C^{A_i})\times(C^*)^k:\ x\in U(P)\}$ be the disjoint union of all U(P). There is the local system $L(\alpha,\beta)$ on U whose restriction on each U(P) is $L(\alpha,\beta,P)$. Its sections are branches of functions of the form $(P,x)\longrightarrow \mathrm{const}.P^\alpha x^\beta$. Let $\pi:U\longrightarrow\prod C^{A_i}$ be the projection. Define $H(\alpha,\beta)=R^k\pi_!L(\alpha,\beta)$. Here $R^k\pi_!$ is the k-th direct image with proper supports, see [Bo]. By definition, the stalk $H(\alpha,\beta)_P$ is $H^k_c(U(P),L(\alpha,\beta,P))$, the *cohomology* with compact supports. By Poincaré duality, one has isomorphisms

$$H^k_c(U(P),L(\alpha,\beta,P))=H^{2k-k}(U(P),L(\alpha,\beta,P)^*)^*=H_k(U(P),L(\alpha,\beta,P)).$$

It is clear that the mapping $E(\alpha,\beta,P)$ from Theorem 3.9 is induced by a morphism of sheaves $E(\alpha,\beta):H(\alpha,\beta)\longrightarrow \mathrm{Hyp}(\gamma)$ by taking stalks. Note that the mapping $G_{P,P'}$ (see remark 3.3) is a special case of "transport" map defined for any constructible sheaf.

3.13. Theorem. *In the conditions of Theorem* 3.9 *the morphism* $E(\alpha,\beta)$ *is an isomorphism of sheaves.*

Consider also the complex $R\pi_*L(\alpha,\beta)$, the full direct image.

3.14. Theorem. $R\pi_*L(\alpha,\beta)$ *is an irreducible perverse sheaf, and the canonical morphism* $R\pi_!L(\alpha,\beta)\longrightarrow R\pi_*L(\alpha,\beta)$ *is an isomorphism.*

3.15. Remark. We have a natural restriction morphism $res: R^k\pi_* L(\alpha,\beta)_P \longrightarrow H^k(U(P),L(\alpha,\beta,P))$. In general, this morphism is not an isomorphism (in contrast with the case of $R^k\pi_!$ and H^k_c). Its image consists of cohomology classes of cocycles which can be extended to $H^k(U(P'),L(\alpha,\beta,P'))$ for all P' close to P. By Poincaré duality, $H^k(U(P),L(\alpha,\beta,P))$ is isomorphic to $H^{lf}_k(U(P),L(\alpha,\beta,P))$, the homology defined by means of locally finite chains, see [VGZ]. For an "extendable" cycle $\sigma\in H^{lf}_k(U(P),L(\alpha,\beta,P))$ we could define the Euler integral as in [VGZ]. But Theorem 3.14 means that the space of extendable cycles is just the image of the canonical map $H_k(U(P),L(\alpha,\beta,P)) \longrightarrow H^{lf}_k(U(P),L(\alpha,\beta,P))$, and various "extensions" of a cycle σ correspond to various compact cycles homologous to σ.

3.16. Example: hypergeometric functions on the Grassmannian $Gr_p(C^{p+q})$ (cf. Example 1.10). These functions were originally defined in [G] in terms of generalized Euler integrals of type (9), where all P_i are (inhomogeneous) linear functions. In this case all A_i $(i=1,\dots,q)$ are equal to $\{0,e_1,\dots,e_{p-1}\}\subset Z^{p-1}$, where the e_j are standard basis vectors. The construction in 3.5 leads to the realization of C^A as the space of polynomials $P(y,x)=\sum_{i=1}^{q} v_{i0}y_i+\sum_{i=1}^{q}\sum_{j=1}^{p-1} v_{ij}y_ix_j$. The polytope Q is the product $\Delta^{p-1}\times\Delta^{q-1}$ of two simplices. It has p+q faces of codimension 1. The corresponding non-resonance conditions have the form $\alpha_i\notin Z$, $i=1,\dots,q$; $\beta_j\notin Z$, $j=1,\dots,p-1$; $(\sum\alpha_i)+(\sum\beta_j)\notin Z$, cf. [VGZ].

The set A can be transformed to a form $\cup(\{e'_i\}\times A'_i)$, $i=1,\dots,m$, $A'_i\subset Z^k$, in three different ways. The first was just described (here $m=q$, $k=p-1$). The second has $m=p$, $k=q-1$, and the transformation is given by transposition of the matrix (v_{ij}). The third has $m=1$, $k=p+q-2$. So we obtain three kinds of Euler integrals for the same hypergeometric system which are taken over cycles of different dimension. The relation between integrals of first two kinds is closely connected with the duality studied in [GG 1]. Both these integrals can be obtained from the integrals of the third kind by an appropriate iterated integration.

REFERENCES

[A] *Aomoto K.* On the structure of integrals of power products of linear functions.-Sci.Papers Coll.Gen. Educ.Univ. Tokyo, 1977, v.27, no.2, 49-61.

[AK] *Appell P., Kampé de Fêriet J.* Fonctions hypergéométriques et hypersphériques; polynomes d'Hermite.-Paris, Gauthier-Villars, 1926.

[BE] *Bateman H.*, *Erdelyi A.* Higher transcendental functions.-vol.1, McGraw-Hill, 1953.

[Bj] *Bjork J.E.* Rings of differential operators.-North-Holland, 1979.

[Bo] *Borel A.* (ed.). Seminar on intersection homology.-Birkhäuser, Boston, 1984.

[G] *Gelfand I.M.* General theory of hypergeometric functions.-Doklady AN SSSR, 1986, v.288, no.1, 14-18 [Sov. Math. Dokl., 1987, v.33, 9-13].

[GGe] *Gelfand I.M*, *Gelfand S.I.* Generalized hypergeometric equations.-Doklady AN SSSR, 1986, v.288, no.2, 279-283 [Sov. Math. Dokl., 1987, v.33, 643-646].

[GG1] *Gelfand I.M.*, *Graev M.I.* A duality theorem for general hypergeometric functions.-Doklady AN SSSR, 1986, v.289, no.1, 19-23 [Sov. Math. Dokl., 1987, v.34, 9-13].

[GG2] *Gelfand I.M.*, *Graev M.I.* Hypergeometric functions associated with the Grassmannian $G_{3,6}$.-Mat.Sbornik, 1989, v.180, no.1, 3-38 [Math. USSR Sb., 1990, v.66, no.1, 1-40].

[GGZ] *Gelfand I.M.*, *Graev M.I.*, *Zelevinsky A.V.* Holonomic systems of equations and series of hypergeometric type.-Doklady AN SSSR, 1987, v.295, no.1, 14-19 [Sov. Math. Dokl., 1988, v.36, 5-10].

[GZK 1] *Gelfand I.M.*, *Zelevinsky A.V.*, *Kapranov M.M.* Equations of hypergeometric type and Newton polytopes.-Doklady AN SSSR, 1988, v.300, no.3, 529-534 [Sov. Math. Dokl., 1988, v.37, 678-683].

[GZK 2] *Gelfand I.M.*, *Zelevinsky A.V.*, *Kapranov M.M.* Hypergeometric functions and toric varieties.-Funkc.Anal., 1989, v.23, no.2, 12-26.

[GZK 3] *Gelfand I.M.*, *Zelevinsky A.V.*, *Kapranov M.M.* A-discriminants and Cayley-Koszul complexes.-Doklady AN SSSR, 1989, v.307, no.6, 1307-1310 [Sov. Math. Dokl., 1990, v.40, no.1, 239-243].

[GZK 4] *Gelfand I.M.*, *Zelevinsky A.V.*, *Kapranov M.M.* Newton polytopes of principal A-determinants.-Doklady AN SSSR, 1989, v.308, no.1, 20-23.

[GZK 5] *Gelfand I.M.*, *Zelevinsky A.V.*, *Kapranov M.M.* On discriminants of polynomials in several variables.-Funkc.Anal., 1990, v.24, no.1, 1-4.

[GZK 6] *Gelfand I.M.*, *Zelevinsky A.V.*, *Kapranov M.M.* Discriminants of polynomials in several variables and triangulations of Newton polytopes.- Algebra i analiz, 1990, v.2,no.3, 1-62.

[GZK 7] *Gelfand I.M.*, *Kapranov M.M.*, *Zelevinsky A.V.* Generalized Euler integrals and A-hypergeometric functions.-to appear in Adv.Math.

[Gi] *Ginsburg V.A.* Characteristic varieties and vanishing cycles.-Invent. Math, 1986, v.84, 327-402.

[H] *Horn J.* Ueber die Konvergenz der hypergeometrischen Reihen zweier und dreier Veranderlichen.-Math.Ann,1889,Bd.34, S.544-600.

[K] *Kashiwara M.* Systems of micro-differential equations. Birkhäuser, Boston, 1983.

[O] *Ore O.* Sur la forme de fonctions hypergeometriques de plusieurs variables.-J.Math.Pures et Appl., 1930, v.9, no.4, 311-327.

[S] *Sato M.* Singular orbits of a prehomogeneous vector space and hypergeometric finctions.-to appear in Nagoya Math.Journal.

[St] *Steenrod N.* Homology with local coefficients.-Ann. Math., 1945, v.44, 610-627.

[VGZ] *Vasiliev V.A.*, *Gelfand I.M.*, *Zelevinsky A.V.* General hypergeometric functions on complex Grassmannians.-Funkc.Anal., 1987, v.21, no.1, 23-38.

On Some Properties of Robinson-Schensted Correspondence

by

Anatoli KIRILLOV
Steklov Mathematical Institute, Leningrad
Fontanka 27, 191011 Leningrad, USSR

Abstract.

Some new results related to C. Green's theorem about the shape of Young tableaux P and Q which correspond to the fixed permutation under Robinson-Schensted correspondence are presented. The results are based on the theory of rigged and filled configurations. We give the explicit description of the first tableaux of the filled configuration for Young tableaux P (or Q) in terms of the structure of descents of the corresponding permutation.

1.

The Robinson-Schensted correspondence establish the relation between permutations σ of N symbols (elements of S_N) and pairs of standard Young tableaux P and Q of the same shape λ ($|\lambda| = N$), [1], [8]. The number of colomns (respectively rows) in the tableax P is equal to the length of maximal increasing (respectively decreasing) subsequence in σ [1]. This theorem was generalized by C. Green [2] (see also [4], [5]), where the description of the shape of the tableu P is given in terms of lengthes of nonselfintersecting increasing subsequences of permutation σ.

2.

Let λ and μ be the pair of partitions with $|\lambda| = |\mu|$. Define the linear functional Q_n over partitions

$$Q_n(\lambda) := \sum_{j\geq 1} \min(n, \lambda_j)$$

Definition [6]. *The configuration ν of the type (λ, μ) is the set of partitions $\nu = \{\nu^{(K)}\}$ such that*

i) $|\nu^{(K)}| = \sum_{j\geq K+1} \lambda_j$
ii) $P_n^{(K)}(\nu \mid \mu) := Q_n(\nu^{(K-1)}) - 2Q_n(\nu^{(K)}) + Q_n(\nu^{(K+1)}) \geq 0$

for any $K, n \geq 1$. We assume that $\nu^{(0)} = \mu$.

Definition [6]. *We will say that the configuration is rigged if with each Young diagramm $\nu^{(K)}$ we affiliate the set $\{J_{\alpha,n}^{(K)}\}$, where $J_{\alpha,n}^{(K)} \in \mathbf{Z}_+$, $1 \leq \alpha \leq m_n(\nu^{(K)})$, such that*

i) $J_{\alpha,n}^{(K)} \leq J_{\beta,n}^{(K)}$, *if* $\alpha \leq \beta$
ii) $J_{\alpha,n}^{(K)} \leq P_n^K(\nu \mid \mu)$, *if* $1 \leq \alpha \leq m_n(\nu^{(K)})$,

for any $K, n \geq 1$. Here $m_n(\lambda)$ is the number of rows of length n in the diagramm λ.

Let us denote the set of all rigged configurations of the type (λ, μ) as $QM(\lambda, \mu)$. It was found in [6], [7] the explicite bijection between the sets of standard Young tableaux of the shape λ and weight μ and rigged configurations of the type (λ, μ):

$$STY(\lambda, \mu) \rightleftharpoons QM(\lambda, \mu). \tag{2.1}$$

Let us formulate results of [3] in terms of rigged configurations. Let $\sigma \in S_N$ and P and Q are the Young tableaux corresponding to this permutation under RS-correspondence. Consider configurations related to these Young tableaux under the bijection (2.1):

$$\{\nu^{(1)}(P), \ldots, \nu^{(l)}(P)\}, \quad \{\nu^{(1)}(Q), \ldots, \nu^{(l)}(Q)\}. \tag{2.2}$$

From the Green's theorem it follows that

$$|\nu^{(K)}(P)| = N - d_K(\sigma), \quad |\nu^{(K)}(Q)| = N - d_K(\sigma^{-1}),$$

where numbers $d_K(\sigma)$ are given by (3.2).

Remark. The number of diagrams in (2.2) is equal to $2l$, where $l+1$ is the length of the maximal decreasing subsequence in σ.

In the last part of the paper we give the explicite combinatorial description of the shape of the diagram $\nu^{(1)}(P)$ and of the corresponding filled configuration (Definition see in [6]).

3.

Let us remind some necessary definitions and results from [3]. We will identify each element of S_N with the sequence $(\sigma(1), \ldots, \sigma(N))$. Let us associate with each permutation σ sets of numbear $\{a_K(\sigma)\}$ and $\{d_K(\sigma)\}$

$$a_K(\sigma) := \max |D_1 \cup D_2 \cup \cdots \cup D_K|, \tag{3.1}$$

where $D_i \cap D_j = \phi$ if $i \neq j$; each subset D_j is a decreasing subsequence in σ;

$$d_K(\sigma) := \max |I_1 \cup I_2 \cup \cdots \cup I_K|, \tag{3.2}$$

where $I_i \cap I_j = \phi$, if $i \neq j$; each subset I_j is an increasing subsequence in σ, $a_0(\sigma) = d_0(\sigma) = 0$.

Let P and Q be the pair of Young tableaux of the shape $\lambda = (\lambda_1 \geq \lambda_2 \geq \cdots \geq \lambda_r)$ corresponding to the permutation $\sigma \in S_N$, and λ' be the Young diagram conjugated to λ.

Theorem [3]. $\lambda_K = d_K(\sigma) - d_{K-1}(\sigma)$, $\lambda'_K = a_K(\sigma) - a_{K-1}(\sigma)$

Corollary. $\sum_{j \geq K} \lambda_j = d_K(\sigma)$. *Other words: the number of boxes in the first K rows of the tableau P is equal to the maximal length of the family of K nonintersecting increasing subsequences in σ (K-family).*

Definition [3]. *Let us say that the set $a_{ij} \in \sigma$, $1 \leq j \leq K$, $1 \leq i \leq \alpha$ be the K-matching in σ if*

i) $a_{i1} > a_{i2} > \cdots > a_{iK}$, $1 \leq j \leq \alpha$

ii) the elements each column $\{a_{1j}, a_{2j}, \ldots, a_{\alpha j}\}$, $1 \leq j \leq K$, are distinct. The set $\{a_{11}, a_{21}, \ldots, a_{\alpha 1}\}$ is called the source of the K-matching. Further, define

$$\varphi_K(\sigma) = \max\{|S| : S \quad \text{is the source of a } K\text{-matching in } \sigma\}.$$

The following result due to C. Green [3] gives the number of elements in configuration $\nu^{(K)}(P)$:

$$|\nu^{(K)}(P)| = \varphi_{K+1}(\sigma).$$

Recall that $\sigma \overset{RS}{\longleftrightarrow} (P, Q)$. We will say that the K-matching in σ is maximal if it has maximal possible number of elements. Let $\{U^{(K)}\}$ be the set of the source of a maximal $(K+1)$-matching in σ, and $U_0^{(K)}$ be the minimal element of $\{U^{(K)}\}$ with respect to lexicographical order on $\{U^{(K)}\}$ (see [3]).

Theorem [3]. *The set $U_0^{(K)}$, $0 \le K \le l-1$, is the union of rows $K+1, K+2, \ldots, l$ in the tableau $P(\sigma)$.*

4.

Let $\{P^{(K)}\}$ be the filled configurations corresponding to tableau P. Here $P^{(K)}$ is the Young tableau of shape $\nu^{(K)}$ and $\{\nu^{(K)}\}$ is the configuration (see [6]).

Consider the set $\mathfrak{X} = \{x_1 < x_2 < \cdots < x_l\}$, $1 \le \bar{\bar{X}}_i \le N$. For each element $x \in \mathfrak{X}$ denote as x^+ the nearest larger element of $\mathfrak{X}$. Let w be the permutation of the set $\mathfrak{X} : w\mathfrak{X} = \{x_{w_1}, \ldots, x_{w_l}\}$.

Definition. *$\mathcal{DES}_{\mathfrak{X}}(w)$ is the set of elements $x \in \mathfrak{X}$ such that x^+ situated on the left from x in $w\mathfrak{X}$. Define $\mathcal{DES}^+_{\mathfrak{X}}(w) = \{x^+ \mid x \in \mathcal{DES}_{\mathfrak{X}}(w)\}$. With each permutation $\sigma \in S_N$ we associate the set $\{\mathcal{D}^+_K\}$, where $\mathcal{D}^+_1 = \mathcal{DES}^+_I(\sigma)$, $\mathcal{D}^+_{m+1} = \mathcal{DES}^+_{I_m}(\sigma_m)$, $m \ge 1$, $I_m = I_{m-1} \backslash \mathcal{D}^+_m$, $I_0 = I = \{1 < 2 < \cdots < N\}$, and the permutation σ_m can be obtained from σ_{m-1} by removing elements from $\mathcal{D}^+_m$, $\sigma_0 = \sigma$.*

Definition. *Let $T^{(1)}(\sigma)$ be the set of columns such that the column with number m is the intersection $U_0^{(1)} \cap \mathcal{D}^+_m(\sigma)$, where $U_0^{(1)}$ is the minimal element in the set of the source of a maximal 2-matching.*

Now we can formulate the main theorem of this paper.

Theorem. *Let $\sigma \in S_N$, $\sigma \overset{RS}{\longleftrightarrow} (P(\sigma), Q(\sigma))$ and $P^{(1)}(\sigma)$ be the first Young tableau from the filled configuration corresponding to $P(\sigma)$.*

i) The set $T^{(1)}(\sigma)$ is the standard tableau of the shape $\nu^{(1)}$

ii) $P^{(1)}(\sigma) = T^{(1)}(\sigma)$.

I would like to thank L. D. Faddeev, N. Yu. Reshetikhin, A. N. Zelevinsky and S. V. Kerov for valuable and stimulating discussions. I also wish to thank Prof. T. Miwa for the opportunity of visiting RIMS.

References

[1] C. Schensted, Longest increasing and decreasing subsequences, *Canad. J. Math.* **13** (1961), 179-191.

[2] C. Greene, An extension of Schensted's theorem, *Advances in Math.*, **14** (1974), 254-265.

[3] C. Greene, Some order-theoretic properties of the Robinson-Schensted correspondence, *Lect. Notes in Math.*, **579** (1976), 114-120.

[4] S. V. Kerov, A. M. Vershik, The characters of the infinite symmetric group and probability properties of the Robinson-Schensted–Knuth algorithm, SIAM *Journ. alg. math.*, **7** (1984), 116-124.

[5] S. V. Fomin, Finite partially ordered sets and Young tableaux, *Soviet Math. Dokl.*, **19** (1978), 1510-1514.

[6] A. N. Kirillov, N. Yu. Reshetikhin, Bethe ansatz and the combinatorics of Young tableaux, *Zap. Nauch. Semin. LOMI*, **155** (1986), 65-115 (in Russian).

[7] A. N. Kirillov, On the Kostka-Green-Foulkes polynomials and Clebsch-Gordan numbers, *JGP*, **5** (1988), 365-389.

[8] M. -P. Schützenberger, La correspondance de Robinson, *Lect. Notes in Math.*, **579** (1976), 59-113.

Standard Monomial Theory for $\hat{Sp}_{2n}$

by
*V. Lakshmibai
Department of Mathematics
Northeastern University
Boston, MA 02115

Introduction:

Let G be a semi-simple algebraic group and B a Borel subgroup. Let X be a Schubert variety in G/B. Let L be an ample line bundle on G/B, as well as its restriction to X. A standard monomial theory for Schubert varieties in G/B is developed in [9],[11],[8],[10] as a generalization of the classical Hodge-Young theory (cf [3],[4]). This theory consists in the construction of a characteristic-free basis for $H^0(X,L)$. This theory is extended to Schubert varieties in the infinite dimensional flag variety $\hat{SL}_n/B$ in [13] (see also [12]). In this paper, we extend the theory to Schubert varieties in the infinite dimensional flag variety $\hat{Sp}_{2n}/B$.

Let A be a symmetrizable, generalized Cartan matrix (cf [5]). Let $\mathfrak{g}$ (resp. $\mathcal{G}$) be the associated Kac-Moody Lie algebra (resp. Kac-Moody group). Let W be the Weyl group. Let U be the universal enveloping algebra of $\mathfrak{g}$ and $U^+_{\mathbb{Z}}$, the $\mathbb{Z}$-subalgebra of U generated by $X^n_\alpha/n!$, α a simple root. Let λ be a dominant, integral weight and V_λ the integrable, highest weight $\mathfrak{g}$ module (over $\mathbb{C}$) with highest weight λ. Let us fix a generator e for the highest weight space (note that e is unique up to scalars). For $\tau \in W$, let $e_\tau = \tau e$, $V_{\mathbb{Z},\tau} = U^+_{\mathbb{Z}} e_\tau$. In [12], we gave a conjectural basis for $V_{\mathbb{Z},\tau}$. Using this, we construct an explicit basis for $V_{\mathbb{Z},\tau}$ for the case $\mathfrak{g} = \hat{sp}(2n,\mathbb{C})$. We briefly describe below the results.

Let us fix a fundamental weight ω_i, $0 \le i \le n$ (notation as in [5]) and denote ω_i by just ω. Let $\mathcal{P}$ be the maximal parabolic subgroup of $\mathcal{G}$ associated to ω. Let V_ω be the irreducible $\mathfrak{g}$-module (over $\mathbb{C}$) with highest

*Partially supported by NSF Grant DMS-8701043

weight ω. Defining e, $e_\tau, V_{\mathbb{Z},\tau}$ etc. as above, for any field k, let $V_\tau = V_{\mathbb{Z},\tau} \otimes k$. Consider the canonical embedding $X(\tau) \hookrightarrow \mathbb{P}(V_\tau)$. Let L be the tautological line bundle on $\mathbb{P}(V_\tau)$. Let us denote $L|_{X(\tau)}$ by just L. We first construct a $\mathbb{Z}$-basis $\{Q_\Lambda\}$ for $V_{\mathbb{Z},\tau}$(cf §3) indexed by "admissible $\hat{Sp}(2n)$-Young tableaux on $X(\tau)$." Let $\{P_\Lambda\}$ be the basis of the $\mathbb{Z}$-dual of $V_{\mathbb{Z},\tau}$ dual to $\{Q_\Lambda\}$ and for any field k, let $p_\Lambda = P_\Lambda \otimes 1$. To an admissible Young tableau Λ on $X(\tau)$, we associate a pair of Weyl group elements $\rho(\Lambda), \delta(\Lambda)$, where $\tau \geq \rho(\Lambda) \geq \delta(\Lambda)$. We then define a monomial $p_{\Lambda_1} \cdot p_{\Lambda_2} \cdot \cdots p_{\Lambda_m}$ to be <u>standard on $X(\tau)$</u> if $\tau \geq \rho(\Lambda_1) \geq \delta(\Lambda_1) \geq \rho(\Lambda_2) \geq \delta(\Lambda_2) \geq \cdots \geq \delta(\Lambda_m)$. Then we prove

Theorem 1: Let $\tau \in W/W_{\mathcal{P}}$ and $X(\tau)$ the associated Schubert variety in $\mathcal{G}/\mathcal{P}$. The standard monomials on $X(\tau)$ of degree m form a basis of $H^0(X(\tau), L^m)$.

Let now $X(\tau)$ be a Schubert variety in $\mathcal{G}/\mathcal{B}$ and $L = \bigotimes_{i=0}^{n} L_i^{a_i}$, $a_i \in \mathbb{Z}^+$ (L_i being the line bundle on $X(\tau)$ defined as above with respect to ω_i (or $\mathcal{P}_i$)). Let $F \in H^0(X(\tau), L)$. Further, let $F = f_0 f_1 f_2 \cdots f_n$, where $f_i = p_{\Lambda_{i1}}, p_{\Lambda_{i2}} \cdots p_{\Lambda_{ia_i}}$. We say, F is <u>standard on $X(\tau)$ of multidegree $\underline{a}$</u> = $(a_0, \cdots, a_n)$ if there exists a sequence in W

$$\tau \geq \theta_{01} \geq \varphi_{01} \geq \theta_{02} \geq \cdots \geq \varphi_{0a_1} \geq \theta_{11} \geq \cdots \geq \varphi_{na_n}$$

such that $\pi_i(\theta_{ij}) = \rho(\Lambda_{ij}), \pi_i(\varphi_{ij}) = \delta(\Lambda_{ij})$, $1 \leq j \leq a_i$, $0 \leq i \leq n$ (here π_i denotes the projection $\pi_i : \mathcal{G}/\mathcal{B} \longrightarrow \mathcal{G}/\mathcal{P}_i$ (or same as $\pi_i : W \longrightarrow W/W_{\mathcal{P}_i}$)). We prove

Theorem 2: Standard monomials on $X(\tau)$ of degree $\underline{a}$ form a basis of $H^0(X(\tau), L)$.

The philosophy of the proof of Theorems 1 and 2 is the same as in [11], namely, given $X = X(\tau)$, we fix a nice Schubert divisor Y in X. We then construct a proper birational morphism $\psi : Z \longrightarrow X$ such that Z is a fiber space over $\mathbb{P}^1$ with fiber Y. By induction, we suppose the results to be true on Y, prove the results for Z and then make the results "go down to X."

As important consequences of the main theorem we obtain

(1) $X(\tau)$ is normal

(2) $H^i(X(\tau),L) = 0$, $i \geq 1$

(3) A character formula for $H^0(X(\tau),L)$.

(4) Character formulae for all integrable, highest weight $\mathfrak{g}$-modules (over $\mathbb{C}$).

§1. **Preliminaries.**

Let k be the base field which we assume to be algebraically closed of arbitrary characteristic. Let V be a 2n-dimensional k-vector space together with a skew-symmetric, non-degenerate bilinear form (,). Let $H = SL(V)$, $G = Sp(V) = \{A \in SL(V) | A \text{ preserves } (,)\}$. With respect to the standard basis $\{e_i, 1 \leq i \leq 2n\}$ of V, let us take the matrix of the form as $E = \begin{pmatrix} 0 & J \\ -J & 0 \end{pmatrix}$; where J = anti-diag (1...1). Then $G = H^{\sigma}$, where σ is the involution $\sigma\colon H \to H$, $\sigma(A) = E({}^tA)^{-1}E^{-1}$. Denoting by T(H) (resp. B(H)) the maximal torus consisting of diagonal matrices (resp. Borel subgroup in H consisting of upper triangular matrices), we have, T(H) and B(H) are stable under σ; further $T(G)(= T(H)^{\sigma})$ and $B(G)(= B(H)^{\sigma})$ are respectively a maximal torus and Borel subgroup in G. In the sequel, we shall denote T(G), B(G), by just T,B respectively. Let $N = N_G(T)$. Let $A = k[t,t^{-1}]$, t being an indeterminate. Let $A^+ = k[t]$. The map $\pi^+\colon A^+ \to k$ sending t to 0 induces a map $\pi^+\colon G(A^+) \to G$. Let $\mathcal{B} = \pi^{+\,-1}(B)$, $W = N(A)/T$, $\mathcal{G} = G(A)$. Let $\mathfrak{g}$ be the Kac-Moody Lie algebra corresponding to the generalized Cartan matrix

$$\begin{pmatrix} 2 & -2 & 0 & \cdots\cdots & 0 & 0 \\ -1 & 2 & -1 & \cdots\cdots & 0 & 0 \\ 0 & -1 & 2 & -1 \cdots & 0 & 0 \\ \vdots & & & & & \\ 0 & \cdots\cdots & & -1 & 2 & -1 \\ 0 & \cdots\cdots & & 0 & -2 & 2 \end{pmatrix}_{n+1 \times n+1}$$

With notation as in [5], let $S = \{\alpha_0, \alpha_1, \ldots, \alpha_n\}$ be the set of simple roots of $\mathfrak{g}$. For $w \in W$, let $X(w) = \bigcup_{\tau \le w} \mathcal{B}\tau\mathcal{B} \pmod{\mathcal{B}}$ be the Schubert variety in $\mathcal{G}/\mathcal{B}$ (see [6] for generalities on the infinite dimensional variety $\mathcal{G}/\mathcal{B}$). Let us fix a fundamental weight $\omega = \omega_i$, $0 \le i \le n$ and let $\mathcal{P}$ be the maximal parabolic subgroup of $\mathcal{G}$ obtained by "omitting α_i." Let $W_{\mathcal{P}}$ be the Weyl group of $\mathcal{P}$, $W^{\mathcal{P}}$ be the set of minimal representatives in W of $W/W_{\mathcal{P}}$. Let ω be the fundamental weight associated to $\mathcal{P}$.

Definition 1.1 (cf [1]): Let $\tau \in W^{\mathcal{P}}$ and let $X(\tau)$ be the associated Schubert variety in $\mathcal{G}/\mathcal{P}$. Let φ in $W^{\mathcal{P}}$ be such that $X(\varphi)$ is a divisor in $X(\tau)$, say $\varphi = \tau s_\beta$, for some positive root β. We define the multiplicity of $X(\varphi)$ in $X(\tau)$ as

$$m(\varphi,\tau) = (\omega,\beta^*)\left(= \frac{2(\omega,\beta)}{(\beta,\beta)}\right).$$

Definition 1.2: With notation as above, if $m(\varphi,\tau) > 1$, then we shall refer to $X(\varphi)$ as a multiple divisor in $X(\tau)$.

Definition 1.3: With notation as above, let $\varphi = s_\gamma\tau$, for some positive root γ. We say $X(\varphi)$ is a moving (resp. nonmoving) divisor in $X(\tau)$ if γ is simple (resp. nonsimple).

Lemma 1.4 (cf [11], Lemma 1.5): Let $X(\varphi)$ be a moving divisor in $X(\tau)$, say $\varphi = s_\alpha\tau$, for some $\alpha \in S$. Then for any $\theta \le \tau$, we have either $\theta \le \varphi$ or $\theta = s_\alpha\theta'$, for some $\theta' \le \varphi$.

Lemma 1.5 ([11]): With notation as in Lemma 1.5, let $X(\theta)$ be a divisor in $X(\tau)$ and let $\theta' = s_\alpha\theta$. Then $m(\theta,\tau) = m(\theta',\varphi)$.

§2. The Bases $\{Q(\lambda,\mu)_N\}$ and $\{P(\lambda,\mu)_N\}$.

We keep the notation of §1.

Let $U = U(\mathfrak{g})$ be the universal enveloping algebra of $\mathfrak{g}$. Let $U_{\mathbb{Z}}$ (resp. $U^+_{\mathbb{Z}}$, $U^-_{\mathbb{Z}}$) denote the $\mathbb{Z}$-subalgebra of U generated by $X^n_{\pm\alpha}/n!$ (resp. $X^n_{\alpha}/n!$, $X^n_{-\alpha}/n!$), $\alpha \in S$, $n \in \mathbb{Z}^+$. In the sequel, we shall denote $X^n_{\alpha}/n!$ by $X^{(n)}_{\alpha}$. Let U_{β} (resp. $U_{\beta,\mathbb{Z}}$) be the $\mathbb{Q}$-vector subspace (resp. $\mathbb{Z}$-submodule) of U (resp. $U_{\mathbb{Z}}$) generated by X^n_{β} (resp. $X^{(n)}_{\beta}$), $\beta = \pm\alpha$, $\alpha \in S$, $n \in \mathbb{Z}^+$. Let $\mathcal{P}, \omega, W^{\mathcal{P}}$ etc. be as in §1. Let V_{ω} or just V be the integrable, highest weight $\mathfrak{g}$-module (over $\mathbb{C}$) with highest weight ω. Let us fix a highest weight vector e (which is unique up to scalars). Let $V_{\mathbb{Z}} = U_{\mathbb{Z}}e$. For $\tau \in W^{\mathcal{P}}$, let $e_{\tau} = \tau e$, $V_{\mathbb{Z},\tau} = U^+_{\mathbb{Z}}e_{\tau}$. For any field k, let $V_{k,\tau}$ (or just V_{τ}) be $V_{\mathbb{Z},\tau} \otimes k$.

A $\mathbb{Z}$-basis $\mathcal{B}_{\mathbb{Z},\tau}$ for $V_{\mathbb{Z},\tau}$:

We first describe an indexing set $I_{\mathbb{Z},\tau}$ or just I_{τ}. The indexing set is going to consist of certain Young tableaux to be called "admissible $\hat{Sp}(2n)$-Young tableaux." Let us first recall an admissible $\hat{SL}(n)$-Young diagram.

Definition 2.1: (cf. [2],[13]). Let $\Lambda = (\lambda_1,\ldots,\lambda_r)$ be a Young diagram with λ_j boxes in the jth row. We say Λ is <u>$\hat{SL}(n)$-admissible</u>, if

(1) $\lambda_1 \geq \lambda_2 \geq \cdots \geq \lambda_r$

(2) Number of rows of same length is $\leq n-1$.

The natural embedding $C^{(1)}_n \hookrightarrow A^{(1)}_{2n-1}$ induces a projection π: weight lattice of $A^{(1)}_{2n-1} \longrightarrow$ weight lattice of $C^{(1)}_n$ given by $\pi(\omega_i) = \omega_i$, $0 \leq i \leq n$, $\pi(\omega_i) = \omega_{2n-i}$, $n+1 \leq i \leq 2n-1$. We use this to realize the admissible $\hat{Sp}(2n)$-Young diagrams as certain admissible $\hat{SL}(2n)$-Young diagrams.

Let Λ be an $\hat{SL}(2n)$-admissible Young diagram. Let $\omega = \omega_d$ for some d, $0 \leq d \leq n$. We fill in Λ with simple roots (or integers $0,1,\ldots,n$) as follows. We fill in the first row (from left to right) with d, d+1, d+2,...

modulo 2n. Then we go down each column decreasing the value by 1 at a time (again modulo 2n). Further, we identify i with 2n-i, if i > n.

Definition 2.2: A $\hat{SL}(2n)$-admissible Young tableau Λ filled with α_i, $0 \le i \le n$ will be referred to as a <u>weakly admissible $\hat{Sp}(2n)$-Young tableau.</u>

Before we could define admissible $\hat{Sp}(2n)$-Young tableaux we need to recall some results from [13].

Let

$$Z = \left\{ (a_1, b_1, a_2, b_2, \cdots, a_s, b_s) \,\middle|\, \begin{array}{l} (1)\ a_i, b_j \in \mathbb{Z}^+ \\ (2)\ a_i, b_j > 0,\ i \ne 1,\ j \ne s \end{array} \right\}$$

Definition 2.3: Let $A \in Z$.

(a) A is called a <u>head</u> if for every b_t, $\sum_{i \le t} b_i \le \sum_{i \le t} a_i$ (in particular, $a_1 \ne 0$)

(b) A is called a <u>tail</u> if for every a_t, $\sum_{i \ge t} a_i \le \sum_{i \ge t} b_i$ (in particular, $b_s \ne 0$).

We recall the following result from [13] (cf. [13], Lemmas 2.2 and 2.3).

Lemma 2.4:

(a) Let A be not a tail. Then there exists a canonical tail associated to A.

(b) Let A be not a head. Then there exists a canonical head associated to A.

Definition 2.5: We call a subset Z_1 of Z <u>complete</u>, if either

1. Z = {A}, where A is both a head and tail (in which case we call it <u>trivial</u>), or

2. $Z_1 = \{A_0, A_1, \cdots, A_t\}$ where A_0 is a head, A_t is a tail and A_i, $1 \le i \le t$ are as in the proof of Lemma 2.2 of [13].

Remark 2.6: Given $A = (a_1, b_1, a_2, b_2, \cdots)$, following the procedure in Lemmas 2.2 and 2.3 of [13], there exists a unique complete subset Z_1 of Z to which

A belongs.

Definition 2.7: Given a weakly admissible $\hat{Sp}(2n)$-Young tableau, Λ, a corner box in Λ is one such that

1. it is a box at the end of a certain row, say the kth row, r_k.
2. $\ell(r_k) > \ell(r_{k+1})$.

Further, if a corner box is filled with the simple root β, we shall refer to it as a corner β box.

Given a weakly admissible $\hat{Sp}(2n)$-Young tableau Λ, let the right most corner box be filled with α_j for some j , $0 \le j \le n$. Let us denote α_j by just α. Consider all the blank boxes filling all of which by α yields a weakly admissible $\hat{Sp}(2n)$-Young tableau. We shall refer to these as blank corner α-boxes. Let $A(\Lambda) = (b_1, a_2, b_2, a_3, b_3, \cdots)$ where

b_1 = # first set of corner α-boxes

a_2 = # first set of blank corner α-boxes

b_2 = # second set of corner α-boxes

and so on.

Let us denote $A(\Lambda)$ by just A, so that $A = (a_1, b_1, a_2, b_2, \cdots)$, where $a_1 = 0$. Let Z_1 be the (unique) complete subset of Z such that $A \in Z_1$ (cf. Remark 2.5). Note that Z_1 is not trivial (note that A is not a head, as $a_1=0$). Let $Z_1 = \{A_0, A_1, \cdots, A_t\}$ (cf. Definition 2.4). Let $A = A_i$. We now associate weakly admissible $\hat{Sp}(2n)$-Young tableaux Λ_j , $0 \le j \le t$ as follows.

Case 1: $j > i$.

Let $m = t - i$. Then $m = \sum_{u=1}^{r} C_u$ (notation as in the proof of Lemma 2.2 of [13]). Let $j-i = C_1 + \ldots + C_{u-1} + k$, where $1 \le k \le c_u$ (here, if $u = 1$, then $j-i = k$). Then Λ_j is obtained by filling in with α, the first D_ℓ blank corner α-boxes in the $(i_\ell-1)^{th}$ set of blank corner α-boxes , $1 \le \ell \le u$, where

$$D_\ell = \begin{cases} C_\ell, & \ell \neq u \\ k, & \ell = u \end{cases}$$

(notation as in the proof of Lemma 2.2 of [13]).

Case 2: $j < i$.

Then the m in the proof of Lemma 2.3 of [13] is simply i. Writing $m = \sum_{u=1}^{r} C_u$ (with notation as in the proof of Lemma 2.3 of [13]), let $j = C_1 + \ldots + C_{u-1} + k$ where $1 \leq k \leq C_u$ (here, if $u = 1$, then $j = k$). Then Λ_j is obtained from Λ by replacing the bottom D_ℓ corner α-boxes by blank corner α-boxes, in the i_ℓ^{th} set (from the bottom) of corner α-boxes, $1 \leq \ell \leq u$, where

$$D_\ell = \begin{cases} C_\ell, & \ell \neq u \\ k, & \ell = u \end{cases}$$

(notation as in the proof of Lemma 2.3 of [13]).

Definition 2.8: Given a weakly admissible $\hat{S}p(2n)$-Young tableau Λ, we define a sequence $\{\Lambda_{ij}\}$ of weakly admissible $\hat{S}p(2n)$-Young tableaux as follows.

Step 1: Carrying out the above discussion for Λ, let us denote Λ_j by Λ_{j1}, $0 \leq j \leq t$.

Step 2: Carrying out the above discussion for Λ_{01} (with respect to the right most corner box in Λ_{01}), we obtain a sequence $\{\Lambda_{j2}\}$.

In Step 3 we work with Λ_{02} and so on. Thus we obtain a sequence $\{\Lambda_{jk}\}$ of weakly admissible $\hat{S}p(2n)$-Young tableaux. Further, if $1 \leq k \leq m$, then Λ_{0m} is the empty Young tableau.

Definition 2.9: Let Λ be a weakly admissible $\hat{S}p(2n)$-Young tableau. We say that Λ satisfies the <u>end-column property</u> if satisfies the following condition:

If the right most column in Λ starts with α_0 or α_n, then its length is 1.

Definition 2.10: Let Λ be a weakly admissible $\hat{Sp}(2n)$-Young tableau and let $\{\Lambda_{jk}\}$ be the sequence of weakly admissible $\hat{Sp}(2n)$-Young tableaux as in Definition 2.8. We say Λ is an <u>admissible $\hat{Sp}(2n)$-Young tableau</u> if all the Λ_{jk}'s have the end-column property. In the sequel, we shall refer to an admissible $\hat{Sp}(2n)$-Young tableau as just an admissible Young tableau.

Definition 2.11: Given an admissible Young tableau Λ, we define inductively the vector Q_Λ in $V_{\mathbb{Z}}$ $(= U_{\mathbb{Z}}e)$ as

$$Q_\Lambda = X_{-\alpha}^{(i)} Q_{\Lambda_0}$$

(with i and Λ_0 as in the discussion above. Note that Λ_0 is simply Λ_{01}).

Remark 2.12(a): With notation as above, let $Q_\Lambda = X_{-\alpha}^{(i)} Q_{\Lambda_0}$. Let $\chi(\Lambda)$ be the weight of Q_Λ (note that $\chi(\Lambda) = \omega - \sum_{i=0}^{n} m_i\alpha_i$, $m_i = \#\alpha_i$-boxes in Λ). It is easily seen that Q_Λ is an extremal weight vector if and only if Q_{Λ_0} is an extremal weight vector and in Λ_0, #{corner α-boxes} = 0 and #{blank corner α-boxes} = i. Conversely, to an extremal weight vector Q_θ, we can associate an admissible Young tableau θ^* (in the obvious way).

(b) Given an extremal weight vector Q_θ and $\alpha \in S$, let y = #{corner α-boxes in θ^*}, x = #{blank corner α-boxes in θ^*}. Then we have, either

(1) $x = 0 = y$, in which case $(\theta(\omega),\alpha^*) = 0$

(2) $x \neq 0$, $y = 0$, in which case $(\theta(\omega),\alpha^*) = x$

(3) $x = 0$, $y \neq 0$, in which case $(\theta(\omega),\alpha^*) = -y$.

(c) Given θ, suppose β is a nonsimple positive root such that $X(\theta)$ is a divisor in $X(s_\beta\theta)$ with $m(\theta,s_\beta\theta) = r$, then we talk about corner β-boxes and it is clear that in θ^*, there are r corner blank β-boxes and no corner β-boxes.

The elements $\rho(\Lambda)$ and $\delta(\Lambda)$:

With notation as above let $Q_\Lambda = X_{-\alpha}^{(i)} Q_{\Lambda_0}$. Then we define $\rho(\Lambda)$ inductively as $\rho(\Lambda) = s_\alpha \rho(\Lambda_0)$. To define $\delta(\Lambda)$, let us write

$$Q_\Lambda = X_{-\beta_t}^{(m_t)} \cdots X_{-\beta_1}^{(m_1)} e$$

where β_i is the right most corner box at each step. Let ℓ, $1 \le \ell \le t$ be the largest integer such that $X_{-\beta_\ell}^{(m_\ell)} \cdots X_{-\beta_1}^{(m_1)} e$ is an extremal weight vector, say Q_θ. Then from the way Q_Λ is defined, it is clear that Λ is obtained from θ^* in $t-\ell$ steps as follows:

We have $(\theta(\omega), \beta_{\ell+1}^*) \ge m_{\ell+1}$. Let Λ_1 be the admissible Young tableau obtained by filling in the first $m_{\ell+1}$ blank corner $\beta_{\ell+1}$ boxes in θ. Next we have $(\chi(\Lambda_1), \beta_{\ell+2}^*) \ge m_{\ell+2}$ and let Λ_2 be the admissible Young tableau obtained by filling in the first $m_{\ell+2}$ blank corner $\beta_{\ell+2}$ boxes in Λ_1 and so on. Now, if in the above process we allow for nonsimple roots also, then it is clear that there exists a unique maximal element $\delta \in W^{\mathcal{P}}$, positive roots γ_k, positive integers $n_k > 0$, $1 \le k \le s$ such that $X(s_k \cdots s_1\delta)$ is a divisor in $X(s_{k+1} \cdots s_1\delta)$ occurring with multiplicity $\ge n_k$ (here, s_i denotes the reflection with respect to γ_i). Further, let Λ_1 be the admissible Young tableau obtained by filling in the first n_1 blank corner γ_1-boxes in δ^* (cf. Remark 2.12(c)) with γ_1. We have $(\chi(\Lambda_1), \gamma_2^*) \ge n_2$. Let Λ_2 be the admissible Young tableau obtained by filling in the first n_2 blank corner γ_2-boxes in Λ_1 and so on. Then $\Lambda = \Lambda_s$. We set

$$\delta(\Lambda) = \delta.$$

Remark 2.13: Starting with Λ, at each step let us delete just one box, namely the right most corner box deleting which yields a weakly admissible Young tableau. This process gives rise to a vector $X_{-\varepsilon_r}^{(a_r)} \ldots X_{-\varepsilon_1}^{(a_1)} e$, $\varepsilon_i \in S$.

Let u be the largest integer such that $X_{-\varepsilon_u}^{(a_u)} \ldots X_{-\varepsilon_1}^{(a_1)} e$ is extremal say e_φ, for some $\varphi \in W^{\mathcal{P}}$. Then it is seen easily that $\varphi = \delta(\Lambda)$.

Notation 2.14: Let us denote $\rho(\Lambda), \delta(\Lambda)$, by just ρ, δ and $s_k \ldots s_1\delta$ by δ_k. We shall in the sequel denote Q_Λ by $Q(\rho,\delta)_N$, also, where

$$N = \{(\delta_0 = \delta < \delta_1 < \cdots < \delta_s = \rho);\ (n_1, \cdots, n_s)\}.$$

We shall have occasion to look at the following special vectors:

(1) Let $s_\alpha\rho = \varphi$, $\alpha \in S$ and let $r = m(\varphi,\rho)$.

If $\delta = \varphi$ and $n_1 = t$ for some $t \le r$, we shall denote Q_Λ by just $Q(\rho,\varphi)_t$ (note that $Q_\Lambda = X_{-\alpha}^{(t)} Q_\varphi$)

(2) Let $\eta = s_\beta\rho$ be such that $X(\eta)$ is a divisor in $X(\rho)$, where $\beta \ne \alpha$. Let $\theta = s_\alpha\eta$. Let $m(\eta,\rho) = m$ (note that $m(\theta,\varphi)$ is also m). Corresponding to $\delta = \eta$, $n_1 = t$ for some $t \le m$, we shall denote Q_Λ by just $Q(\rho,\eta)_t$ (note that $(\rho,\eta)_t = X_{-\alpha}^{(b)}(\varphi,\delta)_t$ where

$$b = \begin{cases} t, & \text{if } (\beta,\alpha^*) = 0 \\ t+r', & \text{if } (\beta,\alpha^*) = 1 \\ r+m-t, & \text{if } (\beta,\alpha^*) = -1 \end{cases}$$

where $r' = m(\theta,\eta)$ and $r = m(\varphi,\rho)$.

§3. First Basis Theorem.

Generalized Demazure character formula (cf [7]).

Let N = {integral weights of $\mathfrak{g}$}. Let $\mathbb{Z}[N]$ denote the group ring of the multiplicative group expN. For a simple root α, let M_{s_α} or just M_α be the linear operator $M_\alpha : \mathbb{Z}[N] \longrightarrow \mathbb{Z}[N]$ defined by

$$M_\alpha(\exp\lambda) = \begin{cases} \exp\lambda + \exp(\lambda-\alpha) + \cdots + \exp(\lambda-m\alpha), & \text{if } m = (\lambda,\alpha^*) \ge 0 \\ -[\exp(\lambda+\alpha) + \cdots + \exp(\lambda+(m-1)\alpha), & \text{if } -m = (\lambda,\alpha^*) < 0 \end{cases}$$

Let $\tau = s_1 \cdots s_r$ (reduced expression) and let $M_\tau = M_{s_1} \circ M_{s_2} \circ \cdots \circ M_{s_r}$.

Let $V_\tau = V_{\mathbb{Z},\tau} \otimes \mathbb{C}$.

Theorem 3.1 (cf [7]):

$$\text{char } V_\tau = M_\tau(\exp(\omega)).$$

Notation 3.2: For τ in W, let

$$I_\tau = \{\text{admissible Young tableau } \Lambda \mid \tau \geq \rho(\Lambda)\}$$

Theorem 3.3:

(a) $\text{char } V_\tau = \sum_{\Lambda \in I_\tau} \exp(\chi(\Lambda))$

(b) $\dim V_\tau = \# I_\tau$

(note that (b) is an immediate consequence of (a)).

Proof: (by induction on $\dim X(\tau)$). If $\dim X(\tau) = 0$, then the result is obvious. Let then $\dim X(\tau) > 0$. Let the right most corner box in τ^* be an α-box (cf §2). Let $\varphi = s_\alpha \tau$. We have

$$\begin{aligned} \text{char } V_\tau &= M_{s_\alpha}(\text{char } V_\varphi) \\ &= M_{s_\alpha} \sum_{\Lambda \in I_\varphi} \exp(\chi(\Lambda)) \end{aligned}$$

Suppose Λ_0 in I_φ is such that $A(\Lambda_0)$ is a head. Let $A(\Lambda_0), A(\Lambda_1), \cdots, A(\Lambda_t)$ be the associated complete subset of Z (cf §2). Then from our discussion in §2, it is clear that either $\Lambda_i \in I_\varphi$, $1 \leq i \leq t$ or $\Lambda_i \notin I_\varphi$, $1 \leq i \leq t$ according as $s_\alpha \rho(\Lambda_0) \leq \varphi$ or $\nleq \varphi$. Hence I_φ breaks up as

$$I_\varphi = \dot{\bigcup_i} B_i \dot{\bigcup_j} C_j$$

where

$$B_i = \left\{ \Lambda_0, \cdots, \Lambda_t \;\middle|\; \begin{array}{l} A(\Lambda_0), \cdots, A(\Lambda_t) \text{ is a} \\ \text{complete subset of Z} \end{array} \right\}$$

and

$$C_j \text{ is a singleton } \{\Lambda\},$$

where $A(\Lambda)$ is a head and $s_\alpha \rho(\Lambda) \nleq \varphi$. We have for a B_i,

$$M_\alpha(\exp(\Lambda_0) + \cdots + \exp(\Lambda_t)) = \exp(\Lambda_0 + \cdots + \exp(\Lambda_t)$$

(note that $s_\alpha(\chi(\Lambda_i)) = \chi(\Lambda_{t-i})$). If a $C_j (= \Lambda)$ is such that $(\chi(\Lambda), \alpha^*) = 0$, again $M_\alpha(\exp(\chi(\Lambda))) = \exp(\chi(\Lambda)))$. Thus the extra terms in char V_τ arise from C_j's such that $(\chi(C_j),\alpha^*) > 0$, and are given by $\Lambda_1,\cdots,\Lambda_t$ (where $A(C_j),A(\Lambda_1),\cdots,A(\Lambda_t)$ form a complete subset of Z). On the other hand any Λ in I_τ arises from an unique Λ_0 in I_τ such that $A(\Lambda_0)$ is a head. The result now follows:

Theorem 3.4 (First Basis Theorem): Let $\mathcal{B}_\tau = \{Q_\Lambda, \Lambda \in I_\tau\}$. Then $\mathcal{B}$ is a $\mathbb{Z}$-basis for $V_{\mathbb{Z},\tau}$.

Proof: (by induction on dim $X(\tau)$). In view of Theorem 3.3, suffices to check that $\mathcal{B}_\tau$ generates $V_{\mathbb{Z},\tau}$. If dim $X(\tau) = 0$, then the result is obvious. Let then dim $X(\tau) > 0$. Let α and φ be as in the proof of Theorem 3.3. We have (cf [11]), $V_{\mathbb{Z},\tau} = U_{-\alpha,\mathbb{Z}}V_{\mathbb{Z},\varphi}$. Hence suffices to check $X_{-\alpha}^{(r)}Q_\Lambda$, $\Lambda \in I_\varphi$ belongs to the $\mathbb{Z}$-span of Q_Δ, $\Delta \in I_\tau$. Again, suffices to consider Q_Λ's such that $s_\alpha\rho(\Lambda) \not\geq \varphi$ (since $X_{-\alpha}^{(r)}Q_\Lambda \in V_{\mathbb{Z},s_\alpha\rho(\Lambda)}$). As in the proof of Theorem 3.3, we have, for such a Λ, $A(\Lambda)$ is a head. Let $A(\Lambda_t)$ be the corresponding tail (note that $\Lambda_t = X_{-\alpha}^{(t)}Q_\Lambda$). Let $A(\Lambda_t) = (a_1b_1a_2b_2 \cdots a_sb_s)$. Consider all elements B_j of Z obtained from $A(\Lambda_t)$ by replacing a_i by $x,1,a_i-(x+1)$, for some i, $1 \leq i \leq s$. (Here $0 \leq x \leq a_i-1$; if x is 0 (resp. a_i-1) then the corresponding B_j has $b_{i-1}+1$, a_i-1 (resp. a_i-1,b_i+1) in the place of b_{i-1},a_i (resp. a_i,b_i).) Let $A(\Delta_j)$ be the corresponding head. Then it is clear that

$$X_{-\alpha}Q_{\Lambda_t} \in \mathbb{Z} \text{ span of } \{X_{-\alpha}^{(r)}Q_{\Delta_j}\}.$$

Thus we may assume Λ is such that $A(\Lambda_t) = (b_1)$. This implies $A(\Lambda) = (a_1)$. In this case, we have obviously

$$X_{-\alpha}^{(i)}Q_\Lambda = Q_{\Lambda_i}\ ,\ 1 \leq i \leq t$$

(note that $t = a_1$) and $X_{-\alpha}^{(i)}Q_\Lambda = 0$, $i > t$. This completes the proof of Theorem 3.4.

Definition 3.5: Let $\{P_\Lambda,\ \Lambda \in I_\tau\}$ be the $\mathbb{Z}$-basis of the $\mathbb{Z}$-dual of $V_{\mathbb{Z},\tau}$, dual to $\{Q_\Lambda, \Lambda \in I_\tau\}$.

Remark 3.6: Let $\theta \le \tau$. Then $P_\Lambda|_{X(\theta)} \equiv 0$ if and only if $\theta \not\ge \rho(\Lambda)$.

§4. Standard monomials and their linear independence.

Let k be the base field, $V_\tau = V_{\mathbb{Z},\tau} \otimes k$, $p_\Lambda = P_\Lambda \otimes 1$. If $\Lambda = \theta^*$, for some $\theta \in W$, then we shall denote p_Λ by just p_θ.

Definition 4.1: A monomial $p_{\Lambda_1} p_{\Lambda_2} \cdots p_{\Lambda_m}$ (in $S^m(V_\tau^*)$) is <u>called standard on X(τ)</u> if

$$\tau \ge \rho(\Lambda_1) \ge \delta(\Lambda_1) \ge \rho(\Lambda_2) \ge \cdots \ge \rho(\Lambda_m) \ge \delta(\Lambda_m).$$

Notation 4.2: Let Λ be given and let $\rho(\Lambda) = \tau$. Let $Q_\Lambda = X_{-\alpha}^{(t)} Q_{\Lambda'}$. We shall denote t by $t(\Lambda)$, δ_{s-1} by $\eta(\Lambda)$, n_s by $n(\Lambda)$ (cf. Notation 2.14). In the sequel, we shall also denote Λ by $(\tau,\eta,\delta)_N$ (where $\eta = \eta(\Lambda), \delta = \delta(\Lambda)$) and p_Λ by $p(\tau,\eta,\delta)_N$. Also, if $\delta = \eta$ (so that s=1), then we shall denote p_Λ by just $p(\tau,\eta)_{n_1}$.

Lemma 4.3: Let $\tau \in W^P$. Let φ, α be as in §3.

(1) Let $\eta \ne \varphi$. Let $m(\eta,\tau) = m$. Then for $\eta' \ne \eta$, we have, on $X(\tau)$,

(a) $p(\tau,\eta',\delta')_{N'} p(\tau,\eta)_{m-s} = p(\tau)p(\tau,\lambda,\lambda')_{M'}$ where either $\eta' \ne \varphi$ or $\eta' = \varphi$ in which case $t(\Lambda') > s$ (here $\Lambda' = (\tau,\eta',\delta')_{N'}$). Further $\lambda = \eta$ or η'.

(b) Let $\Lambda = (\tau,\eta,\delta)_N$. If $n(\Lambda) > s$, then

$$p(\tau,\eta,\delta)_N p(\tau,\eta)_{m-s} = p(\tau)p(\tau,\eta,\delta)_{N'}$$

for some N'.

(c) If $n(\Lambda) = s$, then

$$p(\tau,\eta,\delta)_N p(\tau,\eta)_{m-s} = p(\tau)F$$

where $F|_{X(\eta)} = F(\eta,\lambda,\delta)_{N'}$ for some λ and N'.

(2) Let $m(\varphi,\tau) = r$. Let $\Lambda = (\tau,\eta,\delta)_N$, $\eta \ne \varphi$.

(a) If $t(\Lambda) > a$, then

$$p(\tau,\eta,\delta)_N p(\tau,\varphi)_{r-a} = p(\tau)p(\tau,\lambda,\lambda')_{N'},$$

where $\lambda = \varphi$ or η.

(b) If $t(\Lambda) = a$ and $\delta \le \varphi$, then

$$p(\tau,\eta,\delta)_N p(\tau,\varphi)_{r-a} = p(\tau)F$$

where $F|_{X(\varphi)} = p(\varphi,\varphi',\delta)_{N'}$, for some φ' and N'.

(c) Let $\Lambda = (\tau,\varphi,\delta)_N$. If $t(\Lambda) > s$, then

$$p(\tau,\varphi,\delta)_N p(\tau,\varphi)_{r-a} = p(\tau)p(\tau,\varphi,\delta)_{N'}$$

for some N'.

(d) If $t(\Lambda) = s$, then

$$p(\tau,\varphi,\delta)_N p(\tau,\varphi)_{r-a} = p(\tau)F$$

where $F|_{X(\varphi)} = F(\varphi,\lambda,\delta)_{N'}$ for some λ and N'.

The above assertions are proved in the same spirit as in [10] and we skip the details.

Theorem 4.4: Standard monomials on $X(\tau)$ of degree m are linearly independent.

Proof: (by induction on dim $X(\tau)$). If dim $X(\tau) = 0$, then p_τ^m is the only degree m standard monomial on $X(\tau)$ and the result is obvious. Let then dim $X(\tau) > 1$. Let φ,α be as in §3. Let

$$\text{(I)} \qquad \Sigma a_i F_i = 0 \ , \ a_i \in k$$

be a linear relation among standard monomials of degree m on $X(\tau)$. If there is a F_i such that F_i starts with a p_Λ, with $\rho(\Lambda) < \tau$, then by restricting (I) to $X(\rho(\Lambda))$, we conclude $a_i = 0$ (using induction hypothesis). Hence we may assume that each F_i starts with a p_Λ where $\rho(\Lambda) = \tau$.

Let us then rewrite (I) as

$$\text{(II)} \qquad a_\tau p_\tau F_\tau + \Sigma a_\Lambda p_\Lambda F_\Lambda = 0$$

where F_Λ is a standard monomial of the form $p_{\Lambda_1} p_{\Lambda_2} \cdots$ with $\rho(\Lambda_1) \le \delta(\Lambda)$. Let $Q_\Lambda = X_{-\alpha}^{(t)} Q_{\Lambda'}$ and $t(\Lambda) = t$ (as before). Let $t_0 = \min_\Lambda \{t(\Lambda)\}$.

Case 1: All Λ with $t(\Lambda) = t_0$ are such that $\eta(\Lambda) \ne \varphi$. Let us fix a Λ with

$t(\Lambda) = t_0$. Let $\eta = \eta(\Lambda)$. Consider all Δ (appearing on the L.H.S. of II) with $\eta(\Delta) = \eta$. Let $b = \min_{\Delta} \{n(\Delta)\}$. We now multiply II thruout by $p(\tau,\eta)_{m-b}$ (where $m = m(\eta,\tau)$, use Lemma 4.3, (1), cancel $p(\tau)$, restrict (II) to $X(\eta)$ and conclude $a_\Delta = 0$ for all Δ such that $\eta(\Delta) = \eta$ and $n(\Delta) = b$.

Case 2: There are Λ's, with $\eta(\Lambda) = \varphi$ and $t(\Lambda) = t_0$. We divide this case into the following two subcases:

Subcase 2(a): All Δ's with $t(\Lambda) = t_0$ are such that $\delta(\Delta) \leq \varphi$. In this case, we multiply (II) thruout by $p(\tau,\varphi)_{r-t_0}$, use Proposition 4.3, (2), cancel $p(\tau)$, restrict (II) to $X(\varphi)$ and conclude $a_\Delta = 0$ for all Δ such that $t(\Lambda) = t_0$.

Subcase 2(b): There are Δ's with $t(\Lambda) = t_0$ and $\delta(\Delta) \not\leq \varphi$. We fix such a Δ. Let $\eta(\Delta) = \eta$. Now we consider <u>all</u> Θ's with $\eta(\Theta) = \eta$ and $n(\Theta) \leq n(\Delta)$ and proceed as in Case (1). Then we are reduced to Subcase 2(a). Thus proceeding II reduces to

$$a_\tau p_\tau F_\tau = 0$$

from which we conclude $a_\tau = 0$. Thus II reduces to the trivial relation. This completes the proof of Theorem 4.4.

§5. Filtration for the ideal of $H(\tau)$.

Consider $X(\tau) \hookrightarrow \mathbb{P}(V_\tau)$. Let L denote the tautological line bundle on $\mathbb{P}(V_\tau)$. Let $R(\tau)$ denote the homogeneous coordinate ring of $X(\tau)$ under the above embedding. We shall show (cf Theorem 5.3) that the standard monomials on $X(\tau)$ of degree m form a basis of $H^0(X(\tau),L^m)$. Let $H(\tau)$ = zero set of p_τ in $X(\tau)$ and let $I(H(\tau))$ = the ideal $p_\tau R(\tau)$ in $R(\tau)$. We now take a total ordering on K_τ = {admissible Young diagrams Λ on $X(\tau)$ such that $\rho(\Lambda) = \tau$}.

Fix an integer a, $1 \leq a < r$, (here $r = m(\varphi,\tau)$, $\varphi = s_\alpha\tau$, $\alpha \in S$). Let

$$I_a = \{\Lambda \mid t(\Lambda) = a\}.$$

Let $I = K_\tau - \bigcup_{1\leq a<r} I_a - \{\tau^*\}$. We index K_τ as follows. We keep τ^* as

the first element, followed by I, then $I_{r-1}, I_{r-2}, \cdots, I_1$. We now take a total ordering on I, I_a, $1 \le a < r$ as follows:

1. **Indexing of I:** We fix a Λ in I and denote $\eta = \eta(\Lambda)$. We then arrange all Δ's in I with $\eta(\Delta) = \eta$ with decreasing order of $n(\Delta)$ in such a way that among the Δ's with the same $n(\Delta)$, if $\delta(\Delta_1) < \delta(\Delta_2)$, then Δ_2 will precede Δ_1.

2. **Indexing of I_a:** We first consider $\{\Lambda \in I_a | \eta(\Lambda) = \varphi\}$ and arrange them as in (1). We then consider $\{\Lambda \in I_a | \eta(\Lambda) \ne \varphi$ and $\delta(\Lambda) \le \varphi\}$ and arrange them as in (1). Finally we consider $\{\Lambda \in I_a | \delta(\Lambda) \not\le \varphi\}$ and arrange them as in (1).

We write $K_\tau = \{\Lambda_0, \Lambda_1, \cdots, \Lambda_m\}$ where $\Lambda_0 = \tau^*$ and $m+1 = \#K_\tau$. We now define the ideals I_t, $0 \le t \le m$ in $R(\tau)$ as

$$I_t = \sum_{0 \le i \le t} p_i R(\tau)$$

where $p_i = p_{\Lambda_i}$, $0 \le i \le m$ (note that $I_0 = p_\tau R(\tau)) (= I(H(\tau))$. Let $\underline{I}_t$ be the ideal sheaf in $\mathcal{O}_{X(\tau)}$ associated to I_t. Let us fix a simple root β such that if $w = s_\beta \tau$, then $X(\tau)$ is a divisor in $X(w)$. Proceeding as in [12], we have

Proposition 5.1: Suppose that the standard monomials on $X(\tau)$ of degree s form a basis of $H^0(X(\tau), L^s)$, $s \in \mathbb{N}$.

(1) We have canonical isomorphisms (as $\mathcal{O}_{X(\tau)}$-modules)

$$f_t : \underline{I}_t/\underline{I}_{t-1} \approx \mathcal{O}_{X(\delta)}(-1) , \quad 0 \le t \le m$$

where δ is given by $\delta = \delta(\Lambda), p_\Lambda = p_t$ (here $I_{-1} = (0)$).

(2) Let $B_\beta = \mathcal{B} \cap SL(2,\beta)$, where $SL(2,\beta)$ is the "SL(2)" associated to the simple root β. We have B_β-isomorphisms

$$\underline{I}_t/\underline{I}_{t-1} \approx \chi_t \otimes \mathcal{O}_{X(\delta)}(-1) , \quad 0 \le t \le m$$

where χ_t represents the 1-dimensional B_β-module associated to $-\chi(\Lambda)$ (where recall that $\chi(\Lambda)$ = weight of Q_Λ).

(3) The ideal sheaf $\underline{I}(H(\tau)_{red})$ in $\mathcal{O}_{X(\tau)}$ is precisely $\underline{I}_m$.

With w, τ, β etc. as above, we define (as in [12]) Z_τ (or just Z) as

(i) Z = the fiber product $SL(2,\beta) \times^{B_\beta} X(\tau)$.

(ii) We denote by ψ, the canonical morphism

$$\psi : Z \longrightarrow X(w),$$

defined by

$$(g,y) \longmapsto gy,\ g \in SL(2,\beta),\ y \in X(\tau)$$

(iii) For a coherent $\mathcal{O}_{X(\tau)}$-module F with a B_β action compatible with the action of B_β on $X(\tau)$, we set

$$\tilde{F} = SL(2,\beta) \times^{B_\beta} F$$

(iv) For a bundle N on Z, we set

$$N^{(\ell)} = N \otimes p^*(\mathcal{O}_{\mathbb{P}^1}(\ell))\ ,\ \ell \in \mathbb{Z}$$

where p is the canonical map

$$p : Z \longrightarrow \mathbb{P}^1 (= SL(2,\beta)/B_\beta).$$

(v) Let $H(Z)$ be the closed subscheme of Z with ideal sheaf $\underline{I}(H(Z)) = \underline{I}(H(\tau)) \otimes M$, where $M = \underline{I}(X(\tau))^{\otimes r}$, $r = (\tau(\omega),\beta^*)$ (here, we identify $X(\tau)$ as the closed subscheme $p^{-1}(\bar{e})$ of Z, where $\bar{e}$ = the coset eB_β).

Lemma 5.2: With notation as above, suppose that standard monomials on $X(\tau)$ of degree s form a basis of $H^0(X(\tau),L^s)$, $s \in \mathbb{N}$. Let $(\tau(\omega),\beta^*) = t$. Let us define ideal sheaves K_i, $0 \le i \le m$, M_j, $1 \le j \le t$ on Z as follows:

$$M_j = \tilde{\underline{I}}_0^{(-j)}\ ,\ 1 \le j \le t$$

$$K_j = \tilde{\underline{I}}_j^{(-1)}\ ,\ 0 \le j \le m.$$

(where, recall $m+1 = \#K_\tau$ (cf §4)) so that we have

$$M_t \subset M_{t-1} \subset \cdots \subset M_1 = K_0 \subset K_1 \subset \cdots \subset K_m \subset \mathcal{O}_Z.$$

Then we have

(i) $M_t = \psi^*(L^{-1})$

(ii) $M_j/M_{j+1} \approx L^{-1}\big|_{X(\tau)}\ ,\ 1 \le j \le t-1$

(iii) $K_i/K_{i-1} \approx \psi^*(L^{-1})^{(m_j-1)}\Big|_{Z_\delta}$, $0 \le i \le m$

where δ is given by $\delta = \delta(\Lambda)$, $p_\Lambda = p_i$ and $m_i = (\chi(\Lambda), \beta^*)$.

(iv) $\mathcal{O}_Z/K_m = \mathcal{O}_{H(Z)_{red}}$

Proof: Similar to that of [12], Lemma 5.11.

Theorem 5.3: Let $w \in W^{\mathcal{P}}$. Then

(a) $H^i(X(w), L^s) = 0$, $i \ge 1$, $s \in \mathbb{Z}^+$

(b) $X(w)$ is normal

(c) Standard monomials on $X(w)$ of degree s form a basis for $H^0(X(w), L^s)$

Proof: We prove the theorem by induction on dim $X(w)$, the result being obvious if dim $X(w) = 0$. Let then dim $X(w) > 0$. Let us fix a moving divisor $X(\tau)$ in $X(w)$, say $\tau = s_\beta w$, for some simple root β. Let Z be as before. We now consider the following exact sequences on Z.

$$0 \longrightarrow M_t \longrightarrow \mathcal{O}_Z \longrightarrow \mathcal{O}_Z/M_t \longrightarrow 0$$

$$0 \longrightarrow M_j/M_{j+1} \longrightarrow \mathcal{O}_Z/M_{j+1} \longrightarrow \mathcal{O}_Z/M_j \longrightarrow 0 \text{ , } 1 \le j \le t-1$$

$$0 \longrightarrow K_i/K_{i-1} \longrightarrow \mathcal{O}_Z/K_{i-1} \longrightarrow \mathcal{O}_Z/K_i \longrightarrow 0 \text{ , } 0 \le i \le m$$

(where $K_{-1} = 0$). In view of Lemma 5.2, we have for $s \in \mathbb{N}$,

(1) $\psi^*(L^s) \otimes M_t = \psi^*(L^{s-1})$ (on Z)

(2) $\psi^*(L^s) \otimes M_j/M_{j+1} = L^{s-1}\Big|_{X(\tau)}$

(3) $\psi^*(L^s) \otimes K_i/K_{i-1} = \psi^*(L^{s-1})^{(m_i-1)}\Big|_{Z_\delta}$

(4) $\mathcal{O}_Z/K_m = \mathcal{O}_{H(Z)_{red}}$

Now tensoring the exact sequences in (*) with $\psi^*(L^s)$, we obtain (denoting $\psi^*(L)$ by just L)

$$(**) \quad \begin{cases} \chi(Z, L^s) = \chi(Z, L^{s-1}) + \chi(H(Z)_{red}, L^s) + (t-1)\chi(X(\tau), L^{s-1}) \\ \qquad + \sum\limits_{0 \le i \le m} \chi(Z_\delta, \mathcal{O}_{Z_\delta}^{(m_i-1)} \otimes L^{s-1}) \end{cases}$$

where χ is the Euler-Poincaré characteristic. From (**) we obtain (proceeding as in [12])

(***): $h^0(Z,L^s) = h^0(Z,L^{s-1}) + h^0(H(w)_{red},L^s) + \sum_{\theta<w} c_\theta h^0(X(\theta),L^{s-1})$

where $c_\theta = \#\{\Lambda \in K_w | \delta(\Lambda) = \theta\}$.

For any $\delta \in W^P$, let us denote $S(\delta, s)$ = {standard monomials on $X(\delta)$ of degree s} and $s(\delta, s) = \#S(\delta, s)$. We can write

$$S(w, s,) = S_1 \dot{\cup} S_2 \dot{\cup} S_3,$$

where

$S_1 = \{F \in S(w, s) | F \text{ starts with } p_\Lambda, \text{ where } \rho(\Lambda) < w\}$

$S_2 = \{F \in S(w, s) | F \text{ starts with } p_\Lambda, \text{ where } \rho(\Lambda) = w\}$

$S_3 = \{F \in S(w, s) | F \text{ starts with } p_w\}$

We have

$\#S_1$ = #{standard monomials on $H(w)_{red}$ of degree s} (note that $H(w)_{red}$ = ∪ all divisors $X(w')$ in $X(w)$),

$\#S_2 = \sum_{\theta<w} c_\theta s(\theta, s-1)$, where $c_\theta = \#\{\Lambda \in K_w | \delta(\Lambda) = \theta\}$

and

$\#S_3 = s(w, s-1)$.

Thus we obtain

(****): $s(w, s) = s_3 + s_1 + s_2$

where $s_i = \#S_i$. Now by induction on s, we may assume $H^0(Z,L^{s-1}) = H^0(X(w),L^{s-1})$ and hence we obtain

R.H.S. of (***) = R.H.S. of (****).

Hence we obtain $h^0(Z,L^s) = s(w,s)$. This implies that the canonical inclusions

$$H^0(X(w),L^s) \hookrightarrow H^0(\tilde{X(w)},L^s) \hookrightarrow H^0(Z,L^s)$$

(where $\tilde{X(w)}$ is the normalization of $X(w)$) are in fact isomorphisms. From this we obtain (b) and (c). The assertion (a) is proved as in [12], Theorem 5.14.

§6. Standard monomials for Schubert varieties in $\hat{Sp}(2n)/\mathcal{B}$.

Let $\mathcal{G} = \hat{Sp}(2n)$. Let $\tau \in W$ and let $X(\tau^{(i)})$ be the projection of $X(\tau)$ under $\pi_i: \mathcal{G}/\mathcal{B} \longrightarrow \mathcal{G}/\mathcal{P}_i$, where $\mathcal{P}_i$, $0 \le i \le n$ are the maximal parabolic subgroups of $\mathcal{G}$. Let $V_{\mathbb{Z},\tau^{(i)}}$ be the $\mathbb{Z}$-submodule of $V_{\mathbb{Z},\omega_i}$ as defined in §2. For any field k, let $V_{\tau^{(i)}} = V_{\mathbb{Z},\tau^{(i)}} \otimes k$. Consider the canonical embedding $X(\tau^{(i)}) \hookrightarrow \mathbb{P}(V_{\tau^{(i)}})$. Let L_i be the tautological line bundle on $\mathbb{P}(V_{\tau^{(i)}})$.

Definition 6.1: Let $F \in H^0(X(\tau),L)$, where $L = \bigotimes_{i=0}^{n} L_i^{a_i}$, $a_i \in \mathbb{Z}^+$. Let $F = f_0 f_1 f_2 \cdots f_n$, where $f_i = p_{\Lambda_{i1}} p_{\Lambda_{i2}} \cdots p_{\Lambda_{ia_i}}$ F is said to be <u>standard on $X(\tau)$ of multidegree $\underline{a} = (a_0,\cdots,a_n)$</u>, if there exists a sequence in W

$$\tau \ge \theta_{01} \ge \varphi_{01} \ge \theta_{02} \ge \varphi_{02} \ge\cdots\ge \theta_{0a_1} \ge \varphi_{0a_1} \ge \theta_{11} \ge\cdots\ge \theta_{na_n} \ge \varphi_{na_n}$$

such that $\pi_i(\theta_{ij}) = \rho(\Lambda_{ij})$, $\pi_i(\varphi_{ij}) = \delta(\Lambda_{ij})$, $1 \le j \le a_i$, $0 \le i \le n$. Proceeding as in [11], we have

Theorem 6.2: Standard monomials on $X(\tau)$ of degree $\underline{a}$ form a basis of $H^0(X(\tau),L)$.

Remark 6.3: In Definition 6.1, the order $\{P_0,\cdots,P_n\}$ is immaterial.

REFERENCES

1. C.C. Chevalley - Sur les Décompositions celluaires des espaces G/B (manuscript non publié) 1958.

2. E. Date, M. Jimbo, A. Kuniba, T. Miwa and M. Okado - Paths, Maya diagrams and representations of $\hat{s\ell}(r,\mathbb{C})$ (preprint).

3. W.V.D. Hodge - Some enumerative results in the theory of forms, Proc. Camb. Phil. Soc., 39, 1943, 22-30.

4. W.V.D. Hodge and C. Pedoe - Methods of Algebraic Geometry, Vol. II, Cambridge University Press, 1952.

5. V.G. Kac - Infinite dimensional Lie Algebras, Progress in Math., Vol. 44, Birkäuser, 1983.

6. M. Kashiwara - The flag manifold of Kac-Moody Lie Algebra (preprint)

7. S. Kumar - Demazure character formula in arbitrary Kac-Moody setting, Inventione Math. Vol. 89 (1987), 395-423.

8. V. Lakshmibai - Standard Monomial Theory for G_2, J. Algebra, Vol. 98, 1986, 281-318.

9. V. Lakshmibai, C. Musili and C.S. Seshadri - Geometry of G/P - IV, Proc. Ind. Acad. Sci. 99A, 1979, 279-362.

10. V. Lakshmibai and K.N. Rajeswari - Towards a Standard Monomial Theory for Exceptional Groups, Contemporary Mathematics, Vol. 88, 449-578.

11. V. Lakshmibai and C.S. Seshadri - Geometry of G/P - V, J. Algebra, Vol. 100, 1986, 462-557.

12. ———————————— - Standard Monomial Theory for $\hat{SL}_2$, Infinite Dimensional Lie Algebras and Groups, Advanced Series in Mathematical Physics, Vol. 7, 178-234.

13. ———————————— - Standard Monomial Theory for $\hat{SL}_n$, Volume dedicated to Dixmier's 65th birthday (published by Birkhaüser), 1990.

Quantum Deformations of SL_n/B and its Schubert Varieties

by

V. Lakshmibai
Department of Mathematics
Northeastern University
Boston, MA 02115

and

N. Reshetikhin*
Department of Mathematics
Harvard University
Cambridge, MA 02138

<u>Introduction</u>: In this paper, we prove the results announced in [L-R] for $G = SL_n$. Let G be a simple algebraic group over the base field k. Let M be a maximal torus in G and B, a Borel subgroup, $B \supset M$. Let W be the Weyl group of G. For $w \in W$, let $X(w) = \overline{BwB}$ (mod B) be the Schubert variety in G/B associated to w. Let L be an ample line bundle on G/B. We shall denote the restriction of L to X(w) also by just L. Let $k[X(w)] = \bigoplus_L H^0(X(w),L)$. In this paper, we construct an algebra $k_q[X(w)]$ over k(q), where q is a parameter taking values in k^*, as a quantization of k[X(w)], G being SL_n.

<u>The algebra $k_q[SL_n]$:</u> Let $G = SL_n$. Let $T = (t_{ij})$, $1 \le i, j \le n$. Let

$$R = \sum_{\substack{i \neq j \\ i,j=1}}^{n} e_{ii} \otimes e_{jj} + q \sum_{i \neq i}^{n} e_{ii} \otimes e_{ii} + (q-q^{-1}) \sum_{1 \le j < i \le n} e_{ij} \otimes e_{ji}$$

(here, e_{ij}'s are the elementary matrices). Let A(R) be the associative algebra (with 1) generated by $\{t_{ij}, 1 \le i, j \le n\}$, the relations being given by $RT_1T_2 = T_2T_1R$, where $T_1 = T \otimes Id$, $T_2 = Id \otimes T$ (cf. [F-R-T]). Then A(R) gives a quantization of $k[M_n]$, M_n being the space of $n \times n$ matrices and $k[M_n]$, the coordinate ring of M_n. Now A(R) has a bialgebra structure, the comultiplication being given by Δ: $A(R) \to A(R) \otimes A(R)$, $\Delta(t_{ij}) = \sum_{k=1}^{n} t_{ik} \otimes t_{kj}$. In the sequel, we shall denote A(R) by $k_q[M_n]$. Let

$$D = \sum_{\sigma \in S_n} (-q)^{-\ell(\sigma)} t_{1\sigma(1)} t_{2\sigma(2)} \cdots t_{n\sigma(n)}.$$

We shall refer to D as the <u>q-determinant</u> (or the <u>quantum determinant</u>) of (t_{ij}). We have (cf. [F-R-T]), D is central in $k_q[M_n]$. The algebra $A[\frac{1}{D}]$

---
*On leave of absence from USSR; LOMI, Fontanka 27, Leningrad, 191011

(resp. $A/(D-1)$), where $A = k_q[M_n]$, gives a quantization of $k[GL_n]$ (resp. $k[SL_n]$) (here $k[GL_n]$ (resp. $k[SL_n]$) denotes the coordinate ring of GL_n (resp. SL_n)).

The algebra $k_q[G/P_d]$: Let P_d be the maximal parabolic subgroup of G obtained by omitting the simple root α_d. Let

$$I_{d,n} = \{(i_1,\dots,i_d) \mid 1 \le i_1 < \cdots < i_d \le n\}.$$

For $\tau \in I_{d,n}$, let x_τ denote the q-determinant of the d×d minor of T with column indices $1,\dots,d$ and row indices $i_1,\dots,i_d$. We set $k_q[G/P_d]$ as the subalgebra of $k_q[SL_n]$ generated by x_τ, $\tau \in I_{d,n}$. Let $(k_q[G/P_d])_m$ be the k(q)-span of monomials in x_τ of degree m, $m \in \mathbb{Z}^+$.

One knows that there is a bijection between {Schubert varieties in G/P_d} and $I_{d,n}$. There is a partial order ≥ on $I_{d,n}$ (induced by the partial order on {Schubert varieties in G/P_d} given by inclusion), namely, $(i_1,\dots,i_d) \ge (j_1,\dots,j_d)$, if $i_t \ge j_t$, $1 \le t \le d$.

Definition 1: A monomial $x_{\tau_1} x_{\tau_2} \cdots x_{\tau_m}$ is said to be standard if $\tau_1 \ge \tau_2 \ge \cdots \ge \tau_m$.

Theorem 1: Standard monomials of degree m give a k(q)-basis for $(k_q[G/P_d])_m$.

Theorem 2: (a) The algebra $k_q[G/P_d]$ has a canonical $\mathbb{Z}$-gradation given by $k_q[G/P_d] = \oplus_m (k_q[G/P_d])_m$, $m \in \mathbb{Z}^+$.

(b) $k_q[G/P_d]$ has a canonical left comodule structure over $k_q[G]$, the coaction being given by

$$\Delta: k_q[G/P_d] \to k_q[G] \otimes k_q[G/P_d],\ \Delta(x_\tau) = \sum_\varphi F_{\tau,\varphi} \otimes x_\varphi,$$

where the summation runs over all d-tuples $\varphi = (j_1,\dots,j_d)$ and $F_{\tau,\varphi}$ is the q-determinant of the d × d minor of (t_{ij}) with row indices $i_1,\dots,i_d$ and column indices $j_1,\dots,j_d$ (here, $\tau = (i_1,\dots,i_d)$).

The algebra $k_q[X(\tau)]$: For $\tau \in I_{d,n}$, let $X(\tau)$ denote the associated Schubert variety in G/P_d. Let I_τ be the two-sided ideal in $k_q[G/P_d]$ generated by $\{x_\varphi | \tau \not\geq \varphi\}$. We define $k_q[X(\tau)]$ as $k_q[G/P_d]/I_\tau$. Let $(k_q[X(\tau)])_m$ be the $k(q)$-span (in $k_q[X(\tau)]$) of monomials of degree m.

Definition 2: A monomial $x_{\tau_1} \cdots x_{\tau_m}$ is said to be standard on $X(\tau)$ if $\tau \geq \tau_1 \geq \cdots \geq \tau_m$.

Theorem 3: (a) Standard monomials on $X(\tau)$ of degree m form a $k(q)$-basis for $(k_q[X(\tau)])_m$.

(b) The algebra $k_q[X(\tau)]$ has a canonical $\mathbb{Z}$-gradation given by $k_q[X(\tau)] = \bigoplus_m k_q[X(\tau)]_m$, $m \in \mathbb{Z}^+$.

(c) $k_q[X(\tau)]$ has a canonical left comodule structure over $k_q[B]$ $(= k_q[G]/(t_{ij}, i > j))$.

The algebra $k_q[G/B]$: We define $k_q[G/B]$ as the subalgebra of $k_q[G]$ generated by $\{x_\tau, \tau \in I_{d,n}, 1 \leq d \leq \ell\}$ (here ℓ = rankG = n-1). Let $\underline{a} = (a_1, \ldots, a_\ell)$, $a_i \in \mathbb{Z}^+$ and $(k_q[G/B])_{\underline{a}}$ be the $k(q)$-span of all monomials f such that f has a_i factors $x_{\tau_{ij}}$, $\tau_{ij} \in I_{i,n}$, $1 \leq j \leq a_i$, $1 \leq i \leq \ell$ (the factors appearing in f in some order).

Definition 3: A monomial f in $k_q[G/B]_{\underline{a}}$ is said to be standard if

(a) $f = \prod_i \prod_j x_{\tau_{ij}}$, $\tau_{ij} \in I_{i,n}$, $1 \leq j \leq a_i$, $1 \leq i \leq \ell$ (the factors appearing in that order).

(b) there exists a sequence $\{\theta_{ij}, 1 \leq j \leq a_i, 1 \leq i \leq \ell\}$ in $W(=S_n)$ such that

(1) $\Pi_i(X(\theta_{ij})) = X(\tau_{ij})$ under $\Pi_i: G/B \to G/P_i$.

(2) $X(\theta_{11}) \geq X(\theta_{12}) \geq \cdots \geq X(\theta_{1a_1}) \geq X(\theta_{21}) \geq \cdots \geq X(\theta_{\ell a_\ell})$.

Theorem 4: (a) Standard monomials of multidegree $\underline{a}$ form a $k(q)$-basis of $(k_q[G/B])_{\underline{a}}$.

(b) $k_q[G/B]$ has a canonical $\mathbb{Z}^\ell$-gradation given by $k_q[G/B] = \bigoplus_{\underline{a}} (k_q[G/B])_{\underline{a}}$,

$\underline{a} \in (\mathbb{Z}^+)^{\ell}$.

(c) $k_q[G/B]$ has a canonical left comodule structure over $k_q[G]$.

The algebra $k_q[X(w)]$: Let $w \in W$ and let $w^{(d)}$ be the minimal representative in W of wW_{P_d}, $1 \le d \le \ell$. Let I_w be the two-sided ideal in $k_q[G/B]$ generated by $\{x_{\tau_i} | \tau_i \in W^{P_i}, \tau_i \not\le w^{(i)}, 1 \le i \le \ell\}$. We define $k_q[X(w)] = k_q[G/B]/I_w$.

Let $\underline{a} = (a_1, \dots, a_\ell)$, $a_i \in \mathbb{Z}^+$ and let $(k_q[X(w)])_{\underline{a}}$ be the k(q)-span (in $k_q[X(w)]$) of all monomials f such that f has a_i factors $x_{\tau_{ij}}$, $\tau_{ij} \in I_{i,n}$, $1 \le j \le a_i$, $1 \le i \le \ell$, such that $\tau_{ij} \le w$ (these factors appearing in f in some order).

Definition 4: A monomial f in $(k_q[X(w)])_{\underline{a}}$ is said to be standard on X(w) if it satisfies the conditions (a), (b) of Definition 3 and the condition that $X(w) \ge X(\theta_{11})$.

Theorem 5: (a) Standard monomials on X(w) of multidegree $\underline{a}$ form a k(q)-basis of $(k_q[X(w)])_{\underline{a}}$.

(b) $k_q[X(w)]$ has a canonical $\mathbb{Z}^{\ell}$-gradation given by $k_q[X(w)] = \oplus_{\underline{a}} (k_q[X(w)])_{\underline{a}}$, $\underline{a} \in (\mathbb{Z}^+)^{\ell}$.

(c) $k_q[X(w)]$ has a canonical left comodule structure over $k_q[B]$.

Outline of proof: Linear independence of standard monomials on X(w) in arbitrary characteristic is obtained as a consequence of the linear independence of standard monomials for the case q = 1 (cf. [L-S]). In view of linear independence of standard monomials in arbitrary characteristic, it suffices to prove generation by standard monomials over k(q) when $k = \mathbb{Q}$. This is proved by using certain (qualitative) relations. These relations are obtained by using the Clebsch-Gordan coefficient matrix giving the projection $V^\lambda \otimes V^\mu \to V^\nu$, where V^ν is a factor in the decomposition of $V^\lambda \otimes V^\mu$ as a direct sum of irreducible $U_q(\mathfrak{g})$-modules (here, $U_q(\mathfrak{g})$ denotes the quantized universal enveloping algebra of $\mathfrak{g}$ = Lie G (cf. [D],[J]). For a dominant integral weight λ, V^λ denotes the corresponding irreducible $U_q(\mathfrak{g})$-module (cf. [Lu],[Ro])).

The paper is organized as follows. In §1, we recall results on Schubert varieties in the Grassmannian as well as the Flag variety SL_n/B, Quantum SL_n, $U_q(\mathfrak{g})$, the universal R matrix. In §2, we present quantizations of Grassmannians and its Schubert varieties. In §3, we present quantizations of the Flag variety SL_n/B and its Schubert varieties.

It should be remarked that quantum SL_n/B is constructed in [D-M-M-Z], [S], [T-T]. But our approach is different from the approaches in [D-M-M-Z], [S], [T-T].

§1. Preliminaries:

Schubert varieties in the Grassmannian: Let k be the base field. Let V be an n-dimensional k-vector space. Identifying V with k^n, let $\{e_i, 1 \le i \le n\}$ be the standard basis in V. Let us fix an integer d, $1 \le d \le n$. Let $G_{d,n}$, the Grassmannian of d-dimensional subspaces in V. For the Plücker embedding $G_{d,n} \hookrightarrow \mathbb{P}(\overset{d}{\wedge}V)$, let R be the homogeneous coordinate ring of $G_{d,n}$. Under the map $\pi: \underbrace{V \oplus \cdots \oplus V}_{d} \to \wedge^d V$, $\pi(v_1,\ldots,v_d) = v_1\wedge \cdots \wedge v_d$, we have $\mathrm{Im}\,\pi = \hat{G}_{d,n}$. Thus, we get an identification of R as a subring of the polynomial ring $k[X_{ij}]$, $1 \le i \le n$, $1 \le j \le d$. As a subring of $k[X_{ij}]$, R gets identified with the subring generated by the Plücker coordinates. To make it more precise, let $I_{d,n} = \{(i) = (i_1,\ldots,i_d) \mid 1 \le i_1 < i_2 < \cdots < i_d \le n\}$. For $(i) \in I_{d,n}$, let $P_{(i)}$ = the determinant of the $d \times d$ minor of $k[X_{ij}]$ with row indices given by $i_1,\ldots,i_d$. Then R is the subring of $k[X_{ij}]$ generated by $\{P_{(i)}, (i) \in I_{d,n}\}$.

Now $G_{d,n}$ can also be identified with G/P_d, where $G = SL_n$ and P_d is the maximal parabolic subgroup of SL_n consisting of matrices of the form

$$\begin{pmatrix} * & * \\ \hline O_{n-d\times d} & * \end{pmatrix}$$

Let us denote by B, the Borel subgroup of SL_n consisting of upper triangular matrices and M, the maximal torus of SL_n consisting of diagonal matrices. For the canonical action of G on $G_{d,n}$, the M-fixed points are simply the spans of $\{e_{i_1},\dots,e_{i_d},\ (i_1,\dots,i_d) \in I_{d,n}\}$. Denoting the M-fixed points by e_τ, $\tau \in I_{d,n}$, let $X(\tau)$ be the Schubert variety in $G_{d,n}$ associated to τ, namely, $X(\tau)$ = Zariski closure of the B-orbit Be_τ. Thus, we have a bijection $\{\text{Schubert varieties in } G_{d,n}\} \overset{\text{bij}}{\approx} I_{d,n}$.

Note that $I_{d,n}$ can be identified with W^{P_d}, the set of minimal representatives in W of W/W_{P_d}. To be more precise, we have, $W_{P_d} = S_d \times S_{n-d}$ and an element in W^{P_d} looks like $(i_1,\dots,i_d,j_1,\dots,j_{n-d})$, where $1 \le i_1 < i_2 < \cdots < i_d \le n$, $1 \le j_1 < j_2 < \cdots < j_{n-d} \le n$).

The partial order on {Schubert varieties in $G_{d,n}$} given by inclusion induces a partial order $\ge$ on $I_{d,n}$, namely $(i_1,\dots,i_d) \ge (j_1,\dots,j_d)$ if $i_t \ge j_t$, $1 \le t \le d$.

Definition 1.1: A monomial $P_{\tau_1}P_{\tau_2} \cdots P_{\tau_m}$ is said to be <u>standard</u> if $\tau \ge \tau_1 \ge \tau_2 \ge \cdots \ge \tau_m$.

Definition 1.2: A monomial $P_{\tau_1}P_{\tau_2} \cdots P_{\tau_m}$ is said to be <u>standard on $X(\tau)$</u> if $\tau_1 \ge \tau_2 \ge \cdots \ge \tau_m$. Let $R(\tau)$ be the homogeneous coordinate ring of $X(\tau)$ for the projective embedding $X(\tau) \hookrightarrow G_{d,n} \hookrightarrow \mathbb{P}(\Lambda^d V)$.

Theorem 1.3: (cf. [H], [HO], [M]). Standard monomials on $X(\tau)$ of degree m form a basis for $(R(\tau))_m$, $m \in \mathbb{Z}^+$.

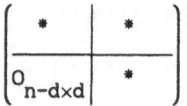

As a consequence, we obtain (cf. [M]) $H^0(X(\tau), L^m) = R(\tau)_m$, for <u>all m</u>, where L denotes the tautological line bundle on $\mathbb{P}(\wedge^d V)$, as well as its restriction to $X(\tau)$. In particular, we obtain

Corollary 1.5: Standard monomials on $X(\tau)$ of degree m form a basis of $H^0(X(\tau),L^m)$, $m \in \mathbb{Z}^+$.

Schubert varieties in the Flag variety SL_n/B: Let $W(=S_n)$ be the Weyl group of G. For $w \in W$, let us denote the Schubert variety $\overline{BwB}$ (mod B) in G/B by $X(w)$. Let L_i be the ample generator of $\text{Pic}(G/P_i)$, $1 \le i \le \ell$, where $\ell = n-1$. For $\underline{a} = (a_1,\ldots,a_\ell)$, $a_i \in \mathbb{Z}^+$, let $L_{\underline{a}} = \bigotimes_{i=1}^{\ell} L_i^{a_i}$. We shall denote the restriction of $L_{\underline{a}}$ to $X(w)$ also by just $L_{\underline{a}}$.

Definition 1.6 [L-S]: A monomial F of multidegree $\underline{a}$,

$$F = \prod_{i=1}^{\ell} \prod_{j=1}^{a_i} P_{\tau_{ij}}, \quad \tau_{ij} \in W^{P_d}$$

is said to be standard on X(w), if there exists a sequence $\{\theta_{ij}\}$ in W such that

(1) $X(w) \supseteq X(\theta_{11}) \supseteq X(\theta_{12}) \supseteq \cdots \supseteq X(\theta_{1a_1}) \supseteq X(\theta_{21}) \supseteq \cdots \supseteq X(\theta_{\ell a_\ell})$.

(2) Under $\pi_i: G/B \to G/P_i$, $\pi_i(X(\theta_{ij})) = X(\tau_{ij})$.

Theorem 1.7 (cf. [L-S]): Monomials of multidegree $\underline{a}$ standard on X(w) form a basis of $H^0(X(w),L_{\underline{a}})$.

Quantum SL_n (cf. [D],[J],[F-R-T]): Let $T = (t_{ij})$, $1 \le i, j \le n$. Let

$$R = \sum_{\substack{i\neq j\\ i,j=1}}^{n} e_{ii} \otimes e_{jj} + q \sum_{i=1}^{n} e_{ii} \otimes e_{ii} + (q-q^{-1}) \sum_{1\le j<i\le n} e_{ij} \otimes e_{ji}$$

where q is a parameter taking values in k^* and e_{ij}'s are the elementary matrices. Let A(R) be the associative algebra with 1 generated by $\{t_{ij}, 1 \le i, j \le n\}$, the relations being given by $RT_1T_2 = T_2T_1R$, where $T_1 = T \otimes Id$, $T_2 = Id \otimes T$ (cf. [D],[J],[F-R-T]). Then A(R) is a quantization of

$k[M_n]$, the coordinate ring of M_n (= space of nxn matrices). Now A(R) has a bialgebra structure, the comultiplication being given by

$$\Delta: A(R) \to A(R) \otimes A(R),\ \Delta(t_{ij}) = \sum_{k=1}^{n} t_{ik} \otimes t_{kj}\ .$$

Let us denote A(R) by just A. Let $D = \Sigma(-1)^{-\ell(\sigma)} t_{1\sigma(1)} t_{2\sigma(2)} \cdots t_{n\sigma(n)}$. We shall refer to D as the <u>quantum determinant</u> (or the <u>q-determinant</u>) of T. We have, D is central in A and A/(D-1) is a quantization of $k[SL_n]$ (cf. [D], [J], [F-R-T]).

<u>The quantized universal enveloping algebra</u>: Let $k = \mathbb{C}$. Let $\mathfrak{g} = \mathrm{Lie}(G)$ and $U(\mathfrak{g})$, the universal enveloping algebra of $\mathfrak{g}$. Let $U_q(\mathfrak{g})$ be the quantized universal enveloping algebra as constructed in [J]. This is a Hopf algebra over $\mathbb{C}(q)$ with generators X_i, Y_i, and invertible K_i, $1 \le i \le \ell$, relations and comultiplication on the generators as described in [J].

<u>The universal R-matrix</u>: Let $U_q(b)$ (resp. $U_q(b^-)$) be the subalgebra of $U_q(\mathfrak{g})$ generated by $\{K_i, X_i\}$ (resp. $\{K_i, Y_i\}$), $1 \le i \le \ell$. There is a duality between $U_q(b)$ and $U_q(b^-)$ (cf. [D]). Let $\{e_s\}$ be a basis of $U_q(b)$ and $\{f_s\}$ the dual basis in $U_q(b^-)$. Then there is an element $R \in U_q(\mathfrak{g})^{\hat{\otimes}2}$ (where $\hat{\otimes}$ is the tensor product completed by formal power series over (q-1)). This element can be identified with $\sum_s e_s \otimes f_s \in U_q(b) \hat{\otimes} U_q(b^-)$).

§2. <u>**Quantum Grassmannians**</u>: In this section as well as in the section that follows, q will denote a parameter taking values in k^*. If $q^r = 1$, then r will be supposed to be sufficiently large. We keep the notation of §1. We set

$$k[G/P_d] = \oplus_m H^0(G/P_d, L^m)\ ,\ m \in \mathbb{Z}^+$$

where L is the tautological line bundle (as well as its restriction to G/P_d)

on $\mathbb{P}(\overset{d}{\wedge}V)$.

The algebra $k_q[G/P_d]$: Let $\tau \in W^{P_d}$, say $\tau = (i_1,\ldots,i_d)$. Let δ_τ be the q-determinant of the d×d minor of $T = (t_{ij})$ with column indices $1,\ldots,d$ and row indices $i_1,\ldots,i_d$ (recall that the q-determinant of a square matrix (a_{ij}), $a_{ij} \in k(q)$ of size r is $\sum_{\sigma\in S_r} (-q)^{-\ell(\sigma)} a_{1\sigma(1)} \cdots a_{r\sigma(r)}$). We define $k_q[G/P_d]$ as the subalgebra of $k_q[G]$ generated by $\{\delta_\tau,\ \tau \in W^{P_d}\}$.

Definition 2.1: A monomial $\delta_{\tau_1} \cdots \delta_{\tau_m}$ is said to be standard if $\tau_1 \geq \cdots \geq \tau_m$.

Proposition 2.2: Standard monomials are linearly independent over k(q).

Proof: Let $\Sigma a_i f_i = 0$, $a_i \in k(q)^*$ be a non-trivial linear relation among standard monomials. Clearing the denominators, we may suppose that $a_i \in k[q]$. Let r be the largest integer such that $(q-1)^r$ divides all the a_i's. Cancelling $(q-1)^r$ and going modulo the ideal (q-1), we obtain a nontrivial relation among standard monomials, for q = 1, which is not possible (cf. [M]). The required result follows from this.

Remark 2.3: In view of Proposition 2.2, to prove generation by standard monomials, over k(q), k being an arbitrary base field, it suffices to prove it for the case $k = \mathbb{Q}$.

Generation by standard monomials for the case $k = \mathbb{Q}$: We have (cf. [Lu], [Ro]), the structure theory of finite dimensional $U_q(\mathfrak{g})$-modules is the same as that of finite dimensional $\mathfrak{g}$-modules. Thus we have,

Proposition 2.4: The finite dimensional representations of $U_q\mathfrak{g}$ are completely reducible.

Proposition 2.5: (a) The finite dimensional, irreducible representations of $U_q\mathfrak{g}$ are parameterized by the dominant, integral weights of $\mathfrak{g}$.

(b) Given a dominant integral weight λ, let V^λ be the corresponding irreducible $U_q\mathfrak{g}$-module. Let $V^\lambda = \oplus_\mu V^\lambda(\mu)$ (direct sum of weight spaces): the dimensions of $V^\lambda(\mu)$'s are the same as those of the corresponding weight spaces of the irreducible $\mathfrak{g}$-module (over $\mathbb{Q}$) with highest weight λ.

Let ω_d, $1 \le d \le \ell$ be the fundamental weights of G. As a consequence of Propositions 2.4 and 2.5, we have

Proposition 2.6: Let $G = SL_n$. Then

(a) $$V^{\omega_d} \otimes V^{\omega_d} = \bigoplus_{p\geq 0}^{\min(d,n-d)} V^{\omega_{d-p}+\omega_{d+p}}$$

(b) V^{ω_d} occurs as a factor in $(V^{\omega_1})^{\otimes d}$.

<u>A Basis for V^{ω_d}</u>: Let us fix a highest weight vector e in V^{ω_d}. Let W^{P_d} be the set of minimal representatives in W of W/W_{P_d}. Let $w \in W^{P_d}$. Further, let $w = s_{\beta_1} \cdots s_{\beta_r}$, ($\beta_i$ simple) be a reduced expression for w. Then it is easily checked (since ω_d is a minuscule weight) that the vector we = $X_{-\beta_1} \cdots X_{-\beta_r} e$ is an extremal weight vector (of weight $w(\omega_d)$). On the other hand, in view of Proposition 2.5(b), there are no non-extremal weights in V^{ω_d}. Thus the set $\{\tau e, \tau \in W^{P_d}\}$ is a basis for V^{ω_d} (we remark that the vectors $X_{-\beta_1} \cdots X_{-\beta_r} e$ and $X_{-\gamma_1} \cdots X_{-\gamma_r} e$ (corresponding to two reduced expressions $s_{\beta_1} \cdots s_{\beta_r}, s_{\gamma_1} \cdots s_{\gamma_r}$ of τ) differ by a scalar and thus τe is uniquely determined up to scalars). Further, we have (cf. §1), $W^{P_d} \overset{\text{bij}}{\approx} I_{d,n}$. Hence if $\tau = (i_1,\dots,i_d)$, then τe is a weight vector of weight $\epsilon_{i_1} + \dots + \epsilon_{i_d}$ (here ϵ_i is the element of h^*, $\epsilon_i(A) = a_i$, where A = $\mathrm{diag}(a_1,\dots,a_n)$). We shall index the basis $\{\tau e, \tau \in I_{d,n}\}$ as $\{v_i, 1 \le i \le N_d\}$ so that v_1 is the highest weight vector.

Definition 2.7:

<u>The elements $x_i^{\omega_d}$</u>: We set

K^{ω_d} = the projection $(V^{\omega_1})^{\otimes d} \longrightarrow V^{\omega_d}$

$\bar{K}^{\omega_d}$ = the inclusion $V^{\omega_d} \hookrightarrow (V^{\omega_1})^{\otimes d}$, such that $\bar{K}^{\omega_d}K^{\omega_d} = id_{V^{\omega_d}}$

$T^{\omega_d} = K^{\omega_d}(T^{\omega_1})^{\otimes d}\bar{K}^{\omega_d}$

(here T^{ω_1} is just $T = (t_{ij})$). Let us write $T^{\omega_d} = (t_{ij}^{\omega_d})$, $1 \le i, j \le N_d$, where $N_d = \binom{n}{d}$ (= dim V^{ω_d}). We define $x_i^{\omega_d}$ (or just x_i) as $x_i = t_{i1}^{\omega_d}$, $1 \le i \le N_d$.

Proposition 2.8: With notation as above, let v_i (in V^{ω_d}) be of weight $\epsilon_{i_1} + \epsilon_{i_2} + \dots + \epsilon_{i_d}$ and let $\tau = (i_1,\dots,i_d)$. Then x_i is simply δ_τ, the q-determinant of the dxd minor of $T(=(t_{ij}))$ with column indices $1,\dots,d$ and row indices $i_1,\dots,i_d$.

Proof: Since the basis $\{v_i, 1 \le i \le N_d\}$ for V^{ω_d} is parametrized by $I_{d,n}$, a typical entry in T^{ω_d} may be denoted by $t_{\tau\sigma}^{\omega_d}$, $\tau,\sigma \in I_{d,n}$. To be very precise, if $\tau = (i_1,\dots,i_d)$ and $\sigma = (j_1,\dots,j_d)$, then we have

$$t_{\tau\sigma}^{\omega_d} = \sum_{s\in S_d} (-q)^{-\ell(s)} t_{s(i_1)j_1}^{\omega_1} \cdots t_{s(i_d)j_d}^{\omega_1}$$

$$= \sum_{s\in S_d} (-q)^{-\ell(s)} t_{i_1 s(j_1)}^{\omega_1} \cdots t_{i_d s(j_d)}^{\omega_1}.$$

The required result follows from this.

The straightening relations: Let $X^d = \Sigma x_i v_i$ $(=T^{\omega_d}v_1)$. Let us denote the action of the universal R-matrix R on $V^{\omega_d} \otimes V^{\omega_d}$ as $R^{\omega_d\omega_d}$. Let $R^{\vee\omega_d\omega_d} = P \circ R^{\omega_d\omega_d}$, where P: $V^{\omega_d} \otimes V^{\omega_d} \longrightarrow V^{\omega_d} \otimes V^{\omega_d}$ is the map $P(u\otimes v) = v \otimes u$. Then we have (cf. [R])

Proposition 2.9: Let $V^{\omega_d} \otimes V^{\omega_d} = \bigoplus_\nu V^\nu$. Let us denote by P_ν the projection

$V^{\omega_d} \otimes V^{\omega_d} \longrightarrow V^{\nu}$. Then

(a) $R^{V\omega_d\omega_d} = \sum_{\nu} (-1)^{\text{parity of } \nu} q^{2c(\omega_d)-c(\nu)} P_{\nu}$

(b) $R^{V\omega_d\omega_d} X^d \otimes X^d = q^{2c(\omega_d)-c(2\omega_d)} X^d \otimes X^d$

where c is the Casimir operator, $c(\lambda) = (\lambda,\lambda) + 2(\rho,\lambda)$, $\rho = \frac{1}{2}$ sum of positive roots. (Here $(-1)^{\text{parity of } \nu}$ = eigenvalue of the restriction to V^{ν} of the operator P).

Corollary 2.10: With notation as in Proposition 2.6, we have

$$P_{\nu}(X^d \otimes X^d) = 0 \ , \ \nu \neq 2\omega_d.$$

Quantum Clebsch-Gordan coefficients: Let $\{e_i^{\nu}\}$ be a basis, consisting of weight vectors, for V^{ν}. The coefficients appearing in the expression for e_i^{ν} in terms of the basis $\{v_j \otimes v_k\}$ for $V^{\omega_d} \otimes V^{\omega_d}$ will be referred to as the quantum Clebsch-Gordan coefficients. Let us consider a typical expression

$$e_i^{\nu} = \sum_{j,k} c_{ijk}^{\nu} v_j \otimes v_k \ , \ c_{ijk}^{\nu} \in k(q).$$

In view of Corollary 2.10, we obtain

$$\sum_{j,k} c_{ijk}^{\nu} x_j x_k = 0 \ , \ \nu \neq 2\omega_d \qquad \text{(I)}$$

More generally, setting $X^{m\omega_d} = X^d \otimes \cdots \otimes X^d$ (m factors), we have

Proposition 2.11: Let V^{ν} be an irreducible factor occuring in the direct sum decomposition of $(V^{\omega_d})^{\otimes m}$. Let P_{ν} be the projection $(V^{\omega_d})^{\otimes m} \longrightarrow V^{\nu}$. Then

$$P_{\nu}(X^{m\omega_d}) = 0 \ , \ \nu \neq m\omega_d.$$

The proof is carried out in the same spirit as for the case m=2 above and we skip the details.

Generation by standard monomials: Let V^{ν} be a factor appearing in the expression for $(V^{\omega_d})^{\otimes m}$ as a direct sum of irreducible $U_q(\mathfrak{g})$-modules. Let

$\{e_i^\nu\}$ be a basis of V^ν consisting of weight vectors. Writing

$$e_i^\nu = \sum_J c_{iJ}^\nu v_{j_1} \otimes \cdots \otimes v_{j_m}, \quad J = (j_1, \ldots, j_m), \quad c_{iJ}^\nu \in k(q),$$

we have (in view of Proposition 2.11)

$$\sum_J c_{iJ}^\nu x_{j_1} \cdots x_{j_m} = 0. \qquad \text{(I)}$$

(here, $\{v_j\}$ as before is the basis of V^{ω_d} consisting of extremal weight vectors). There are $N_d^m - s(d,m)$ such linear equations among the N_d^m monomials of degree m, where $N_d = \dim V^{\omega_d} (= \binom{n}{d})$ and $s(d,m) = \dim V^{m\omega_d}$ (= # {standard monomials of degree m}). Also, the coefficient matrix of (I) has maximal rank (= $N_d^m - s(d,m)$), in view of linear independence of $\{e_i^\nu\}_{i,\nu}$. Hence taking the standard monomials of degree m as the free variables of (I) (in view of linear independence of standard monomials), we obtain that each nonstandard degree-m monomial has an expression as a linear combination of degree-m standard monomials. Combining this with linear independence of standard monomials (cf. Proposition 2.2) we obtain

Theorem 2.12: Let $(k_q[G/P_d])_m$ be the k(q)-span of monomials in x_τ's of degree m. Standard monomials of degree m form a k(q)-basis for $(k_q[G/P_d])_m$ (k being an arbitrary field).

We next prove

Theorem 2.13: (a) The algebra $k_q[G/P_d]$ has a canonical $\mathbb{Z}$-gradation.

(b) $k_q[G/P_d]$ has a canonical left comodule structure over $k_q[G]$.

Proof: We have $k_q[G/P_d] = \bigoplus_m (k_q[G/P_d])_m$, $m \in \mathbb{Z}^+$ which gives a canonical $\mathbb{Z}$-gradation for $k_q[G/P_d]$. Now, under Δ: $k_q[G] \to k_q[G] \otimes k_q[G]$, (where, recall $\Delta(t_{ij}) = \sum_{k=1}^{n} t_{ik} \otimes t_{kj}$), we have $\Delta(x_i^{\omega_d})$ $(= \Delta(T_{i1}^{\omega_d})) = \sum_k T_{ik}^{\omega_d} \otimes T_{k1}^{\omega_d} = \sum_k T_{ik}^{\omega_d} \otimes x_k^{\omega_d}$. Thus if i corresponds to the d-tuple $(i_1, \ldots, i_d)$, then $\Delta(x_i^{\omega_d}) = \sum_j F_{ij} \otimes x_j^{\omega_d}$, where the summation runs over all d-tuples $j = (j_1, \ldots, j_d)$ and

F_{ij} is the q-determinant of the d×d minor of (t_{ij}) with row indices $i_1,\dots,i_d$ and column indices $j_1,\dots,j_d$. Thus Δ induces a map $k_q[G/P_d] \to k_q[G] \otimes k_q[G/P_d]$ giving a canonical left $k_q[G]$-comodule structure to $k_q[G/P_d]$.

In the sequel, we shall denote $x_i^{\omega_d}$ by just x_i.

Quantum Schubert schemes: Let us fix a $w \in W^{P_d}$, say $w = (i_1,\dots,i_d)$. Let I_w be the two-sided ideal in $k_q[G/P_d]$ generated by $\{x_\tau | w \not\geq \tau\}$. Set $k_q[X(w)]$ or just $R_q(w)$ as $R_q(w) = k_q[G/P_d]/I_w$.

Definition 2.14: A monomial $x_{\tau_1} x_{\tau_2} \cdots x_{\tau_m}$ is said to be standard on $X(w)$ (or in $R(w)$) if $w \geq \tau_1 \geq \cdots \geq \tau_m$.

Proposition 2.15: Standard monomials in $R(w)$ are linearly independent over $k(q)$.

Proof: If possible, let $\Sigma a_i f_i = 0$ (in $R_q(w)$), where $a_i \in k(q)^*$ and each f_i is a monomial which is standard on $X(w)$. Clearing the denominators of a_i, we may assume $a_i \in k[q]$. Let r be the largest integer such that $(q-1)^r$ divides all a_i's. Cancelling $(q-1)^r$ and going modulo $(q-1)$, we obtain a nontrivial linear relation among standard monomials in $k[X(w)]$, which is a contradiction (cf. [LS]). Hence a relation as above cannot exist in $R_q(w)$ and the required result follows from this.

Corollary 2.16: The standard monomials in $R_q(w)$ form a $k(q)$-basis for $R_q(w)$.

Proof: That standard monomials in $R_q(w)$ generate $R_q(w)$ follows in view of the facts that standard monomials in $k_q[G/P_d]$ generate $k_q[G/P_d]$, and $R_q(w) = k_q[G/P_d]/I_w$. This together with linear independence (cf. Proposition 2.15) implies the required result.

Combining Proposition 2.15 and Corollary 2.16, we obtain

Theorem 2.17: Let $(R_q(w))_m$ be the k(q)-span of monomials of degree m in $R_q(w)$. Standard monomials in $R_q(w)$ of degree m form a k(q)-basis for $(R_q(w))_m$.

Let $k_q[B]$ be the quantization of B given by $k_q[G]/(t_{ij}, i>j)$ (cf. §1). We have

Theorem 2.18: (a) The algebra $R_q(w)$ has a canonical $\mathbb{Z}$-gradation.

(b) $R_q(w)$ has a canonical left $k_q[B]$-comodule structure.

Proof: The canonical $\mathbb{Z}$-gradation is simply $R_q(w) = \oplus_m (R_q(w))_m$, $m \in \mathbb{Z}^+$. To exhibit a canonical $k_q[B]$-comodule structure, we first observe that I_w is $U_q(b^+)$-stable; for, let β be a positive root. Then for any τ, $X_\beta x_\tau$ is either 0 or is a scalar multiple of $x_{s_\beta \tau}$. Further, we have $\tau < s_\beta \tau$ (by weight considerations). This implies the $U_q(b^+)$-stability for I_w and hence we obtain (dually) that under

$$\Delta: k_q[G/P_d] \longrightarrow k_q[G] \otimes k_q[G/P_d]$$

$$\Delta(I_w) \subset J \otimes k_q[G/P_d] + k_q[G] \otimes I_w,$$

where J is the ideal defining $k_q[B]$. The required result follows from this.

§3. Quantum Flag Manifold

We preserve the notation of the previous sections.

The algebra $k_q[G/B]$: We define $k_q[G/B]$ as the subalgebra of $k_q[G]$ generated by

$$\{x_\tau^{\omega_d}, \tau \in W^{P_d}, 1 \le d \le \ell\}.$$

We fix a total ordering on the set of maximal parabolic subgroups.

Definition 3.1 (cf. [LS]: Let $\underline{a} = (a_1, \ldots, a_\ell)$, $a_i \in \mathbb{Z}^+$. A Young diagram of type $\underline{a}$ is a sequence $\Lambda = \{\lambda_{ij}, \lambda_{ij} \in W^{P_i}, 1 \le j \le a_i, 1 \le i \le \ell\}$.

Definition 3.2: A Young diagram $\Lambda = \{\lambda_{ij}\}$ is said to be standard if there exists a sequence (called a defining sequence)

$$\Theta = \{\theta_{ij},\ \theta_{ij} \in W\ ,\ 1 \le j \le a_i\ ,\ 1 \le i \le \ell\}$$

such that

1. $X(\theta_{11}) \ge X(\theta_{12}) \ge \cdots \ge X(\theta_{1a_1}) \ge X(\theta_{21}) \ge \cdots \ge X(\theta_{\ell a_\ell})$ (in G/B)

2. $\pi_i(X(\theta_{ij})) = X(\lambda_{ij})$ under π_i: $G/B \to G/P_i$

Definition 3.3: A Young diagram $\Lambda = (\lambda_{ij})$ is said to be standard on $X(w)$, $w \in W$, if there exists a sequence $\Theta = \{\theta_{ij}\}$ satisfying conditions (1) and (2) of Definition 3.2 and the additional condition that $X(w) \ge X(\theta_{11})$ (in G/B).

Definition 3.4: Given $\Lambda = (\lambda_{ij})$, we define x_Λ (in $k_q[G/B]$) as $x_\Lambda = \prod_i \prod_j x_{\lambda_{ij}}$. Further, we say x_Λ is standard (resp. standard on $X(w)$) if Λ is standard (resp. standard on $X(w)$).

Proposition 3.5: Standard monomials are linearly independent over $k(q)$.

Proof: The proof is the same as that of Proposition 2.2.

As in §2, to prove generation by standard monomials, it suffices to prove it for the case $k = \mathbb{Q}$.

Generation by standard monomials when $k = \mathbb{Q}$: As in §2, we write $V^{\omega_d} \otimes V^{\omega_{d'}} = \bigoplus_\nu V^\nu$ ($= \bigoplus_p^{\min(r,n-s)} V^{\omega_{d+p}+\omega_{d'-p}}$ where $r = \min(d,d')$, $s = \max(d,d')$) and denote $X^d = \sum_i x_i^{\omega_d} v_i^d$, $\{v_i^d\}$ being the basis of V^{ω_d} consisting of extremal weight vectors. We have, similar to Proposition 2.9 (cf. [R])

Proposition 3.6: Let P_ν be the projection $V^{\omega_d} \otimes V^{\omega_{d'}} \longrightarrow V^\nu$. Then

(a) $R^{V\omega_{d'},\omega_d} = \sum_\nu q^{2(c(\omega_d)+c(\omega_{d'})-c(\nu))} P_\nu$

(b) $R^{V\omega_{d'},\omega_d}(X^d \otimes X^{d'}) = q^{2(c(\omega_d)+c(\omega_{d'})-c(\omega_d+\omega_{d'}))} X^d \otimes X^{d'}$

(c) $P_\nu(X^d \otimes X^{d'}) = 0\ ,\ \nu \ne \omega_d + \omega_{d'}$

(d) $P_\nu(X^d \otimes X^{d'}) = P_\nu(X^{d'} \otimes X^d)\ ,\ \nu = \omega_d + \omega_{d'}$,

(here, c is the Casimir operator $c(\lambda) = (\lambda,\lambda) + 2(\rho,\lambda)$. In (d), the P_ν on the R.H.S. is the projection $V^{\omega_{d'}} \otimes V^{\omega_d} \longrightarrow V^{\nu}$).

More generally, given $\underline{a} = (a_1,\dots,a_\ell)$, $a_i \in \mathbb{Z}^+$, denoting $X^{\underline{a}} = \bigotimes_{i=1}^{\ell} (X^i)^{\otimes a_i}$, we have (similar to Proposition 2.11)

Proposition 3.7: Let V^{ν} be an irreducible factor in the direct sum decomposition of $\bigotimes_{i=1}^{\ell} (V^{\omega_i})^{\otimes a_i}$. Let P_{ν} be the projection

$$\bigotimes_{i=1}^{\ell} (V^{\omega_i})^{\otimes a_i} \longrightarrow V^{\nu}.$$

Then $P_{\nu}(X^{\underline{a}}) = 0$, $\nu \neq \sum_{i=1}^{\ell} a_i\omega_i$.

Generation by standard monomials: Let $N_{\underline{a}}$ = #{all monomials (in $x_{\lambda_{ij}}$'s) of type $\underline{a}$ or multidegree $\underline{a}$} and let $s_{\underline{a}}$ = #{standard monomials of type $\underline{a}$} (note that $s_{\underline{a}} = \dim V^{\nu}$, $\nu = \sum_{i=1}^{\ell} a_i\omega_i$).

Let V^{ν} be a factor appearing in the expression for $\bigotimes_{i=1}^{\ell} (V^{\omega_i})^{\otimes a_i}$ as a direct sum of irreducible $U_q(\mathfrak{g})$-modules. Let $\{e_i^{\nu}\}$ be a basis of V^{ν} consisting of weight vectors. Writing

$$e_i^{\nu} = \sum_{J_{\underline{a}}} c^{\nu}_{iJ_{\underline{a}}} v_{J_{\underline{a}}} \ , \quad c^{\nu}_{iJ_{\underline{a}}} \in k(q),$$

where $J_{\underline{a}} = \{J_{mj}\ ,\ 1 \le J_{mj} \le N_m\ ,\ 1 \le j \le a_m\ ,\ 1 \le m \le \ell\}$ (recall that $N_m = \dim V^{\omega_m}$) and $v_{J_{\underline{a}}} = \bigotimes_{m=1}^{\ell} \bigotimes_{j=1}^{a_m} v_{J_{mj}}$. We have, in view of Proposition 3.7,

$$\sum_{J_{\underline{a}}} c^{\nu}_{iJ_{\underline{a}}} x_{J_{\underline{a}}} = 0 \qquad \text{(I)}$$

where $x_{J_{\underline{a}}} = \prod_{m=1}^{\ell} \prod_{j=1}^{a_m} x_{J_{mj}}$. There are $N_{\underline{a}} - s_{\underline{a}}$ such linear equations among the $N_{\underline{a}}$ monomials of type $\underline{a}$. Further, the coefficient matrix of (I) has maximal rank ($= N_{\underline{a}} - s_{\underline{a}}$), in view of linear independence of $\{e_i^{\nu}\}_{i,\nu}$. Hence taking the standard monomials of type $\underline{a}$ as the free variables of (I) (in view of linear independence of standard monomials), we obtain that each nonstandard monomial of type $\underline{a}$ has an expression as a linear combination of standard monomials of type $\underline{a}$. Combining this with linear independence of standard monomials, we obtain

<u>**Theorem 3.8**</u>: Let $(k_q[G/B])_{\underline{a}}$ be the k(q)-span of monomials of type $\underline{a}$. Standard monomials of type $\underline{a}$ form a k(q)-basis for $(k_q[G/B])_{\underline{a}}$ (k being an arbitrary field).

Using the comultiplication $\Delta: k_q[G] \to k_q[G] \otimes k_q[G]$, we obtain

<u>**Theorem 3.9**</u>: (a) The algebra $k_q[G/B]$ has a canonical $\mathbb{Z}^{\ell}$-gradation.

(b) $k_q[G/B]$ has a canonical left comodule structure over $k_q[G]$.

<u>Proof</u>: We have, $k_q[G/B] = \underset{\underline{a}}{\otimes}(k_q[G/B])_{\underline{a}}$, $\underline{a} \in (\mathbb{Z}^+)^{\ell}$ which gives a canonical $\mathbb{Z}^{\ell}$-gradation for $k_q[G/B]$. As in the proof of Theorem 2.13, $\Delta: k_q[G] \to k_q[G] \otimes k_q[G]$ induces a map $\Delta: k_q[G/B] \to k_q[G] \otimes k_q[G/B]$, given by $\Delta(x_i) = \sum_j F_{ij} \otimes x_j$ (notation as in the proof of Theorem 2.13), giving a left $k_q[G]$-module structure for $k_q[G/B]$.

<u>Quantum Schubert schemes</u>: Let $w \in W$ and let I_w be the two-sided ideal in $k_q[G/B]$ generated by $\{x_\tau | \tau \in W^{P_d}, 1 \le d \le \ell, \tau \not\le w^{(d)}\}$ (here, for $1 \le d \le \ell$, $w^{(d)}$ is the minimal representative in W of wW_{P_d}). Set <u>$k_q[X(w)]$</u> or just <u>$R_q(w)$</u> as

$$R_w = k_q[G/B]/I_w.$$

<u>**Remark 3.10**</u>: In the case of G/P_d, a monomial $x_{\tau_1} \cdots x_{\tau_m}$ standard on G/P_d, either remains standard on $X(\tau) \subset G/P_d$, namely if $\tau \ge \tau_1$, or belongs to I_τ, namely if $\tau \not\ge \tau_1$. The situation is quite different in the case of G/B. A monomial $x_{\tau_1} \cdots x_{\tau_m}$ standard on G/B could be nonzero in $k_q[X(w)]$, for some $w \in W$ and be not standard on X(w). For example, for G = SL(3), the monomial x_2x_{13} which is standard on G/B is not standard on X(w), w = (312). In $k_q[X(w)]$, we have the relation $x_2x_{13} = q^{-1}x_3x_{12}$. This relation arises from

the quantum Plücker relation $x_1x_{23} = q^{-1}x_2x_{13} - q^{-2}x_3x_{12}$ (in $k_q[G/B]$). (Note that $x_{23} \in I_w$.)

<u>**Theorem 3.11**</u>: Let $\underline{a} = (a_1, \ldots, a_\ell)$, $a_i \in \mathbb{Z}^+$. Let $(R_q(w))_{\underline{a}}$ be the k(q)-span (in $R_q(w)$) of monomials of type (or multidegree) $\underline{a}$. Standard monomials on X(w) of type $\underline{a}$ form a k(q)-basis for $(R_q(w))_{\underline{a}}$.

<u>**Proof**</u>: The proof of linear independence of standard monomials of type $\underline{a}$ on X(w) is carried out on the same lines as that of Proposition 2.15.

<u>Generation by standard monomials</u>: We first observe that all relations in $R_q(w)$ are consequences of relations of the following type. Let f be a nonstandard monomial of type $\underline{a}$ in $k_q[G/B]$. Writing

$$(*) \qquad f_i = \Sigma a_i f_i \ , \ a_i \in k(q)$$

where f_i are standard monomials of type $\underline{a}$ in $k_q[G/B]$. Now going modulo I_w, some of the f_i's in the R.H.S. of (*) may not be standard in $R_q(w)$, while the L.H.S. of (*) is nonzero or zero in $R_q(w)$ according as f does not (resp. does) contain a factor x_τ, $\tau \in W^{P_d}$ (for some d) such that $w^{(d)} \not\geq \tau$. When q=1, these relations in k[X(w)] give rise to expressions for nonstandard monomials of type $\underline{a}$ on X(w) as a sum of standard monomials on X(w) of type $\underline{a}$ (cf. [LS]). From this, it follows that a nonstandard monomial (in $R_q(w)$) of type $\underline{a}$ has an expression as a linear combination of standard monomials of type $\underline{a}$ in $R_q(w)$. To make it more precise, considering all the above relations (in $R_q(w)$) as a linear system of equations in monomials of type $\underline{a}$ in $R_q(w)$, let us denote the corresponding coefficient matrix by $A_q(w)$. We have (in view of linear independence of standard monomials in $R_q(w)$), denoting $s_{\underline{a}}(w)$ = #{standard monomials of type $\underline{a}$ in $R_q(w)$},

$$\begin{aligned} s_{\underline{a}}(w) &\leq \#\{\text{free variables of the above system}\} \\ &\leq \#\{\text{free variables of the system for } q = 1\} \\ &= s_{\underline{a}}(w) \end{aligned}$$

Hence we obtain that #{free variables of the system} = $s_{\underline{a}}(w)$. This together with the linear independence of standard monomials in $R_q(w)$ implies the required result.

Theorem 3.12: $R_q(w)$ has a canonical $\mathbb{Z}^{\ell}$-gradation and a canonical left $k_q[B]$-comodule structure.

Proof: The canonical $\mathbb{Z}^{\ell}$-gradation is simply $R_q(w) = \underset{\underline{a}}{\oplus}(R_q(w))_{\underline{a}}$, $a \in (\mathbb{Z}^+)^{\ell}$. We have (as in the proof of Theorem 2.18), I_w is $U_q(b^+)$-stable from which the required result follows (as in the proof of Theorem 2.18).

References

[D] V. Drinfeld, Quantum Groups, Proc. of the ICM, Berkeley, 1986.

[D-M-M-Z] E.E. Demidov, Yu. I. Manin, E.E. Mukhin, and D.V. Zhdanowich, Nonstandard Quantum Deformations of GL(n) and Constant Solutions of the Yang-Baxter Equation (preprint).

[F-R-T] L. Faddeev, N. Reshetikhin, and L. Takhtajan, Quantization of Lie Groups and Lie Algebras, preprint, LOMI -14-87, 1987; Algebra and Analysis, vol.1, no:1 (1989).

[H] M. Hochster, Grassmannians and their Schubert varieties are arithmetically Cohen-Macaulay, J. Alg., Vol. 25 (1973), 40-57.

[Ho] W.V.D. Hodge, Some enumerative results in the theory of forms, Proc. Camb. Phil. Soc., 39 (1943), 22-30.

[J] M. Jimbo, A q-difference analogue of U($\mathfrak{g}$) and the Yang-Baxter equation, Lett. Math. Phys. 10, (1985), 63-69.

[K-R] A. Kirillov and N. Reshetikhin, q-Weyl group and multiplicative formula for R-matrices, preprint of Harvard University, January 1990.

[L-R] V. Lakshmibai and N. Reshetikhin, Quantum deformations of Flag and Schubert schemes, to appear in Comptes Rendus, Paris.

[L-S] V. Lakshmibai and C.S. Seshadri, Geometry of G/P-V, J. Alg. 100 (1986), 462-557.

[Lo] G. Lusztig, Quantum deformations of certain simple modules over enveloping algebras, Adv. in Math., 70, 237-249 (1988).

[M] C. Musili, Postulation formula for Schubert varieties, J. Indian Math. Soc. 36 (1972), 143-171.

[R] N. Reshetikhin, Quantized universal enveloping algebras, Yang-Baxter equation and invariants of links, LOMI-preprint, E-4-87, E-17-87.

[Ro] M. Rosso, Finite Dimensional Representations of Quantum Analog of the Enveloping Algebra of a Complex Simple Lie Algebra, Comm. Math. Phys. 117 (1988), 581-593.

[S] Ya. Soibelman, Algebra of functions on compact quantum group and its applications, Alg. and Anal., Vol. 2, No: 1 190-212 (1990).

[T-T] E. Taft, J. Towber, "Quantum deformation of flag schemes and Grassmann schemes," 1989-preprint.

Supernumary Polylogarithmic Ladders and Related Functional Equations

L. LEWIN*

Professor Emeritus, University of Colorado, Boulder, CO 80309, U.S.A.

1. INTRODUCTION

As discussed in several recent papers(1-7) the polylogarithmic function of order n and argument z can be defined through the series

$$\mathrm{Li}_n(z) = \sum_{r=1}^{\infty} z^r/r^n \quad , \; |z| \leq 1 \tag{1.1}$$

It satisfies the recursion formula

$$\mathrm{Li}_n(z) = \int_0^z \mathrm{Li}_{n-1}(z')\, dz'/z' \tag{1.2}$$

and this, together with the elementary relation

$$\mathrm{Li}_1(z) = -\log(1-z) \tag{1.3}$$

extends the definition throughout the complex z-plane.

The ladder, of order n, of a real algebraic number u, $(0 < u < 1)$, can be defined through the sum of terms:

$$L_n(N, u) = \frac{\mathrm{Li}_n(u^N)}{N^{n-1}} - \left\{ \sum_r \frac{A_r \mathrm{Li}_n(u^r)}{r^{n-1}} + \frac{A_o \log^n(u)}{n!} \right\} \tag{1.4}$$

Here, N, called the index, is the highest power of u occurring in the expression, and the powers r run through a limited set ranging from 1 to (N-1). But in practice they are confined, in many cases, to the factors of N, though there are a few exceptions as will be mentioned later.

The coefficients A_r, $0 \le r \le N-1$, are determined implicitly through the polynomial equation defining u.

There is also a set of logarithmic terms

$$L_n = \sum_{m=2}^{n} \frac{B_m \zeta(m) \log^{n-m}(u)}{(n-m)!} \tag{1.5}$$

to be subtracted from (1.4), and which, according to convenience, may or may not be explicitly included in the definition of the ladder.

If (1.4) and (1.5) are equated the resulting expression becomes a set of equations for determining the B_m, which are, in general, irrational. But there is an important subset of u for which B_2, and in many cases some of the other B_m, are rational. When this is the case the ladder is said to be <u>valid</u>.

If the ladder is valid at n=2 then it has been found, with absolutely no exceptions, that it is also valid at n=1, where it takes, in logarithmic form, the structure of a <u>cyclotomic equation</u>:

$$(1 - u^N) = u^{-A_o} \prod_r (1 - u^r)^{A_r} \tag{1.6}$$

This equation can be written down by inspection from (1.4); and, in reverse, (1.4) can be written down from (1.6). A partial proof for (1.6), applying for those ladders that can be deduced in a finite number of algebraic steps from Kummer's second-order functional equation[3], is given in reference 2. It has also been shown[4], again for those ladders that can be deduced from Kummer's functional equations[3] of order n, ($n \leq 5$), that if the ladder is valid for a certain value of n it is also valid for all lower values of n down to n=1.

It may be noticed that this result ensues if u in (1.4) is treated as if it were a variable, and the equation is differentiated with respect to u. We call this process pseudo-differentiation. The inverse process, which can be designated pseudo-integration, is very useful for generating expressions at higher orders, which can then be investigated for validity. All of the currently known ladders for $6 \leq n \leq 9$ were found in this way.

Ladders that can be directly deduced algebraically from Kummer's equations are said to be accessible; otherwise they are inaccessible. Many valid results, known at the present time either from numerical evaluation or from other branches of analysis, appear to be of this inaccessible character. It is not currently known whether this feature is genuine or merely an artefact due simply to a failure to find out how to deduce the corresponding ladder.* However, Wechsung[8] has shown that no functional equations of Kummer's type exist for $n > 5$. Now there are in existence at least 19 ladders, numerically verified, in this "trans-Kummer" region ($n \geq 6$) whose structure is no different from those of lower order, suggesting that, in some

* D. Zagier, in recent findings, claims that all results at n=2 and n=3 are deducible from Kummer's equations. In many cases, additional variables appear at intermediate stages of the calculations.

cases, at least, the appellation "inaccessible" is not entirely inappropriate.

The subset of algebraic numbers u for which the ladders are valid at n=2 is apparently rather limited. An extensive discussion of some of the few usable families of generating polynomial equations is given in reference 2.

2. GENERAL RESULTS FROM THE EQUATION-FAMILY $u^p + u^q = 1$

2.1 Kummer's functional equations involve two variables x and y, together with (1-x) and (1-y), all raised to various powers. The only way in which all the arguments can finish up as powers of u, as required by (1.4), is if u satisfies the two-term equation

$$u^p + u^q = 1 \tag{2.1}$$

and x and y are taken as u^p or u^q or harmonic group variants thereof ($1-u^p$, $-u^p/(1-u^p)$ and inversions). No other equation than (2.1) will achieve this; though, for n=2 and 3, Kummer's equations involve (1-x) and (1-y) in only their ratio (1-x)/(1-y), making a somewhat broader three-term equation[1] usable for $n \leq 3$, as examined also in section 8.2.

2.2 The present discussion is concerned only with u determined by (2.1), in which p and q are given positive integers and, as will be explained later, they are also mutually prime:

$$(p, q) = 1 \tag{2.2}$$

It is, at the present time, an unproven conjecture that only redundant results ensue if (2.2) is not satisfied.

2.3 For general integers p and q, subject to (2.2), there are eight independent generic cyclotomic equations, with different indices N. Since these equations can be multiplied together in arbitrary ways there is no unique set, and the ones given in the following constitute merely one convenient and simple selection. They are all easily proved by straightforward manipulation of (2.1). There would appear to be no others, in the general case, though there does not seem to be any current proof of this. These eight generic cyclotomic equations are:

$$1 - u^q = u^p \qquad (2.3)$$

$$1 - u^p = u^q \qquad (2.4)$$

$$1 - u^{2(p-q)} = (1 - u^{p-q})\, u^{-q} \qquad (2.5)$$

$$1 - u^{2(2p-q)} = (1 - u^{2p-q})(1 - u^{2(p-2q)})(1 - u^{p-2q})^{-1} u^q \qquad (2.6)$$

$$1 - u^{3(p-q)} = (1 - u^{p-q})(1 - u^{2(p-2q)})(1 - u^{p-2q})^{-1} \qquad (2.7)$$

$$1 - u^{p+q} = (1 - u^{2(p-2q)})(1 - u^{p-2q})^{-1} u^{2q} \qquad (2.8)$$

$$1 - u^{6q} = (1 - u^{3q})(1 - u^{2q})(1 - u^{2(p-2q)})(1 - u^{p-2q})^{-1} u^{2q-p} \qquad (2.9)$$

$$1 - u^{6p} = (1 - u^{3p})(1 - u^{2p})(1 - u^{2(p-2q)})(1 - u^{p-2q})^{-1} u^q \qquad (2.10)$$

It will be noticed that (p - 2q) occurs in many places in the above formulas. For the smaller values of p and q this will often be a factor of the indices appearing on the left-hand sides of these equations, giving rise to the observed result that, in many cases, all the powers r in (1.4) are factors of the index. For the record, these generic indices are:

$$N = \{q,\ p,\ 2(p-q),\ 2(2p-q),\ 3(p-q),\ p+q,\ 6q,\ 6p\} \qquad (2.11)$$

2.4 Apart from the special case $p = q = 1$, $(u = 1/2)$, we take p as the greater of p and q; and for the interim, $p > 2q$ also. This ensures that all the powers in (2.3) to (2.10) are positive. Modifications for the ladders with $q < p < 2q$ are found by inverting those arguments with negative exponent. The special case $p = 2q$ is discussed in more detail in sections 4.1 and 4.2.

2.5 Corresponding to the eight cyclotomic equations, eight ladders can be written down. Thus, from (2.3), one can define, at n=2, the ladder

$$L_2(q, u) = \frac{Li_2(u^q)}{q} + \frac{p}{2} \log^2 u \qquad (2.12)$$

It is only in certain very special cases that the ladders constructed directly in this way are valid, and it may be preferable to think of them more as component-ladders, i.e., pieces of what will eventually be a valid ladder. The two known exceptions in this family are i) $p = q = 1$; $u = 1/2$, for which the indices degenerate into only two surviving values, namely $N = 1$ and 6; and ii) $q = 1$, $p = 2$; $u \equiv \rho = (5^{1/2} - 1)/2$, for which the indices degenerate to N=1, 2, 6 and 12. In both these cases the ladders pseudo-integrate directly[3] to give valid results as far as n=5 (for additional results on ρ see later), and this property is called transparency. The only other known transparent results stem from the family $u^2 - mu + 1 = 0$, $2 \leq m \leq 10$, $(m \neq 9)$ as discussed in references 2 and 3; and a new result, mentioned later, stemming from research on cyclotomic equations generated by quadratic equations[9].

2.6 The first three cyclotomic equations, (2.3) to (2.5) can be grouped together. Algebraically they are almost trivial re-arrangements of (2.1) and correspond to three component-ladders which, at n=2, combine in pairs to yield two valid ladders. These two then combine to give a single valid ladder at n=3, and which involves $\zeta(3)$ via B_3 in (1.5). These are the well-established results of Euler and Landen[10,11], and are given in (2.21), (2.22) and (2.31).

This tendency for ladders to combine in pairs, resulting in one fewer at the next higher order, in going from even to odd orders, is very marked.

2.7 The remaining five cyclotomic equations give five component-ladders at n=2 which combine in pairs to give four valid ladders at n=2 and three at n=3. At this juncture we encounter the effect of the arbitrariness of the combinations used to construct the cyclotomic equations. Thus, in (2.10), if the final factor u^q is replaced by $(1 - u^p)$ from (2.4), the effect is to add a component-ladder in $L_2(p, u)$ into the definition of $L_2(6p, u)$. This affects the ensuing ladder at n=3, altering the multiplier of $\zeta(3)$ therein. Now the valid ladders at n=4 contain no net $\zeta(3)$ term. It is a conjecture that, quite generally, zeta functions of odd argument less than n do not appear in ladders of order n; and this hypothesis was critical in enabling the trans-Kummer results to n=9 to be generated. Thus we have two options: either to use Landen's result at n=3 to cancel out any $\zeta(3)$ term before proceeding to n=4;

or, with some hind-sight, to choose combinations of component-ladders so that the $\zeta(3)$ term is absent from the beginning. The outcome is the same, of course, but the form taken by the results is simpler if the second alternative is taken. Now it happens that a very convenient functional equation, (6.108) of ref. 12, is absent any $\zeta(3)$ term, and if the valid ladders at n=3 are generated from it, combinations of component-ladders free of $\zeta(3)$ are produced quite naturally. Thus the component-ladder from (2.7) less two-thirds of that from (2.5) is used; that from (2.9) plus one-third that from (2.3); and (2.10) plus one-third of (2.4). The resulting component-ladders, for general n, are given in (2.13) to (2.20) and the valid formulas at n=2 in (2.21) to (2.26).

Component-Ladders

$$L_n(q, u) = \frac{Li_n(u^q)}{q^{n-1}} + \frac{p \log^n(u)}{n!} \tag{2.13}$$

$$L_n(p, u) = \frac{Li_n(u^p)}{p^{n-1}} + \frac{q \log^n(u)}{n!} \tag{2.14}$$

$$L_n[2(p-q), u] = \frac{Li_n(u^{2(p-q)})}{[2(p-q)]^{n-1}} - \left\{\frac{Li_n(u^{p-q})}{(p-q)^{n-1}} + \frac{q \log^n(u)}{n!}\right\} \tag{2.15}$$

$$L_n[2(2p-q), u] = \frac{Li_n(u^{2(2p-q)})}{[2(2p-q)]^{n-1}} - \left\{\frac{Li_n(u^{2p-q})}{(2p-q)^{n-1}} + \frac{Li_n(u^{2(p-2q)})}{[2(p-2q)]^{n-1}} - \frac{Li_n(u^{p-2q})}{(p-2q)^{n-1}} - \frac{q \log^n(u)}{n!}\right\} \tag{2.16}$$

$$L_n[3(p-q), u] = \frac{Li_n(u^{3(p-q)})}{[3(p-q)]^{n-1}} - \left\{ \frac{2}{3} \frac{Li_n(u^{2(p-q)})}{[2(p-q)]^{n-1}} + \frac{Li_n(u^{p-q})}{3(p-q)^{n-1}} \right.$$

$$\left. + \frac{Li_n(u^{2(p-2q)})}{[2(p-2q)]^{n-1}} - \frac{Li_n(u^{p-2q})}{(p-2q)^{n-1}} - \frac{2q}{3} \frac{\log^n(u)}{n!} \right\} \quad (2.17)$$

$$L_n(p+q, u) = \frac{Li_n(u^{p+q})}{(p+q)^{n-1}} - \left\{ \frac{Li_n(u^{2(p-2q)})}{[2(p-2q)]^{n-1}} - \frac{Li_n(u^{p-2q})}{(p-2q)^{n-1}} \right.$$

$$\left. - \frac{2q \log^n(u)}{n!} \right\} \quad (2.18)$$

$$L_n(6q, u) = \frac{Li_n(u^{6q})}{(6q)^{n-1}} - \left\{ \frac{Li_n(u^{3q})}{(3q)^{n-1}} + \frac{Li_n(u^{2q})}{(2q)^{n-1}} - \frac{Li_n(u^{q})}{3q^{n-1}} \right.$$

$$\left. + \frac{Li_n(u^{2(p-2q)})}{[2(p-2q)]^{n-1}} - \frac{Li_n(u^{p-2q})}{(p-2q)^{n-1}} + \frac{2(p-3q)}{3} \frac{\log^n(u)}{n!} \right\} \quad (2.19)$$

$$L_n(6p, u) = \frac{Li_n(u^{6p})}{(6p)^{n-1}} - \left\{ \frac{Li_n(u^{3p})}{(3p)^{n-1}} + \frac{Li_n(u^{2p})}{(2p)^{n-1}} - \frac{Li_n(u^{p})}{3p^{n-1}} \right.$$

$$\left. + \frac{Li_n(u^{2(p-2q)})}{[2(p-2q)]^{n-1}} - \frac{Li_n(u^{p-2q})}{(p-2q)^{n-1}} - \frac{4q}{3} \frac{\log^n(u)}{n!} \right\} \quad (2.20)$$

Valid Ladders at n=2

From Kummer's functional equations it is found that:

$$q\, L_2\, (q,\, u) + p\, L_2\, (p,\, u) = \zeta(2) \tag{2.21}$$

$$p\, L_2\, (p,\, u) + (p-q)\, L_2\, [2(p-q),\, u] = 0 \tag{2.22}$$

$$3(p-q)\, L_2\, [3(p-q),\, u] - (2p-q)\, L_2\, [2(2p-q),\, u] = 0 \tag{2.23}$$

$$(p+q)\, L_2\, (p+q,\, u) - (2p-q)\, L_2\, [2(2p-q),\, u] = 0 \tag{2.24}$$

$$3q\, L_2\, (6q,\, u) - (2p-q)\, L_2\, [2(2p-q),\, u] = -\zeta(2) \tag{2.25}$$

$$3p\, L_2\, (6p,\, u) - 2\, (2p-q)\, L_2\, [2(2p-q),\, u] = 0 \tag{2.26}$$

As already mentioned, (2.21) and (2.22) are essentially due to Euler and Landen.

As can be readily seen, there are six relations between the eight component-ladders, so two of them are independent; they are conveniently taken to be $L_2(q,\, u)$ or $L_2(p,\, u)$, and $L_2[2(2p-q),\, u]$; and these are candidates for combination with the supernumary component-ladders considered in section 3.

2.8 At n=3 certain combinations of component-ladders give rise to valid structures, as found from the functional equations. Again there is a certain amount of arbitrariness, but the following second-degree ladders are convenient to define:

$$L_n{}^{(2)}(p,\, u) = p^2\, L_n\, (p,\, u) + q^2\, L_n\, (q,\, u) + (p-q)^2\, L_n\, [2(p-q),\, u] \tag{2.27}$$

$$L_n{}^{(2)}[3(p-q),\, u] = 9(p-q)^2\, L_n\, [3(p-q),\, u] + (p+q)^2\, L_n\, (p+q,\, u) - 2(2p-q)^2\, L_n\, [2(2p-q),\, u] \tag{2.28}$$

$$L_n{}^{(2)}(6q,\, u) = 9q^2\, L_n\, (6q,\, u) - 2(p+q)^2\, L_n\, (p+q,\, u) + (2p-q)^2\, L_n\, [2(2p-q),\, u] \tag{2.29}$$

$$L_n^{(2)}(6p, u) = 9p^2 L_n(6p, u) - 2(p+q)^2 L_n(p+q, u) - 2(2p-q)^2 L_n[2(2p-q), u] \quad (2.30)$$

Then at n=3 we have

$$L_3^{(2)}(p, u) = \zeta(3) + \zeta(2)\, q \log(u) \quad (2.31)$$

$$L_3^{(2)}[3(p-q), u] = 0 \quad (2.32)$$

$$L_3^{(2)}(6p, u) = 0 \quad (2.33)$$

$$L_3^{(2)}(6q, u) = -3\, \zeta(2)\, q \log(u) \quad (2.34)$$

Landen's formula reduces to (2.31) and, as already discussed, the remaining results are free of $\zeta(3)$. In extending results to n=4, (2.31) is therefore not needed, in the general case, and only (2.32) to (2.34) are used.

2.9 At n=4 these remaining three equations combine in pairs to give, from Kummer's functional equation,

$$p\, L_4^{(2)}(6p, u) + (p-q)L_4^{(2)}[3(p-q), u] = 0 \quad (2.35)$$

$$q\, L_4^{(2)}(6q, u) - (p-q)L_4^{(2)}[3(p-q), u] = \frac{19}{12}\zeta(4) - \frac{3}{2} q^2 \log^2(u) \quad (2.36)$$

Were it not that (2.32) to (2.34) are already free of $\zeta(3)$, (2.31) would be needed to eliminate it first, so the equations really combine in threes to produce two results from four initial ladders. In the supernumary cases at higher orders we see this type of combining more clearly in effect: ladders combine in threes when going from odd to even orders, resulting in a loss of two results as the order is increased.

At n=5 there is only one valid ladder coming from combining the forms in (2.35) and (2.36):

$$3p^2L_5^{(2)}(6p, u) + 3q^2L_5^{(2)}(6q, u) + 3(p-q)^2L_5^{(2)}[3(p-q), u] =$$

$$7\,\zeta(5) + \frac{19}{4}\,\zeta(4)\,q\,\log(u) - \frac{3}{2}\,\zeta(2)\,q^3\log^3(u) \qquad (2.37)$$

All these results come from Kummer's functional equations, or by pseudo-integration of the earlier results. However, the latter method does not give the B_m coefficients, the multipliers of $\zeta(m)$, and these have to be found numerically or by other means.

This is as far as one can go in the general case. One can hypothesize that <u>at least</u> one more result containing $\zeta(5)$, and, more likely, two such, would be required to reach a result at n=6, where, as previously stated, no equations of Kummer's type exist. In fact, at the present time, no non-trivial functional equations of any kind are known for n > 5.

2.10 Figure 1 is a flow-chart showing how the ladders at one order combine to give results at the next. The three elementary cyclotomic equations (order 1) give two valid ladders at order two and one at order 3, where this sequence stops. The remaining five cyclotomic equations give four results at the second order, decreasing by one each time the order increases, and terminating with a single result at n=5. It is believed that this is as far as the sequence goes in the general case (arbitrary p and q).

FIGURE 1
FLOW CHART FOR GENERAL CASE

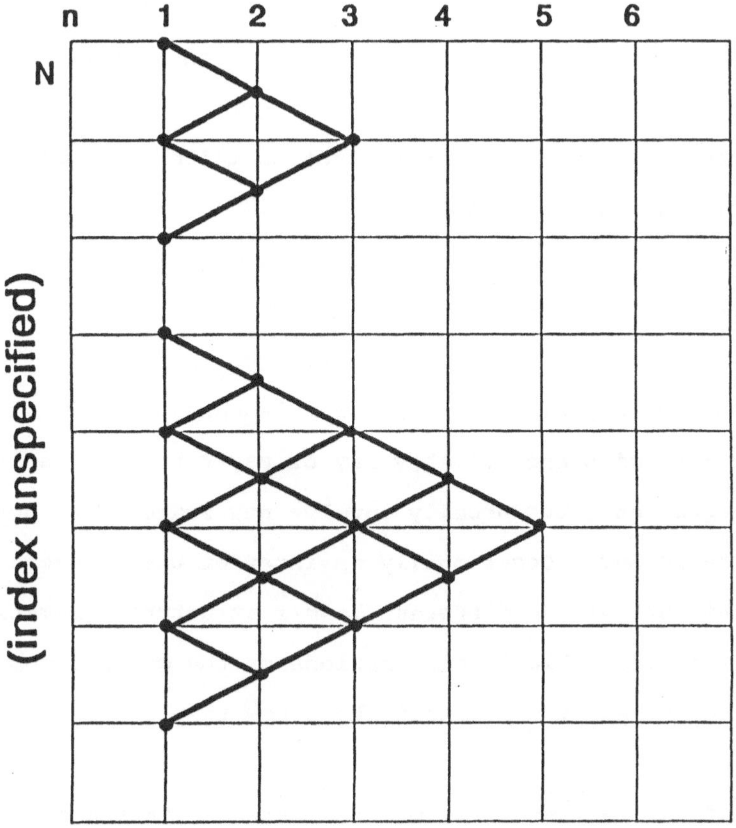

One interesting consequence that can be drawn from these results concerns the special case $p = q = 1$, $u = 1/2$. Euler's and Landen's formulas give $Li_2(1/2)$ and $Li_3(1/2)$ in terms of $\log(1/2)$, $\zeta(2)$ and $\zeta(3)$. But according to figure 1 the sequence stops there (because of the term in $\zeta(3)$), so there is no corresponding formula for $Li_4(1/2)$ in terms of, say, $\log(1/2)$, $\zeta(2)$ and $\zeta(4)$. Some researchers have sought (in vain) to find such a result via numerical computations. According to figure 1, no such result is to be expected, though this demonstration falls short of constituting a rigorous proof.

3. SUPERNUMARY LADDERS

3.1 The supernumary ladders arise from combinations of the results of section 2 with additional component-ladders occurring for special values of p and q. They may or may not be accessible from Kummer's equations, but usually involve new indices not contained in the generic set. Occasionally an index of the generic set may be repeated, but with a different ladder structure. Sometimes this ladder may be reducible to combinations of the generic component-ladders; sometimes it may constitute a new result.

It had earlier appeared, at least at $n=2$, that a supernumary component-ladder always led to a new valid result. More recent work has cast a doubt on this conjecture, however, but the matter is currently unresolved. The difficulty comes partly from practical limitations on the needed numerical computations, and partly from not knowing whether further, undiscovered, supernumary component-ladders exist and need to be combined into the formula. As mentioned later, systematic procedures for the generation of cyclotomic equations now exist. Although the verification of a cyclotomic equation is a matter of straightforward algebraic manipulation, finding them in the first place is more difficult. For this reason some of the details are given in the later sections.

3.2 Since, at n=2, component-ladders in the form (1.4) are assumed to be associated with the cyclotomic equation (1.6), the supernumary ladders are constructed, in the first place, by generating a supernumary cyclotomic equation, if this can be done, and then determining the coefficient B_2 in (1.5) by requiring $L_2(N, u) = L_2$. If B_2 as thus determined (numerically, to a sufficient accuracy) is rational - in practice only very simple rationals are encountered - then the new ladder is valid, and, incidentally, is also transparent. If not, one can try combining it with rational amounts of component-ladders from the generic set (2.13) to (2.20). However, only two of these are independent, so essentially only three integer multipliers need to be sought. If there exists a set to give a rational B_2 then a new valid ladder, outside the generic set, has been found. Failure to find one can mean i) no such result exists; ii) a programming error has been made; iii) the necessary integers lie outside the search range of the program; or iv) a further supernumary component-ladder is needed. In any particular case it is difficult to decide between these possibilities. It should be noted that multi-precision computations are normally required, both to avoid artefacts and to prevent missing genuine results.

If a valid result is obtained at n=2 then its extension to higher orders can be sought. If a new formula at n=5 can be reached then elimination of the $\zeta(5)$ term with (2.37) enables the trans-Kummer region to be explored. This has been done successfully, so far, with three different bases.

3.3 Systematic procedures for constructing cyclotomic equations have now been developed by Browkin, Ray and Zagier. Those developed here have arisen from examining the forms taken by $(1 - u^r)$ for various values of r. In many cases a partial formula can be found involving prime factors 2, 3 and/or 5. If further such formulas can be found, so that these factors can be eliminated, then a cyclotomic equation emerges. If this does not reduce to some combination from the generic set, and it will not if a new index is involved, then a supernumary cyclotomic equation has been generated. This method was employed in ref. 2, and a further example is illustrated here by a quadratic equation due to Professor Browkin(9). It is apparently the only remaining quadratic equation giving real roots not already considered in the previous studies.

3.4 Browkin's quadratic is

$$3u^2 + u - 1 = 0 \tag{3.1}$$

It has roots u = U, -V where both

$$U = (13^{1/2} - 1)/6 \text{ and } V = (13^{1/2} + 1)/6 \qquad (3.2)$$

are real and positive and lie in the interval (0,1).

From (3.1) we have

$$1 - u = 3u^2 \qquad (3.3)$$

$$\begin{aligned} 1 - u^3 &= (1 - u)(1 + u + u^2) = (1 - u)[2u(1 + 2u)] \\ &= 2u(1 - u)(3u + 3u^2) \\ &= 6u^2(1 - u^2) \qquad (3.4) \end{aligned}$$

$$\begin{aligned} 1 - u^6 &= (1 + u)(1 - u + u^2)(1 - u^3) \\ &= (1 - u^2)(1 - u)^{-1}(4u^2)(1 - u^3) \qquad (3.5) \end{aligned}$$

Eliminating factors 2 and 3 between (3.3), (3.4) and (3.5) gives a single cyclotomic equation (there do not appear to be any others):

$$1 - u^6 = u^2(1 - u^3)^3(1 - u^2)^{-1}(1 - u)^{-3} \qquad (3.6)$$

From this the ladder for U can be written down and the coefficient B_2 determined by numerical computation, leading to the valid ladder

$$\frac{\mathrm{Li}_2(U^6)}{6} - \left\{ \mathrm{Li}_2(U^3) - \frac{1}{2}\mathrm{Li}_2(U^2) - 3\mathrm{Li}_2(U) - \log^2 U \right\} = \frac{4}{3}\zeta(2) \quad (3.7)$$

The corresponding structure for V comes by replacing U by $-V$ in (3.7) followed by minor re-arrangements, to give

$$\frac{\mathrm{Li}_2(V^6)}{3} - \left\{ \mathrm{Li}_2(V^3) + 2\mathrm{Li}_2(V^2) - 3\mathrm{Li}_2(V) + \log^2 V \right\} = \frac{2}{3}\zeta(2) \quad (3.8)$$

These two results are "transparent" in the sense used earlier, but do not extend to the third order.

There is no known prescription about (3.1) that determines that it should possess valid ladders. However, it has always been observed that if one root of an equation gives a valid result then so do the others: this feature is exemplified here.

It has not proved feasible to deduce (3.7) or (3.8) from Kummer's equations; nor, so far, has it been possible to construct a dilogarithmic functional equation from which these results could be obtained analytically.

4. THE TRANS-KUMMER RESULTS FOR $6 \le n \le 9$

4.1 The quantity $\rho = (5^{1/2} - 1)/2$ is the solution (p=2, q=1) of $u^2 + u = 1$, and is treated extensively in references 2 and 3. There are four accessible ladders with indices 1, 2, 6 and 12. Of these, the ladders with indices 1 and 6 contain powers ρ^r with r both even and odd; and the resulting ladder of index 6 extends transparently to the fourth order but no further. The ladders with indices 2 and 12 involve only even powers of r, so they really can be put in terms of $\rho_2 \equiv \rho^2 = (3 - 5^{1/2})/2$. The first ladder extends to the third order (Landen's result) and the other to the fifth order, where it involves $\zeta(5)$. All these results are accessible from Kummer's equations, and this is as far as the equations can be taken analytically.

There are two supernumary ladders of indices 20 and 24 and they can be found via the easily verifiable relations

$$1 + \rho^{10} = (1 + \rho^2)^3 \rho^2 \qquad (4.1)$$

$$1 + \rho^{12} = (1 + \rho^6)(1 + \rho^4)^2(1 + \rho^2)^{-1} \tag{4.2}$$

The n=2 ladders, due, respectively to Coxeter(13) and Philips(14) come from a consideration of a certain infinite series, and from a manipulation of multiple integrals. These methods do not seem capable of extension beyond n=2, but both ladders transparently extend to the fifth order where they involve $\zeta(5)$. The elimination of the $\zeta(5)$ term with the index-12 ladder then produces two ladder sequences, each of which extends to the seventh order. The elimination of the $\zeta(7)$ term then provides a single ladder expression which further extends transparently to the ninth order, where the process halts due to the lack of any other relevant result containing $\zeta(9)$. It is believed that the extension cannot be taken any further. To be able to do so would require, at the very least, a further supernumary cyclotomic equation as a starting point, but none has come to light. A flow chart* for these ρ-ladders is shown in figure 2. There is a total of six results for $n \geq 6$.

Although the results for indices 20 and 24 were originally obtained by pseudo-integration, followed by numerical determination of the B_m coefficients, it has since proved possible(5,7) to generate two families of single-variable functional equations, up to the fifth order, from which these results can now be proven analytically. The arguments are all of the form $\pm z^m(1-z)^r(1+z)^s$, where z is the variable and m, r and s are certain various integral exponents; but these equations cannot be extended beyond the fifth order, so all the trans-Kummer results currently depend on

* Figure 3 of ref.4 presents the same information in a slightly different form.

FIGURE 2

FLOW CHART FOR ρ-LADDERS

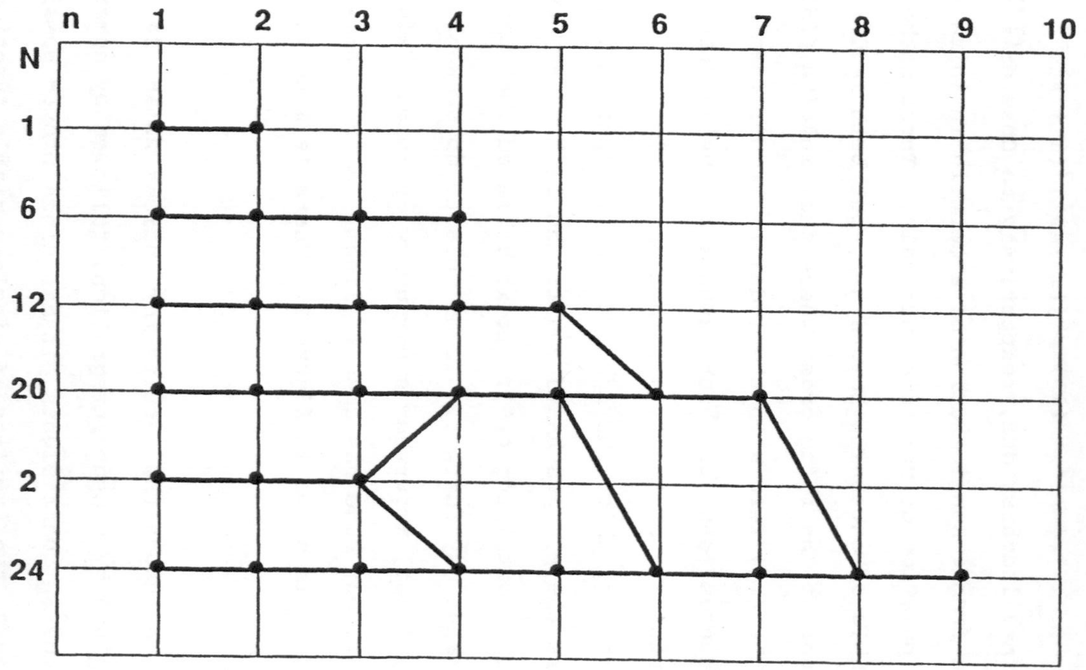

numerical determination. In fact it can be plausibly demonstrated, though a rigorous proof is currently lacking, that no functional equation exists at the sixth order with arguments limited to the above form in z, whatever the selection of integers m, r and s.

4.2 We shall touch briefly here on the assertion of section 2.2 that only redundant results ensue if $(p, q) \neq 1$ by a consideration of the equation

$$u^4 + u^2 = 1 \tag{4.3}$$

Clearly $u = \rho^{1/2}$, so any ρ-ladder becomes a corresponding u-ladder involving, however, only even powers of u. For a novel result to emerge, odd powers would also be required. It has been possible to find only one such result, obtained by adding u^3 to both sides of (4.3) to give, after minor manipulation,

$$1 - u^6 = (1 - u^3)^2(1 - u)^{-1}u^2 \tag{4.4}$$

The corresponding ladder is

$$L_2(6, u) = \frac{Li_2(u^6)}{6} - \left\{ \frac{2}{3} Li_2(u^3) - Li_2(u) - \log^2 u \right\} \tag{4.5}$$

where $u = \rho^{1/2}$.

If this ladder is equated to $C\,\zeta(2)$ then a numerical calculation gives

$$C = 0{\cdot}46928439504570001 \; . \; . \; . \; . \; . \tag{4.6}$$

which clearly is not a simple rational.

It is not known if there is a further result that could be combined with (4.5) to produce a rational coefficient. Subtracting rather than adding u^3 produces only a trivial change to (4.5). Tentatively one concludes that there are no new ladders of this character in the present case. To extend this conclusion to all cases where $(p, q) \neq 1$ is clearly a big leap. Currently it is just a conjecture with (4.5) and (4.6) as its sole backing.

4.3 The quantity ω is treated in detail in reference 7. It is the solution in (0, 1) of

$$u^3 + u^2 = 1 \tag{4.7}$$

and also of

$$u^5 + u = 1 \tag{4.8}$$

It can be shown that ω is the only base satisfying more than one equation of the form $u^p + u^q = 1$ with different integer values of p and q.

As a consequence of both (4.7) and (4.8), ω has an extraordinarily rich sequence of results. Its generic indices are comprised of

$$N = \{1, 2, 3, 5, 7, 8, 12, 18, 30\} \tag{4.9}$$

In addition there are four supernumary indices:

$$N = \{14, 20, 28, 42\} \tag{4.10}$$

of which the first three give ladders accessible from Kummer's equations, and the last comes from[7]

$$1 - \omega^{42} = (1 - \omega^{21})(1 - \omega^{14})(1 - \omega^7)^{-1}(1 - \omega^6)^2\omega^{-1} \tag{4.11}$$

It was originally investigated numerically but the functional equations subsequently developed, as far as the fifth order, now enable the ladders to be found analytically[7].

The flow chart, shown in figure 3 , is an ideal demonstration of the contentions that i) ladders combine in pairs when the order increases from even to odd; ii) ladders combine in threes when the order increases from odd to even; and iii) zeta functions of odd argument less than the order are always absent from the ladders. (This observation comes from the equations that underlie the flow chart.) The trans-Kummer results give 5 equations at the sixth order, 4 at the seventh, 2 at the eighth, and a single one, recently found by colleagues at the University of New South Wales, at the ninth; a total of twelve for $n \geq 6$. These equations have since been determined analytically, apart from the coefficients of $\zeta(n)$, by D. Zagier.

5. SUPERNUMARY RESULTS FOR p=4

5.1 The case of p=4, q=2 was addressed in section 4.2. It comes into the category $(p, q) \neq 1$, and has yielded only redundant results.

The cases p=4, q=1 and p=4, q=3 are closely related. We define ϕ and ψ to be, respectively, the solutions in (0, 1) of the equations

$$u^4 + u = 1 \qquad (5.1)$$

$$u^4 + u^3 = 1 \qquad (5.2)$$

Each equation possesses two real roots, the second being negative and numerically greater than unity. The transformation $u \rightarrow -1/u$ transforms (5.1) into (5.2) or *vice versa*. As a consequence, any ψ-ladder goes over into a related ϕ-ladder of corresponding

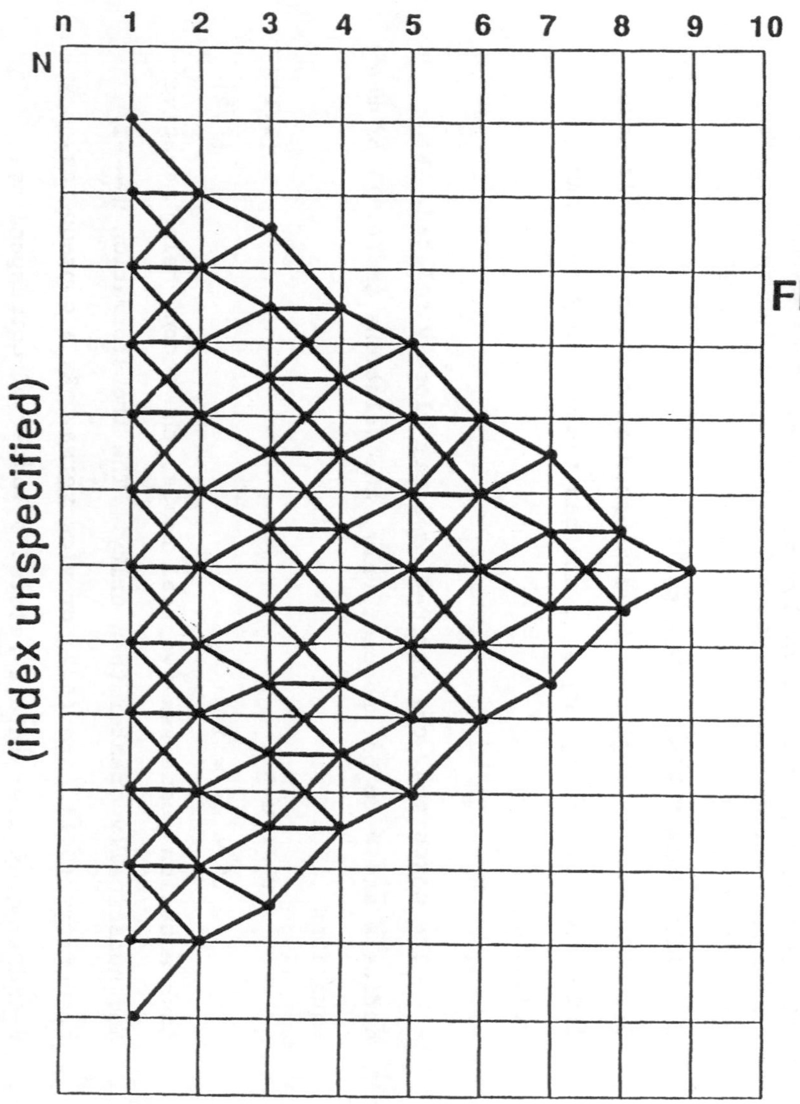

FIGURE 3
FLOW CHART FOR ω-LADDERS

structure. The duplication and inversion formulas are needed to put it in standard form, but essentially the one can be written down from the other. Only the coefficients B_m are different, and need to be evaluated separately.

Each base possesses seven surviving generic cyclotomic equations and a corresponding quota of generic ladders up to the fifth order. Each also possesses two supernumary cyclotomic equations, one of which leads to valid ladders up to the fourth order, and the other to the second order only. Thus no trans-Kummer result can be found for these bases based on the present data, despite the presence of the needed pair of additional ladders at n=2.

We shall deal only briefly with (5.1) and give the results for (5.2) in more detail.

5.2 A cyclotomic equation of index N=21 can be found as follows:

$1 + \phi^7 = 1 + \phi^3 \cdot \phi^4 = 1 + \phi^3 - \phi^4 = \phi(1 + \phi^2)$.

Squaring,

$$(1 + \phi^7)^2 - \phi^7 = \phi^2(1 + 2\phi^2 + \phi^4 - \phi^5) = \phi^2[(1 + \phi^2 + \phi^4) + \phi^2(1 - \phi^3)].$$

$$1 + \phi^7 + \phi^{14} = \phi^2\left[\frac{1 - \phi^6}{1 - \phi^2} + \phi^2(1 - \phi^3)\right] = \phi^2\,\frac{(1 - \phi^3)}{(1 - \phi^2)}\,[1 + \phi^3 + \phi^2(1 - \phi^2)]$$

$$= \phi\,\frac{(1 - \phi^3)}{(1 - \phi^2)}\,(1 + \phi^3 - \phi^5) = \frac{1 - \phi^3}{1 - \phi^2}\,(1 - \phi^6).$$

Hence,

$$1 - \phi^{21} = (1 - \phi^7)(1 - \phi^3)(1 - \phi^6)(1 - \phi^2)^{-1} \tag{5.3}$$

This can be combined with the readily verified equations

$$1 - \phi^6 = (1 - \phi^3)\phi^{-1} \tag{5.4}$$

$$1 - \phi = \phi^4 \tag{5.5}$$

to give the preferred combination

$$1 - \phi^{21} = (1 - \phi^7)(1 - \phi^6)^2(1 - \phi^2)^{-1}(1 - \phi)^{-1}\phi^5 \tag{5.6}$$

The corresponding ladder is

$$L_2(21, \phi) = \frac{Li_2(\phi^{21})}{21} - \left\{\frac{Li_2(\phi^7)}{7} + \frac{Li_2(\phi^6)}{3} - \frac{Li_2(\phi^2)}{2} - Li_2(\phi) - \frac{5}{2}\log^2\phi\right\} \tag{5.7}$$

This can now be combined with the index-one ladder

$$L_2(1, \phi) = Li_2(\phi) + 2 \log^2 \phi \tag{5.8}$$

coming from (5.5).

A numerical search yields

$$42 L_2(21, \phi) - 37 L_2(1, \phi) = 11 \zeta(2) \tag{5.9}$$

This result seems incapable of extension to n=3.

An index-30 cyclotomic equation can be generated as follows: $1 - \phi^5 = 1 - \phi(1 - \phi) = 1 - \phi + \phi^2 = \phi^2(1 + \phi^2)$. Squaring gives $1 - 2\phi^5 + \phi^{10} = \phi^4(1 + 2\phi^2 + \phi^4) = \phi^4(2 - \phi + 2\phi^2)$. Hence

$1 - \phi^5 + \phi^{10} = 2\phi^4(1 + \phi^2)$,

$1 + \phi^5 + \phi^{10} = 2\phi^4(1 + \phi + \phi^2)$.

Eliminating the factor 2 between the last two equations and re-arranging leads to

$$1 - \phi^{30} = (1 - \phi^{15})^2(1 - \phi^{10})(1 - \phi^5)^{-2}(1 - \phi^3)^{-1}(1 - \phi^2)^{-1}\phi^5 \quad (5.10)$$

The index-30 ladder corresponding to this is

$$L_2(30, \phi) = \frac{Li_2(\phi^{30})}{30} - \left\{\frac{2}{15} Li_2(\phi^{15}) + \frac{Li_2(\phi^{10})}{10} - \frac{2}{5} Li_2(\phi^5) - \frac{Li_2(\phi^3)}{3} - \frac{Li_2(\phi^2)}{2} - \frac{5}{2} \log^2\phi\right\} \quad (5.11)$$

This can be combined with (5.8), and a numerical search gives

$$30\, L_2(30, \phi) - 2\, L_2(1, \phi) = 13\, \zeta(2) \quad (5.12)$$

This equation can be extended numerically as far as the fourth order, but apparently no further. Neither (5.9) nor (5.12) have any current analytical proof. It is not known if there are yet further cyclotomic equations which could be incorporated into these results.

5.3 If we replace ϕ by $-1/\psi$ in (5.6), or alternatively go through a corresponding algebraic development starting with $1 - \psi^7 = \psi^4(1 + \psi^2)$, the index-42 cyclotomic equation so generated leads to

$$1 - \psi^{42} = (1 - \psi^{21})(1 - \psi^{14})(1 - \psi^7)^{-1}(1 - \psi^6)^2(1 - \psi^2)^{-2}(1 - \psi) \quad (5.13)$$

The related ladder is

$$L_2(42, \psi) = \frac{Li_2(\psi^{42})}{42} - \left\{\frac{Li_2(\psi^{21})}{21} + \frac{Li_2(\psi^{14})}{14} - \frac{Li_2(\psi^7)}{7} + \frac{Li_2(\psi^6)}{3} - Li_2(\psi^2) + Li_2(\psi)\right\} \quad (5.14)$$

(One can note the limitation of the powers r to factors of 42 in this result.)

From the equation $(1 - \psi^2) = (1 - \psi)\psi^{-3}$ one can define an index-2 ladder

$$L_2(2, \psi) = \frac{Li_2(\psi^2)}{2} - \left\{Li_2(\psi) + \frac{3}{2}\log^2\psi\right\} \qquad (5.15)$$

A numerical search then yields the counterpart of (5.9) in the form

$$42\, L_2(42, \psi) - 37\, L_2(2, \psi) = 8\, \zeta(2) \qquad (5.16)$$

Again, this result has not been extended to the next order. The equivalent of (5.10), found in a similar manner is

$$1 - \psi^{30} = (1 - \psi^{15})^2(1 - \psi^{10})(1 - \psi^5)^{-2}(1 - \psi^6)(1 - \psi^3)^{-1}(1 - \psi)\psi^{-3} \qquad (5.17)$$

The corresponding ladder, and the generic ladders up to the fifth order, follow. The generic indices are comprised of

$$N = \{2, 3, 4, 7, 10, 18, 24\} \qquad (5.18)$$

with the index 3 repeated. Along with the supernumary index 30 the following component-ladders can be defined (these are convenient forms; other legitimate variants exist):

$$L_n(2, \psi) = \frac{Li_n(\psi^2)}{2^{n-1}} - Li_n(\psi) - \frac{3\log^n\psi}{n!} \qquad (5.19)$$

$$L_n(3, \psi) = \frac{Li_n(\psi^3)}{3^{n-1}} + \frac{4\log^n\psi}{n!} \qquad (5.20)$$

$$L_n(4, \psi) = \frac{Li_n(\psi^4)}{4^{n-1}} + \frac{3\log^n\psi}{n!} \qquad (5.21)$$

$$L_n(7, \psi) = \frac{Li_n(\psi^7)}{7^{n-1}} + Li_n(\psi) + \frac{10 \log^n \psi}{n!} \tag{5.22}$$

$$L_n(10, \psi) = \frac{Li_n(\psi^{10})}{10^{n-1}} - \frac{Li_n(\psi^5)}{5^{n-1}} + Li_n(\psi) + \frac{7 \log^n \psi}{n!} \tag{5.23}$$

$$L_n(18, \psi) = \frac{Li_n(\psi^{18})}{18^{n-1}} - \frac{Li_n(\psi^9)}{9^{n-1}} - \frac{Li_n(\psi^6)}{6^{n-1}} + \frac{Li_n(\psi^2)}{2^{n-1}} + \frac{3 \log^n \psi}{n!} \tag{5.24}$$

$$L_n(24, \psi) = \frac{Li_n(\psi^{24})}{24^{n-1}} - \frac{Li_n(\psi^{12})}{12^{n-1}} - \frac{Li_n(\psi^8)}{8^{n-1}} + \frac{Li_n(\psi^2)}{2^{n-1}} + \frac{4 \log^n \psi}{n!} \tag{5.25}$$

$$L_n(30, \psi) = \frac{Li_n(\psi^{30})}{30^{n-1}} - \frac{2\, Li_n(\psi^{15})}{15^{n-1}} - \frac{Li_n(\psi^6)}{6^{n-1}} + \frac{Li_n(\psi^5)}{5^{n-1}} \tag{5.26}$$

These give rise to six accessible ladders, and one of index 30, numerically determined:

$$L_n^{(2)}(3, \psi) = 3\, L_n(3, \psi) - L_n(2, \psi) - \zeta(2) \frac{\log^{n-2} \psi}{(n-2)!} \tag{5.27}$$

$$L_n^{(2)}(4, \psi) = 4\, L_n(4, \psi) + L_n(2, \psi) \tag{5.28}$$

$$L_n^{(2)}(7, \psi) = 7\, L_n(7, \psi) + 6\, L_n(2, \psi) - 3\, \zeta(2) \frac{\log^{n-2} \psi}{(n-2)!} \tag{5.29}$$

$$L_n^{(2)}(10, \psi) = 10\, L_n(10, \psi) + 7\, L_n(2, \psi) - 4\, \zeta(2)\, \frac{\log^{n-2}\psi}{(n-2)!} \quad (5.30)$$

$$L_n^{(2)}(18, \psi) = 18\, L_n(18, \psi) + L_n(2, \psi) - 4\, \zeta(2)\, \frac{\log^{n-2}\psi}{(n-2)!} \quad (5.31)$$

$$L_n^{(2)}(24, \psi) = 24\, L_n(24, \psi) - 7\, L_n(2, \psi) - 10\, \zeta(2)\, \frac{\log^{n-2}\psi}{(n-2)!} \quad (5.32)$$

$$L_n^{(2)}(30, \psi) = 30\, L_n(30, \psi) + 3\, L_n(2, \psi) + \zeta(2)\, \frac{\log^{n-2}\psi}{(n-2)!} \quad (5.33)$$

These ladders are all zero at n=2. At n=3 the following combinations are found:

$$L_n^{(3)}(4, \psi) = 4\, L_n^{(2)}(4, \psi) + 3\, L_n^{(2)}(3, \psi) \quad (5.34)$$

$$L_n^{(3)}(10, \psi) = 10\, L_n^{(2)}(10, \psi) - 14\, L_n^{(2)}(7, \psi) - 9\, L_n^{(2)}(3, \psi) \quad (5.35)$$

$$L_n^{(3)}(18, \psi) = 6\, L_n^{(2)}(18, \psi) - 14\, L_n^{(2)}(7, \psi) + 17\, L_n^{(2)}(3, \psi) \quad (5.36)$$

$$L_n^{(3)}(24, \psi) = 8\, L_n^{(2)}(24, \psi) - 28\, L_n^{(2)}(7, \psi) - 19\, L_n^{(2)}(3, \psi) \quad (5.37)$$

$$L_n^{(3)}(30, \psi) = 10\, L_n^{(2)}(30, \psi) + 17\, L_n^{(2)}(3, \psi) \quad (5.38)$$

At n=3 these are equal to $\zeta(3)$ multiplied, respectively, by 1, -1, 1, -5, -2. Combinations with the $\zeta(3)$ term cancelled are

$$L_n^{(4)}(10, \psi) = 10\, \{5\, L_n^{(2)}(10, \psi) - 7\, L_n^{(2)}(7, \psi) + 2\, L_n^{(2)}(4, \psi) - 3\, L_n^{(2)}(3, \psi)\} \quad (5.39)$$

$$L_n^{(4)}(18, \psi) = 18\, \{3\, L_n^{(2)}(18, \psi) - 7\, L_n^{(2)}(7, \psi) - 2\, L_n^{(2)}(4, \psi) + 7\, L_n^{(2)}(3, \psi)\} \quad (5.40)$$

$$L_n^{(4)}(24, \psi) = 24\, \{2\, L_n^{(2)}(24, \psi) - 7\, L_n^{(2)}(7, \psi) + 5\, L_n^{(2)}(4, \psi) - L_n^{(2)}(3, \psi)\} \quad (5.41)$$

$$L_n^{(4)}(30, \psi) = 30\, \{10\, L_n^{(2)}(30, \psi) + 8\, L_n^{(2)}(4, \psi) + 23\, L_n^{(2)}(3, \psi)\} \quad (5.42)$$

Combinations of these, required either by Kummer's equations at n=4, or by numerical computation for the index-30 formula, are

$$L_n^{(5)}(18, \psi) = L_n^{(4)}(18, \psi) + L_n^{(4)}(10, \psi) - \frac{61}{3} \zeta(4) \frac{\log^{n-4}\psi}{(n-4)!} \quad (5.43)$$

$$L_n^{(5)}(24, \psi) = L_n^{(4)}(24, \psi) - L_n^{(4)}(10, \psi) + \frac{7}{2} \zeta(4) \frac{\log^{n-4}\psi}{(n-4)!} \quad (5.44)$$

$$L_n^{(5)}(30, \psi) = L_n^{(4)}(30, \psi) + 14\, L_n^{(4)}(10, \psi) - \frac{41}{3} \zeta(4) \frac{\log^{n-4}\psi}{(n-4)!} \quad (5.45)$$

These are all zero at n=4. At n=5 we have

$$9\, L_5^{(5)}(18, \psi) + 24\, L_5^{(5)}(24, \psi) = 28\, \zeta(5) \quad (5.46)$$

obtained from Kummer's equation at n=5. No corresponding result for $L_5^{(5)}(30, \psi)$ has so far been located.

Results exist for the base ϕ of section 5.2 corresponding to these equations.

6. SUPERNUMARY RESULTS FOR p=5

6.1 For q=1 the base is ω, as discussed in section 4.3. For q=3 the equation is $u^5 + u^3 = 1$. No novel cyclotomic formula has been found for this equation. The results obtained by adding (or subtracting) u^4 reduce to the generic ones. It is not known if there are any further equations, but in view of the tortuous routes that are often needed to generate new cyclotomic equations it is very difficult to know for sure that no further relevant results exist.

The cases q=2 and 4 do give valid ladders but the two sets have no obvious connection.

6.2 A cyclotomic equation for the case q=2 comes by squaring the equation $u^2 = 1 - u^5$ to give $u^4 = 1 - 2u^5 + u^{10}$. Multiply by u, replace u^5 by $1 - u^2$ and add u^{10} to both sides to get, on re-arrangement, $1 - u + u^{10} - u^{11} = u^2 - 2u^6 + u^{10}$ or $(1 + u^{10})(1 - u) = u^2(1 - u^4)^2$.

Hence

$$1 - u^{20} = (1 - u^{10})(1 - u^4)^2(1 - u)^{-1}u^2 \qquad (6.1)$$

The related ladder is

$$L_2(20, u) = \frac{Li_2(u^{20})}{20} - \left\{\frac{Li_2(u^{10})}{10} + \frac{Li_2(u^4)}{2} - Li_2(u) - \log^2 u\right\} \qquad (6.2)$$

This can be combined with the generic ladder of index p=5,

$$L_2(5, u) = \frac{Li_2(u^5)}{5} + \log^2 u, \qquad (6.3)$$

and the generic ladder of index 2(2p-q) = 16,

$$L_2(16, u) = \frac{Li_2(u^{16})}{16} - \left\{\frac{Li_2(u^8)}{8} + \frac{Li_2(u^2)}{2} - Li_2(u) - \log^2 u\right\}, \qquad (6.4)$$

to give, via numerical computation,

$$20\, L_2(20, u) - 32\, L_2(16, u) + 55\, L_2(5, u) = \zeta(2) \qquad (6.5)$$

This is the only known supernumary result for this base.* It is not known if it extends to n=3.

6.3 For q=4 we proceed by adding u^7 to both sides of $u^4 + u^5 = 1$ and multiplying by $1 - u$ to get

$$(1 + u^7)(1 - u) = u^4(1 - u^2 + u^3 - u^4) = u^4(-u^2 + u^3 + u^5)$$
$$= u^6(-1 + u + u^3) = u^6(u + u^3 - u^4 - u^5)$$
$$= u^7(1 + u^2 - u^3 - u^4) = u^7(u^2 - u^3 + u^5)$$
$$= u^8(u - u^2 + u^4) = u^8(1 + u - u^2 - u^5)$$
$$= u^8[1 + u(1 - u - u^4)] = u^8[1 - u(u - u^5)]$$
$$= u^8(1 - u^7)$$

Hence

$$1 - u^{14} = (1 - u^7)^2(1 - u)^{-1}u^8 \qquad (6.6)$$

The related ladder is

$$L_2(14, u) = \frac{Li_2(u^{14})}{14} - \left\{\frac{2\ Li_2(u^7)}{7} - Li_2(u) - 4\log^2 u\right\} \qquad (6.7)$$

This can be combined with generic ladders of index q=4 and $2(2p-q) = 12$:

$$L_2(4, u) = \frac{Li_2(u^4)}{4} + \frac{5}{2}\log^2 u \qquad (6.8)$$

$$L_2(12, u) = \frac{Li_2(u^{12})}{12} - \left\{\frac{Li_2(u^6)}{6} - \frac{Li_2(u^3)}{3} - \frac{1}{2}\log^2 u\right\} \qquad (6.9)$$

A numerical search produced

$$14\ L_2(14, u) + 12\ L_2(12, u) + 44\ L_2(4, u) = 17\ \zeta(2) \qquad (6.10)$$

---

* A further supernumary ladder of index 24 has recently been found by G. Ray.

Again, it is not known if this result extends to n=3, or whether there are further supernumary equations.

7. SUPERNUMARY RESULTS FOR p=6

7.1 The only cases for which $(p, q) = 1$ are $q = 1$ and 5. As with ϕ and ψ of section (5.1) these are related, the base equations transforming into each other on replacing u by $-1/u$.

It is not known whether the cases q=2, 3 and 4 give other than redundant results, but it is conjectured that this is always so when $(p, q) \neq 1$.

In all the analyses up to now a supernumary cyclotomic equation has always been found to be associated with a <u>valid</u> ladder result at n=2, and it was conjectured that this would be so in all cases where a novel cyclotomic equation could be generated. The results at p=6 confirm this.

7.2 The base equation for $p,q = 6,1$ is

$$u^6 + u = 1 \qquad (7.1)$$

Starting with $1 + u^{10}$ and factorizing gives

$$1 + u^{10} = (1 + u^2)(1 - u^2 + u^4 - u^6 + u^8) = (1 + u^2)(u - u^3 + u^4)$$
$$= u(1 + u^2)(1 + u^3 - u^2)$$
$$= u(1 + u^2)(1 - u^3)^{-1} [1 - u^6 - u^2(1 - u^3)]$$
$$= u(1 + u^2)(1 - u^3)^{-1}(u - u^2 + u^5)$$
$$= u^2(1 + u^2)(1 - u^3)^{-1}(1 - u + u^4)$$
$$= u^2(1 + u^2)(1 - u^3)^{-1} (u^4 + u^6) = u^6(1 + u^2)^2(1 - u^3)^{-1}$$

Hence

$$1 - u^{20} = (1 - u^{10})(1 - u^4)^2(1 - u^3)^{-1}(1 - u^2)^{-2}u^6 \qquad (7.2)$$

with a corresponding ladder

$$L_2(20, u) = \frac{Li_2(u^{20})}{20} - \left\{\frac{Li_2(u^{10})}{10} + \frac{Li_2(u^4)}{2} - \frac{Li_2(u^3)}{3} - Li_2(u^2) - 3\log^2 u\right\} \qquad (7.3)$$

According to the recipe adopted successfully so far, this must be combined with two independent generic results, up till now taken to be of index p or q and 2(2p-q). In the present case the choice q=1 gives

$$L_2(1, u) = Li_2(u) + 3\log^2 u, \qquad (7.4)$$

The second component-ladder used comes from the index 2(2p-q) = 22, and is

$$L_2(22, u) = \frac{Li_2(u^{22})}{22} - \left\{\frac{Li_2(u^{11})}{11} + \frac{Li_2(u^8)}{8} - \frac{Li_2(u^4)}{4} - \frac{1}{2}\log^2 u\right\} \qquad (7.5)$$

A numerical determination then produces

$$60\, L_2(20, u) + 352\, L_2(22, u) - 7\, L_2(1, u) = 51\, \zeta(2) \qquad (7.6)$$

A further supernumary result of index 8 has recently been found by G. Ray.

7.3 The transformation $u \longrightarrow -1/u$ changes the base equation into

$$u^6 + u^5 = 1 \qquad (7.7)$$

The corresponding change in the ladders requires the use of the inversion formula, and leads to defining

$$L_2(20, u) = \frac{Li_2(u^{20})}{20} - \left\{ \frac{Li_2(u^{10})}{10} - \frac{Li_2(u^6)}{6} + \frac{Li_2(u^4)}{2} \right.$$

$$\left. + \frac{Li_2(u^3)}{3} - Li_2(u^2) - \frac{3}{2} \log^2 u \right\} \qquad (7.8)$$

$$L_2(11, u) = \frac{Li_2(u^{11})}{11} - \left\{ \frac{Li_2(u^8)}{8} - \frac{Li_2(u^4)}{4} - 3 \log^2 u \right\} \qquad (7.9)$$

$$L_2(2, u) = \frac{Li_2(u^2)}{2} - \left\{ Li_2(u) + \frac{5}{2} \log^2 u \right\} \qquad (7.10)$$

The structure of the ladder is determined by (7.6); only the new coefficient of $\zeta(2)$ has to be found. The resulting formula is

$$60\, L_2(20, u) + 352\, L_2(11, u) - 7\, L_2(2, u) = 64\, \zeta(2) \qquad (7.11)$$

Incidentally, this result argues forcibly that (7.6) and (7.11) cannot be artefacts; the probability that the same ladder structure could give "nearly" rational coefficients for $\zeta(2)$ in both cases must be quite remote. Thirty significant figures were used in both cases.

8. THE FAMILY $u^{6m+1} + u^{6r-1} = 1$

8.1 The difficulty in developing any systematic analysis stems from the lack, until recently, of a synthesis procedure for the cyclotomic equations. In section 4.3 the equation $\omega^3 + \omega^2 - 1 = 0$ is converted into $\omega^5 + \omega - 1 = 0$ by multiplying by $\omega^2 - \omega + 1$. Since the roots of the latter are at $\exp(\pm i\pi/3)$ it follows that if either of the above terms ω^5 or ω are multiplied by any power of ω^6 the resulting equation will still be divisible by $\omega^2 - \omega + 1$. The polynomial quotient so developed can then be examined for the production of cyclotomic equations by algebraic re-arrangement. This leads to a consideration of the equation

$$u^{6m+1} + u^{6r-1} = 1 \qquad (8.1)$$

for integers $r > 0$, $m \geq 0$.

It is fairly easy to reproduce many of the generic formulas this way. In some cases new results ensue. Apparently the pairs $(p, q) = (11, 1)$, $(13, 5)$ and $(13, 11)$ are unproductive, but $(7, 5)$ and $(11, 7)$ give new results by this method. It may be noted that one can always produce a novel cyclotomic equation of index $N = (p+q)/2$ from (8.1) by subtracting $u^{(p+q)/2} = u^{3(m+r)}$ from each side; in fact this method works whenever both p and q are odd. For small values of p and q the resulting equations tend to be contained within the generic results, but otherwise this process doesn't seem to be very useful in producing valid ladders.

8.2 For the second and third orders, Kummer's equations consist of polylogarithmic arguments of products of powers of x, y and $(1-x)/(1-y)$ only. Hence one can generate powers of u by taking

$$x = \pm u^m;\ y = -u^{r-s};\ (1-x)/(1-y) = u^s \qquad (8.2)$$

Eliminating x and y gives the three-term equations for the base u:

$$u^r + u^s \pm u^m = 1 \qquad (8.3)$$

There are three generic forms that the two-term equation $u^p + u^q = 1$ takes when re-arranged in three-term form. They are

$$u^p + u^{q+p} + u^{2q} = 1 \tag{8.4}$$

$$u^q + u^{q+p} + u^{2p} = 1 \tag{8.5}$$

$$u^q + u^{p-q} - u^{2p-q} = 1 \tag{8.6}$$

The corresponding dilogarithmic equations are (23) to (25) and (28) to (30) of reference 1 (a change of notation has been made to avoid confusion with n, p and q as used in the present study). If it is possible to generate a further, non-generic, three-term equation in a particular case, then the use of these equations will generate an accessible supernumary ladder. The method will be illustrated for the case p=7, q=5.

8.3 From the identity $x^7 + x^5 - 1 = (x^2 - x + 1)(x^5 + x^4 + x^3 - x - 1)$ we get, from the quotient polynomial, with $x = u$: $1 + u = u^3(1 + u + u^2)$, or $1 - u^2 = u^3(1 - u^3)$, whence

$$u^2 + u^3 - u^6 = 1 \tag{8.7}$$

This is of the form (8.3) and is non-generic for p=7, q=5. Before taking m=6, r=3, s=2 in the equations, (8.7) must first be put in cyclotomic form to generate the necessary ladders. The equation

$$1 - u^2 = u^3(1 - u^3) \tag{8.8}$$

suggests indices 2 and 3, neither of which is generic. Now the index 6 comes from both the generic 3(p-q) and the non-generic (p+q)/2. These both happen to give the same result which degenerates, in this case, to an index-3 form. Therefore, the index 2 is the new one. To isolate it from (8.8), square the equation to get

$$(1 - u^2)^2 = u^6(1 - u^3)^2 = u^6(1 - u)(1 + u + u^2)(1 - u^3)$$
$$= u^6(1 - u)(1 + u + u^2 - u^3 - u^4 - u^5)$$
$$= u^8(1 - u), \qquad (8.9)$$

since $u^5 + u^4 + u^3 - u - 1 = 0$. Hence the supernumary component-ladder, on taking the square root of (8.9), is

$$L_2(2, u) = \frac{Li_2(u^2)}{2} - \left\{\frac{Li_2(u)}{2} - 2\log^2 u\right\} \qquad (8.10)$$

This has to be combined with, as it happens, only one generic component-ladder to produce a valid result. The index-4 formula suffices, and is given by

$$L_2(4, u) = \frac{Li_2(u^4)}{4} - \left\{\frac{Li_2(u^2)}{2} + \frac{5}{2}\log^2 u\right\} \qquad (8.11)$$

The resulting supernumary, accessible result is found to be

$$4\, L_2(4, u) + 2\, L_2(2, u) + \zeta(2) = 0 \qquad (8.12)$$

There is a corresponding accessible result at n=3, but apparently none at n=4. G. Ray has recently produced a further supernumary equation of index 24.

8.4 For p=11, q=7 we have the identity

$$x^{11} + x^7 - 1 = (x^2 - x + 1)(x^9 + x^8 - x^6 + x^4 + x^3 - x - 1) \quad (8.13)$$

Hence, on multiplying by u^2, we get, from the quotient polynomial:

$$u^{11} + u^{10} - u^8 + u^6 + u^5 - u^3 - u^2 = 0 \qquad (8.14)$$

This can be re-arranged, on putting $u^{11} = 1 - u^7$, as $u^5(1 - u) = (1 + u^5)(1 - u^2 - u^3 + u^5)$ to give a non-generic index-10 formula

$$1 - u^{10} = (1 - u^5)(1 - u^3)^{-1}(1 - u^2)^{-1}(1 - u)u^5 \qquad (8.15)$$

There is also an index-9 result from the power $(p+q)/2$, obtained as in section 8.3. If this is combined with the generic $3(p-q)$ formula one finds

$$1 - u^9 = (1 - u^3)^{-1}(1 - u^2)u^4 \tag{8.16}$$

The corresponding component-ladders are readily generated, but despite a lengthy numerical search, no valid ladders have been found for them. This is the first apparent exception encountered to the rule that supernumary cyclotomic equations give rise to corresponding valid ladders at n=2.

9. Cyclotomic Equations for a Salem Number

9.1 The Salem number x, satisfying the base equation

$$x^{10} + x^9 - x^7 - x^6 - x^5 - x^4 - x^3 + x + 1 = 0 \tag{9.1}$$

was studied by Ray[15] in relation to its dilogarithmic properties. From a rather involved functional equation he was able to derive the analytic result

$$\frac{1}{2}\,\mathrm{Li}_2(x^{18}) - \mathrm{Li}_2(x^{17}) + \frac{1}{2}\,\mathrm{Li}_2(x^{16}) - \mathrm{Li}_2(x^{13}) + 2\,\mathrm{Li}_2(x^{11})$$

$$+ \mathrm{Li}_2(x^{10}) - 2\,\mathrm{Li}_2(x^9) - \frac{1}{2}\,\mathrm{Li}_2(x^8) + \mathrm{Li}_2(x^7) - \mathrm{Li}_2(x^5) - \mathrm{Li}_2(x^4)$$

$$+ \mathrm{Li}_2(x^3) + \frac{3}{2}\,\mathrm{Li}_2(x^2) - \mathrm{Li}_2(x) - \frac{1}{2}\log^2(x) = 0 \tag{9.2}$$

where x = 1.1762808 . . . is a real root of (9.1). Because the equation is symmetric it possesses roots in inverse pairs, the other real root of (9.1) being x = 0.8501371 . . .; and it may be verified, using the inversion formula for the dilogarithm, that (9.2), with no change, is also satisfied for the smaller root. All the other roots occur in conjugate pairs on the unit circle, the one with the smallest argument being

$$u = e^{i\theta}; \ \theta = 62.81495 \ldots^{\circ} \qquad (9.3)$$

It is a conjecture, with no known exceptions, that, under these circumstances, if the dilogarithm $Li_2(x^m)$ is replaced by $Cl_2(m\theta)$, an equation of the form (9.2), with all non-Claussen terms suppressed, will also be satisfied. It is straightforward to verify numerically that this is indeed the case for θ given by (9.3). An attempt to justify this close linkage between Claussen and dilogarithmic ladders is provided in section 9.6.

9.2 Corresponding to (9.2) a cyclotomic equation can be written down, containing 14 factors. Manipulations of (9.1) might be expected to yield some additional results, and in fact a total of 50 cyclotomic equations have been thus determined. This is by far and away the largest number known for any base, and they are listed in section 9.8. They have indices N ranging from 6 to 132, the distribution thinning out considerably at the upper end.* Because of the arbitrary way in which the cyclotomic equations can be combined, a construction was selected so that, in the vast majority of cases, the powers involved were 1, 2, 3, 4, 5 and N. In a few cases, when N is even, a power of N/2 needs to be included; and in a very few other cases, in the interests of simplicity, some other powers are also used. The cyclotomic equation corresponding to (9.2) in fact is a combination of nine of these more "basic" equations.

---

* D. Zagier has subsequently found a further 21 indices, the highest being 360.

It is not intended to give here the derivation for each case, but some representative examples are shown. Mostly, they are easy to derive.

9.3 If the negative terms in (9.1) are taken to the right-hand side of the equation and summed the result is

$$x^{10} + x^9 + x + 1 = (1 + x)(1 + x^9) = x^3(1 - x^5)(1 - x)^{-1}, \quad (9.4)$$

which can be put in the form

$$1 - x^{18} = (1 - x^9)(1 - x^5)(1 - x^2)^{-1}x^3, \quad (9.5)$$

a cyclotomic equation of index 18. Dividing by the factor $(1 - x^9)$, (15.68) is $(1 - x^2)(1 + x^9) = x^3(1 - x^5)$ which can be multiplied out and re-arranged as either

$$1 - x^{11} = (1 - x^6)(1 - x^2)(1 - x)^{-1}x^2 \quad (9.6)$$

or

$$1 - x^{16} = (1 - x^8)(1 - x^7)(1 - x^3)^{-1}x^2 \quad (9.7)$$

Many combinations of factors can be deduced in this way. A somewhat different sequence comes by multiplying (9.1) by x^2-x+1. Some terms cancel and the resulting formula is

$$1 + x^{12} = x^5(1 + x + x^2) = x^5(1 - x^3)(1 - x)^{-1} \quad (9.8)$$

Hence

$$1 - x^{24} = (1 - x^{12})(1 - x^3)(1 - x)^{-1}x^5 \quad (9.9)$$

Equation (15.71) can also be written as $1 - x^6 + x^{12} = x^5(1 + x^2)$, whence

$$1 + x^{18} = (1 + x^6)(1 + x^2)x^5 \quad (9.10)$$

which can be expressed as an index-36 cyclotomic equation:

$$1 - x^{36} = (1 - x^{18})(1 - x^{12})(1 - x^6)^{-1}(1 - x^4)(1 - x^2)^{-1}x^5 \quad (9.11)$$

Equation (9.8) can also be re-arranged and factorized as

$$(1 - x^7)(1 - x^{11}) = (1 - x^{16})(1 - x^8)^{-1}x^5 \quad (9.12)$$

Combining with (9.7) this gives

$$(1 - x^{11})(1 - x^3) = x^7 \qquad (9.13)$$

and this can be combined with (9.6) to give an index-6 result. Equation (9.8) can also be re-arranged as

$$(1 - x^7)(1 - x^5) = x^6 \qquad (9.14)$$

9.4 In the course of generating these and many other similar equations it was noticed that when apparently disparate indices, for example, 11 and 16 in (9.12), occurred in one formula, they also occurred in a different combination in another, enabling them to be separated out. The resulting equations then exhibited the straightforward type of factorization on which the $r|N$ ladder property was first conjectured. (However, any or all of the indices 1, 2, 3, 4 and 5 may also be involved.) The only apparent exception encountered was a formula combining indices 63 and 132:

$$(1-x^{132}) = (1-x^{66})(1-x^{63})(1-x^{24})(1-x^{21})^{-1}(1-x^6)^{-1}x^3 \qquad (9.15)$$

No further equations with these two indices could be found, but in the belief that the two should indeed be separable the following direct factorization of $(1 - x^{63})$ was generated. Write

$$1 - x^{63} = (1 - x^{18}) + x^{18}(1 - x^{45}) \qquad (9.16)$$

Using the already developed factorization of $(1 - x^{18})$ and $(1 - x^{45})$, (see section 9.8), this reduces to

$$1 - x^{63} = [1 - x^2 + x^{25} - x^{40}](1 - x^3)^{-1}(1 - x)^{-1}x^{10} \qquad (9.17)$$

with

$$1 - x^2 + x^{25} - x^{40} = (1 - x^{40}) - x^2(1 - x^{23})$$

$$= [x^5 + x - 1 - x^9](1 - x^6)^2(1 - x)^{-2}x^8, \qquad (9.18)$$

again on using formulas for $(1 - x^{40})$ and $(1 - x^{23})$.

Now $x^5 + x - 1 - x^9 = -(1 - x^5) + x(1 - x^8)$, and with the equation for $(1 - x^8)$ this reduces to

$$x^5+x-1-x^9=(1-x^5)(1-x)^{-1}x^{-3}[(1-x^5)(1-x^4)(1-x^3)-x^3(1-x)] \quad (9.19)$$

This result is multiplied by $1 - x^7$ and use made of (9.14) to give

$$x^5+x-1-x^9=(1-x^5)(1-x^7)^{-1}(1-x)^{-1}[-1+x+x^3-x^6-x^8+x^{10}] \quad (9.20)$$

From (9.8), multiplied by x, $x - x^6 - x^8 = x^7 - x^{13}$; and from (9.13), $-1 + x^3 = -x^7 - x^{11} - x^{14}$. Substituting in (9.20) gives

$$-1+x+x^3-x^6-x^8+x^{10}=x^{10}-x^{11}-x^{13}+x^{14}=x^{10}(1-x)(1-x^3) \quad (9.21)$$

Insertion back into (9.9) and (9.8) finally yields

$$1 - x^{63} = (1 - x^7)^{-2}(1 - x^6)^2(1 - x)^{-3}x^{34} \quad (9.22)$$

This was by far the most difficult index to track down.

9.5 If a cyclotomic equation is written in the form

$$\prod_m^N (1 - x^m)^{A_m} = x^n \quad (9.23)$$

then, since this has to be an algebraic derivation from the symmetric base equation (9.1), the substitution of 1/x for x must leave the formula unaltered. This has, as a direct consequence,

$$\left.\begin{array}{ll} \text{a)} & \sum_m^N A_m \text{ is even} \\ & \\ \text{b)} & n = \frac{1}{2}\sum_m^N mA_m \end{array}\right\} \quad (9.24)$$

The second part of this equation was found useful for checking the results in section 9.8, and it also appears in a different guise in relation to the conjecture in the next section dealing with the linkage of dilogarithmic and Clausen ladders.

9.6 Writing z for x in the base equation (9.1) it may be conjectured that, if the structural information contained in this equation is utilized in some dilogarithmic functional equations so that all the arguments can be expressed as powers of z, then the resulting relation can be put in the form

$$\sum_{m}^{N} D_m[\mathrm{Li}_2(z^m) - \mathrm{Li}_2(z^{-m})] = C_o\zeta(2) + i\pi C_1 \log z + C_2 \log^2 z \qquad (9.25)$$

All known relations are of a similar form, with D_m, C_o, C_1 and C_2 simple, real rationals. The form of the left-hand side of (9.25) is taken so as to facilitate the replacement $z \to 1/z$, which must leave the equation unaltered. This immediately requires $C_o = C_2 = 0$. On the unit circle take $z = e^{i\theta}$ to give

$$2i\sum_{m}^{N} D_m \,\mathrm{Cl}_2(m\theta) = -\pi C_1 \theta \qquad (9.26)$$

Hence C_1 (which is real) is also zero, and (9.26) gives the associated Clausen ladder.

Now let $z = x$, with $0 < x < 1$, the smaller real root of the base equation. Terms in $\mathrm{Li}_2(z^{-m})$ are complex, and can be put in terms of $\mathrm{Li}_2(z^m)$ on using the inversion formula

$$\mathrm{Li}_2(y) + \mathrm{Li}_2(1/y) = 2\zeta(2) - i\pi \log y - \frac{1}{2}\log^2 y; \; y > 1 \qquad (9.27)$$

Hence (9.25) becomes

$$\sum_{m}^{N} D_m \mathrm{Li}_2(x^m) = \zeta(2)\sum_{m}^{N} D_m + \frac{i\pi}{2}\log x \sum_{m}^{N} mD_m - \frac{1}{4}\log^2 x \sum_{m}^{N} m^2 D_m \qquad (9.28)$$

This equation, if true, would give the constant associated with the corresponding dilogarithmic ladder. Despite the fact that it happens to hold for Ray's result (9.2), it is not true in general. The coefficient of $\zeta(2)$ is not always equal to $\sum_{m}^{N} D_m$, neither is $\sum_{m}^{N} mD_m$ always zero, as it would have to be since all other terms are real. However, the formula does correctly give the coefficient of $\log^2 x$. To see this, note that the cyclotomic equation associated with (9.28) is

$$\prod_{m}^{N} (1 - x^m)^{mD_m} = x^{\frac{1}{2}\sum_{m}^{N} m^2 D_m} \qquad (9.29)$$

Comparing with (9.24), with $A_m = mD_m$, we see that the power of x is indeed $\frac{1}{2}\sum_{m}^{N} mA_m$, as required.

It may be noted that if the logarithm of (9.29) is taken, and x replaced by 1/x, the accumulated phase would appear to be simply $i\pi \sum_{m}^{N} mD_m$. The fact that this is not, in general, equal to zero, indicates that the flaw in the hypothesis, as given, relates to branch cuts; and the extensive work needed by Ray[15] to prove his functional equations shows that the conjecture, in its present form, is too simplistic. C_0 and C_1, for instance, need not be true constants, but could depend on the location of z with respect to the unit circle.

One possible, though arbitrary, way of meeting the above objections is to take the right-hand side of (9.25) in the form $C_0 \mathrm{sgn}(\log|z|) + i\pi C_1 \log|z|$. No C_2 term is required, because of

(9.29), and (9.28) gives $c_1 = -\sum_{m}^{N} mD_m$, with c_o undetermined.

Thus this modified conjecture fails to provide the constant in the dilogarithmic ladder, though it does succeed in indicating the linkage between it and the associated Claussen ladder.

9.7 Corresponding to (9.23) is a dilogarithmic component-ladder

$$L_2(N, x) = \sum_{m}^{N} \frac{A_m}{m} Li_2(x^m) + \frac{n}{2} \log^2(x) \qquad (9.30)$$

The leading term has a factor 1/N and it is desirable to clear this by defining a modified component-ladder by

$$M_2(N, x) = N\, L_2(N, x) \qquad (9.31)$$

In terms of it, (9.2) can be written

$$\frac{1}{2} M_2(18,x) - M_2(17,x) + \frac{1}{2} M_2(16,x) - M_2(13,x) + 2\, M_2(11,x)$$

$$+ M_2(10,x) - M_2(9,x) + \frac{1}{2} M_2(8,x) + M_2(7,x) = 0 \qquad (9.32)$$

where the component-ladders are defined as in the next section. The terms for $N < 6$ are all absorbed, but the remaining terms do not quite follow the coefficients in (9.2) because, as here defined, $L_2(18,x)$ and $L_2(16,x)$ contain, respectively, terms in $Li_2(x^9)$ and $Li_2(x^8)$. (A minor re-definition would amend this, which is simply a consequence of the arbitrariness inherent in the ladder definition.) With 50 component-ladders given in section 9.8 it might be expected that there would exist many more relations of the form (9.32). D.Zagier has predicted the the component-ladders should combine in fives, making a total of 46 valid ladders at $n = 2$. These are shown in section 9.9. The additional 21 indices he has found should therefore lead to a total of 67 such results.

9.8 As a shorthand in this section we use the notation

$$(m) \equiv (1 - x^m) \tag{9.33}$$

TABLE I 50 Cyclotomic Equations

Index N	Cyclotomic Equation
6.	$(6) = (3)^{-1}(2)^{-1}(1)x^5$
7.	$(7) = (5)^{-1}x^6$
8.	$(8) = (5)^2(4)(3)(1)^{-1}x^{-4}$
9.	$(9) = (5)^{-1}(3)^{-1}(2)^2(1)^{-1}x^7$
10.	$(10) = (5)^{-1}(3)(2)^{-1}x^7$
11.	$(11) = (3)^{-1}x^7$
12.	$(12) = (6)(5)(4)^2(1)^{-1}x^{-3}$
13.	$(13) = (5)(3)(1)^{-1}x^3$
16.	$(16) = (8)(5)^{-1}(3)^{-1}x^8$
17.	$(17) = (5)^{-1}(3)^{-1}(2)^{-2}(1)x^{14}$
18.	$(18) = (9)(5)(2)^{-1}x^3$
20.	$(20) = (5)^{-1}(4)(1)^{-1}x^{11}$
21.	$(21) = (7)^{-3}(3)^2(1)^{-2}x^{19}$
23.	$(23) = (6)^2(1)^{-1}x^6$
24.	$(24) = (12)(3)(1)^{-1}x^5$
27.	$(27) = (9)(5)^{-1}(3)(2)^{-2}x^{12}$
28.	$(28) = (14)(5)(4)(3)^{-1}(2)^{-1}x^5$
29.	$(29) = (5)(3)^{-1}(1)^{-1}x^{14}$
30.	$(30) = (15)(5)^{-1}(2)^{-1}x^{11}$
34.	$(34) = (17)(5)^2(3)(2)(1)^{-2}x^2$
36.	$(36) = (18)(12)(6)^{-1}(4)(2)^{-1}x^5$

37. $(37) = (5)^{-2}(3)^{2}(2)^{-2}(1)^{-1}x^{23}$

38. $(38) = (19)(9)(2)^{-1}x^{6}$

40. $(40) = (6)^{2}(4)(1)^{-2}x^{13}$

42. $(42) = (21)(14)(3)^{-1}x^{5}$

44. $(44) = (22)(4)(3)(2)^{-1}(1)^{-1}x^{9}$

45. $(45) = (15)(3)^{-1}(1)^{-1}x^{17}$

47. $(47) = (5)^{-1}(1)^{-2}x^{27}$

50. $(50) = (25)(5)(3)(2)^{-1}(1)^{-1}x^{10}$

52. $(52) = (26)(4)(3)^{-2}(2)^{-1}x^{15}$

56. $(56) = (28)(7)(3)^{-1}(2)^{2}(1)^{-2}x^{11}$

60. $(60) = (30)(5)^{2}(4)^{2}(1)^{-2}x^{7}$

62. $(62) = (31)^{2}(6)x^{-3}$

63. $(63) = (7)^{-2}(6)^{2}(1)^{-3}x^{34}$

64. $(64) = (32)(16)(8)^{-1}(5)^{-1}(3)(2)^{-2}x^{15}$

65. $(65) = (5)^{-2}(3)^{-3}(2)^{4}(1)^{-4}x^{40}$

66. $(66) = (33)(22)(11)^{-1}(5)(1)^{-1}x^{9}$

70. $(70) = (35)(14)(1)^{-1}x^{11}$

74. $(74) = (37)(5)^{-1}(3)^{-3}(2)(1)^{-1}x^{25}$

76. $(76) = (38)(5)^{3}(4)(3)(2)^{-1}(1)^{-2}x^{10}$

78. $(78) = (39)(26)(13)^{-1}(5)^{-1}(1)^{-1}x^{16}$

84. $(84) = (42)(7)(4)^{2}(2)^{-2}(1)^{-1}x^{16}$

86. $(86) = (43)(6)^{2}(5)(3)^{-1}(2)^{-1}(1)^{-1}x^{16}$

92. $(92) = (46)(5)(4)(2)(1)^{-3}x^{19}$

96. $(96) = (48)(32)(16)^{-1}(3)^{-1}(1)^{-1}x^{18}$

98. $(98) = (49)(14)(3)^{-1}(2)^{-2}x^{21}$

110. $(110) = (55)(22)(5)^{-1}(3)^{-1}(1)^{-1}x^{21}$

118. $(118) = (59)(5)^{2}(3)^{2}(2)^{-1}(1)^{-3}x^{24}$

124. $(124) = (62)(31)(4)(2)^{-1}(1)^{-1}\, x^{15}$

132. $(132) = (33)(22)(12)(6)(1)^{-3}\, x^{31}$

The additional indices recently found by D.Zagier are

N = 57,75,105,108,130,138,144,154,160,165,175,182,186,195,204,212, 240,246,270,286,360.

9.9 The following valid dilogarithmic ladders were found using numerical methods to 33 decimal places, as calculated by M.Abouzahra, and a multi-component search algorithm by G.Szekeres; to both of whom I am grateful for their contributions to these computations.

We use the shorthand notation

$$[N] \equiv M_2(N,x) \tag{9.34}$$

where M is the modified component-ladder of (9.31). The results read across the table. Thus, the first ladder is

$$3[6] = 20\,\zeta(2) + 8[7] + 0[8] - 4[9] - 6[10] \tag{9.35}$$

All the formulas are expressed in the somewhat arbitrary choice of component-ladders of indices 7,8,9, and 10.

TABLE II 46 Dilogarithmic Valid Ladders

	$\zeta(2)$	[7]	[8]	[9]	[10]
3[6]	20	8	0	-4	-6
3[11]	17	2	0	-1	-3
[12]	9	9	2	-2	-4
[13]	-2	-3	1	1	2
3[16]	35	17	0	-4	-9
[17]	23	9	0	-4	-5
3[18]	23	5	3	-4	-3
3[20]	44	26	6	-4	-9
3[21]	-38	-14	9	15	24
9[23]	-28	-13	9	20	30

	$\zeta(2)$	[7]	[8]	[9]	[10]
3[24]	-26	-26	3	10	18
3[27]	32	5	3	-4	6
3[28]	125	59	9	-22	-36
3[29]	89	14	9	-4	-12
3[30]	61	25	3	-8	-9
3[34]	-46	-46	12	20	27
3[36]	122	56	9	-22	-33
3[37]	35	-7	9	5	27
9[38]	123	15	9	-14	0
9[40]	181	85	54	16	6
[42]	31	11	1	-4	-8
3[44]	80	44	18	-10	-9
3[45]	139	31	12	-5	-18
3[47]	113	5	18	11	12
3[50]	49	-23	15	4	18
3[52]	328	154	18	-50	-90
3[56]	53	-31	12	26	9
3[60]	232	112	45	-26	-57
9[62]	-125	-14	-9	16	24
3[63]	104	8	39	32	42
3[64]	86	8	9	-10	15
3[65]	224	20	30	50	-3
[66]	17	-11	5	4	5
3[70]	97	7	15	4	3
3[74]	379	121	15	-20	-78
3[76]	227	41	57	-16	-18
3[78]	107	11	15	8	9
3[84]	541	247	54	-80	-120
9[86]	743	137	81	-28	-42
3[92]	229	37	63	22	-3
[96]	104	22	8	-4	-13
3[98]	535	157	27	-68	-90
3[110]	355	109	27	-14	-45
[118]	48	-54	28	20	42
3[124]	491	212	57	-58	-93
[132]	101	-19	25	22	28

ACKNOWLEDGEMENTS

The early part of this paper is taken from "Supernumery Polylogarithmic Ladders and Related Functional Equations" by L.Lewin and M.Abouzahra, recently published in Aequationes Mathematicae, 39 (1990), 210 - 253.

Many of the numerical calculations and subsequent searches were performed by M.Abouzahra. The determination of the Salem number ladders is due to G.Szekeres, University of New South Wales.

REFERENCES

1. L. Lewin, "The Dilogarithm in Algebraic Fields", J. Austral. Math. Soc. A 33(1982), 302-330.
2. L. Lewin, "The Inner Structure of the Dilogarithm in Algebraic Fields", J. Number Theory 19(1984), 345-373.
3. M. Abouzahra and L. Lewin, " The Polylogarithm in Algebraic Number Fields", J. Number Theory 21(1985), 214-244.
4. L. Lewin, "The Order-independence of the Polylogarithmic Ladder Structure - Implications for a New Category of Functional Equations", Aequationes Math. 30(1986), 1-20.
5. L. Lewin and E. Rost, "Polylogarithmic Functional Equations: A New Category of Results Developed with the Help of Computer Algebra (MACSYMA)", Aequationes Math. 31(1986), 223-242.
6. M. Abouzahra and L. Lewin, "The Polylogarithm in the Field of Two Irreducible Quintics", Aequationes Math. 31(1986), 315-321.
7. M. Abouzahra, L. Lewin and Xiao Hongnian, "Polylogarithms in the Field of Omega (a Root of a Given Cubic): Functional Equations and Ladders", Aequationes Math. 33(1987), 23-45. Addendum in 35(1988), 304.

8. G. Wechsung, "Über die Unmöglichkeit der Vorkommens von Funktionalgleichungen gewisser Struktur für Polylogarithmen", Aequationes Math. 5(1970), 54-62.

9. J. Browkin, personal communication.

10. L. Euler, "Institutiones Calculi Integralis", Bd. 1 (1768), 110-113. (see also 12).

11. J. Landen, "Mathematical Memoirs", Vol. 1 (1780), 112. (see also 12).

12. L. Lewin, "Polylogarithms and Associated Functions", Elsevier/North-Holland, New York, 1981.

13. H. S. M. Coxeter, "The Functions of Schläfli and Lobatschefsky", Quart. J. Math. Oxford Ser. 6. (1935), 13-29.

14. M. J. Phillips and D. J. Whitehouse, "Two-dimensional Discrete Properties of Random Surfaces", Phil. Trans. Roy. Soc. London Ser. A 305(1982), 441-468.

15. G.Ray, "Linear Relations Involving the Dilogarithm Function", submitted to the Journal of Number Theory.

Resurgent Equations and Stokes Multipliers for the Generalized Confluent Hypergeometric Differential Equations of the Second Order*

Hideyuki Majima
Department of Mathematics
Faculty of Science
Ochanomizu University

1. Introduction.

In this paper, we consider the generalized confluent hypergeometric differential equations of the second order. We shall show that the Borel transforms of formal solutions to these equations can be represented by the hypergeometric functions and that the coefficients of resurgent equations of those differential equations can be calculated explicitly by using connection formulae of the hypergeometric differential equations. Moreover, we shall show that the Stokes multipliers are almost equal to the resurgent constants and that the invariants can be calculated explicitly from them. The idea of these calculations are known classically(cf. [7,8,12]). However, the calculations are seen through by using explicitly resurgent equations. It seems that the explicit formula of invariants will be useful for the resurgent calculus in the future.

2. Review on the Hypergeometric Differential Equation.

The hypergeometric series

$$F(a,b\ ;c;\xi)=\sum_{k=0}^{\infty}\frac{\Gamma(a+k)\Gamma(b+k)\Gamma(c)}{\Gamma(a)\Gamma(b)\Gamma(c+k)k!}\xi^k$$

converges on the unit disc at the origin in the complex plane, and represents an analytic function which satisfies the hypergeometric differential equation,

$$(1-\xi)\xi\frac{d^2}{d\xi^2}F+(c-(a+b+1)\xi)\frac{d}{d\xi}F-abF=0.$$

If c is not an integer, then

$$F_0=(F(a,b;c;\xi),\ \xi^{1-c}F(a-c+1,b-c+1;2-c;\xi))$$

and

$$F_1=(F(a,b;1+a+b-c;1-\xi),(1-\xi)^{c-a-b}F(c-a,c-b;1+c-a-b;1-\xi))$$

*Dedicated to Professor Yasutaka Sibuya on the occasion of his sixtieth birthday.

are bases of solutions to the hypergeometric differential equation near the origin and near the point 1, repectively. Further, there exists a linear relation , namely we have the connection matrix P between F_0 and F_1 as follows(cf. [7,8,12]):

$$F_0 = F_1 P,$$

where

$$P = \begin{pmatrix} \dfrac{\Gamma(c)\Gamma(c-a-b)}{\Gamma(c-a)\Gamma(c-b)} & \dfrac{\Gamma(2-c)\Gamma(c-a-b)}{\Gamma(1-a)\Gamma(1-b)} \\ \dfrac{\Gamma(c)\Gamma(a+b-c)}{\Gamma(a)\Gamma(b)} & \dfrac{\Gamma(2-c)\Gamma(a+b-c)}{\Gamma(a-c+1)\Gamma(b-c+1)} \end{pmatrix},$$

and

$$P^{-1} = \begin{pmatrix} \dfrac{\Gamma(1-c)\Gamma(1-c+a+b)}{\Gamma(a-c+1)\Gamma(b-c+1)} & \dfrac{\Gamma(1-c)\Gamma(1-a-b+c)}{\Gamma(1-a)\Gamma(1-b)} \\ \dfrac{\Gamma(c-1)\Gamma(1-c+a+b)}{\Gamma(a)\Gamma(b)} & \dfrac{\Gamma(c-1)\Gamma(1-a-b+c)}{\Gamma(c-a)\Gamma(c-b)} \end{pmatrix}.$$

3. Generalized Confluent Hypergeometric Differential Equations.

3.1. Formal Solutions and Their Borel Transforms to the Generalized Confluent Hypergeometric Differential Equations of the Second Order.

Consider the differential equation of the following type

$$\frac{d^2}{dz^2}w + (A_0 + \frac{A_1}{z})\frac{d}{dz}w + (B_0 + \frac{B_1}{z} + \frac{B_2}{z^2})w = 0.$$

This equation, called the generalized confluent hypergeometric differential equation of the second order, is the most general one which has a regular singular point at the origin and an irregular singular point of rank 1 at the infinity on the Riemann sphere. Supposes that the equation

$$\rho^2 + A_0\rho + B_0 = 0$$

has two distinct roots, which are denoted by ρ_1 and ρ_2, and so we have

$$\rho_1 + \rho_2 = -A_0, \ \rho_1\rho_2 = B_0.$$

Then we have formal solutions to the differential equation of the form,

$$\exp(\rho z)\phi(\rho; z), \phi(\rho; z) = z^{-\kappa}\sum_{k=0}^{\infty} c_k(\rho)z^{-k},$$

where

$$\kappa = \frac{A_1\rho + B_1}{2\rho + A_0}$$

and

$$c_{k+1}(\rho) = \frac{(\kappa+k)(\kappa+k+1) - A_1(\kappa+k) + B_2}{(2\rho + A_0)(\kappa+k+1) - (A_1\rho + B_1)}c_k(\rho),$$

for $\rho = \rho_j \ (j = 1, 2)$. Denote by α and β the two roots of the equation

$$t^2 - (3 - A_1)t + (2 - A_1 + B_2) = 0,$$

and put $\gamma = 2 - \kappa$. Then, we see

$$c_{k+1}(\rho) = \frac{(\alpha-\gamma+k+1)(\beta-\gamma+k+1)}{(2\rho+A_0)(k+1)} c_k(\rho)\,(k = 0,1,2,...),$$

namely

$$c_k(\rho) = \frac{\Gamma(\alpha-\gamma+k+1)\Gamma(\beta-\gamma+k+1)}{\Gamma(\alpha-\gamma+1)\Gamma(\beta-\gamma+1)(2\rho+A_0)^k k!} c_0(\rho)\,(k = 0,1,2,...).$$

Therefore, if κ is not an integer, the (minor representation of) Borel transform of $\phi(\rho;z)$

$$\Phi(\rho;\zeta) = \sum_{k=0}^{\infty} \frac{c_k(\rho)}{\Gamma(k+\kappa)} \zeta^{k+\kappa-1},$$

is equal to

$$\frac{(2\rho+A_0)^{\kappa-1}}{\Gamma(\kappa)} c_0(\rho)\left(\frac{\zeta}{(2\rho+A_0)}\right)^{1-\gamma} F(\alpha-\gamma+1,\beta-\gamma+1;2-\gamma;\frac{\zeta}{2\rho+A_0}),$$

which is a solution to the hypergeometric differential equation with parameters $(\alpha,\beta;\gamma)$ and with the variable $\xi = \dfrac{\zeta}{2\rho+A_0}$ at the origin.

3.2. Resurgent Equations for the Generalized Confluent Hypergeometric Differential Equations of the Second Order.

Now, notice that

$$\rho_1 - \rho_2 = 2\rho_1 + A_0,\ \rho_2 - \rho_1 = 2\rho_2 + A_0,$$

$$1 - \frac{\zeta}{\rho_1-\rho_2} = \frac{\zeta-(\rho_1-\rho_2)}{\rho_2-\rho_1},$$

$$1 - \frac{\zeta}{\rho_2-\rho_1} = \frac{\zeta-(\rho_2-\rho_1)}{\rho_1-\rho_2},$$

$$\gamma_1-\alpha-\beta = 1-\gamma_2,\ \gamma_1-\alpha = \beta-\gamma_2+1,\ \gamma_1-\beta = \alpha-\gamma_2+1.$$

Then, by the connection formula, we see that

$$\left(\frac{(2\rho_1+A_0)^{\kappa_1-1}}{\Gamma(\kappa_1)} c_0(\rho_1)\right)^{-1}\Phi(\rho_1;\zeta)$$

$$= a_{11}F(\alpha_1,\beta_1;1+\alpha_1+\beta_1-\gamma_1;1-\frac{\zeta}{\rho_1-\rho_2})$$

$$+a_{12}\left(1-\frac{\zeta}{\rho_1-\rho_2}\right)^{\gamma_1-\alpha_1-\beta_1} F(\gamma_1-\alpha_1,\gamma_1-\beta_1;1+\gamma_1-\alpha_1-\beta_1;1-\frac{\zeta}{\rho_1-\rho_2}),$$

namely,

$$\left(\frac{(2\rho_1+A_0)^{\kappa_1-1}}{\Gamma(\kappa_1)} c_0(\rho_1)\right)^{-1}\Phi(\rho_1;\zeta)$$

$$= a_{11}F(\alpha_2,\beta_2;\gamma_2;1+\frac{\zeta}{\rho_2-\rho_1})$$

$$+a_{12}(\frac{(\rho_2-\rho_1)^{\kappa_2-1}}{\Gamma(\kappa_2)}c_0(\rho_2))^{-1}\Phi(\rho_2;\zeta-(\rho_1-\rho_2)),$$

where

$$\alpha_1=\alpha,\ \beta_1=\beta,\ \alpha_2=\beta,\ \beta_2=\alpha,$$

$$a_{11}=\frac{\Gamma(2-\gamma_1)\Gamma(\gamma_1-\alpha-\beta)}{\Gamma(1-\alpha)\Gamma(1-\beta)},$$

$$a_{12}=\frac{\Gamma(2-\gamma_1)\Gamma(\alpha+\beta-\gamma_1)}{\Gamma(\alpha-\gamma_1+1)\Gamma(\beta-\gamma_1+1)}.$$

Hence, we obtain the relation (R-1):

$$\Phi(\rho_1;\zeta\exp(2i\pi))-\Phi(\rho_1;\zeta)$$

$$=(\exp(2i\pi\kappa_1)-1)\Phi(\rho_1;\zeta)$$

$$=(\exp(2i\pi\kappa_1)-1)ca_{12}\Phi(\rho_2;\zeta-(\rho_1-\rho_2))$$

$$+(\exp(2i\pi\kappa_1)-1)\frac{(\rho_1-\rho_2)^{\kappa_1-1}}{\Gamma(\kappa_1)}c_0(\rho_1)a_{11}F(\alpha_2,\beta_2;\gamma_2;1+\frac{\zeta}{\rho_2-\rho_1}),$$

where

$$c=\frac{(\rho_1-\rho_2)^{\kappa_1-1}\Gamma(\kappa_2)c_0(\rho_1)}{(\rho_2-\rho_1)^{\kappa_2-1}\Gamma(\kappa_1)c_0(\rho_2)}.$$

In the same way, we also obtain the relation (R-2):

$$\Phi(\rho_2;\zeta\exp(2i\pi))-\Phi(\rho_2;\zeta)$$

$$=(\exp(2i\pi\kappa_2)-1)\Phi(\rho_2;\zeta)$$

$$=(\exp(2i\pi\kappa_2)-1)c^{-1}a_{21}\Phi(\rho_1;\zeta-(\rho_2-\rho_1))$$

$$+(\exp(2i\pi\kappa_2)-1)\frac{(\rho_2-\rho_1)^{\kappa_2-1}}{\Gamma(\kappa_2)}c_0(\rho_2)a'_{11}F(\alpha_1,\beta_1;\gamma_1;1+\frac{\zeta}{\rho_1-\rho_2}),$$

where

$$a_{11}{}'=\frac{\Gamma(2-\gamma_2)\Gamma(\gamma_2-\alpha-\beta)}{\Gamma(1-\alpha)\Gamma(1-\beta)},$$

$$a_{21}=\frac{\Gamma(2-\gamma_2)\Gamma(\alpha+\beta-\gamma_2)}{\Gamma(\alpha-\gamma_2+1)\Gamma(\beta-\gamma_2+1)}.$$

These relations are nothing but the resurgent equations of Ecalle[6] for the generalized confluent hypergeometric differential equation of the second order.

Now consider the generalized Laplace transforms of $\Phi(\rho_j;\zeta)$,

$$\mathcal{L}(C;\Phi(\rho_j;\zeta);z)=\int_C\exp(-z\zeta)\Phi(\rho_j;\zeta)d\zeta,\ (j=1,2),$$

where C is one of the following paths of integral;

- $C(\rho_1-\rho_2;\theta)$:the path on which $\arg(\zeta-(\rho_1-\rho_2))$ is taken to be initially θ and finally $\theta+2\pi$,

- $C(0;\theta)$:the path on which $\arg(\zeta)$ is taken to be initially θ and finally $\theta+2\pi$,
- $C(\rho_2-\rho_1;\theta)$:the path on which $\arg(\zeta-(\rho_2-\rho_1))$ is taken to be initially θ and finally $\theta+2\pi$.

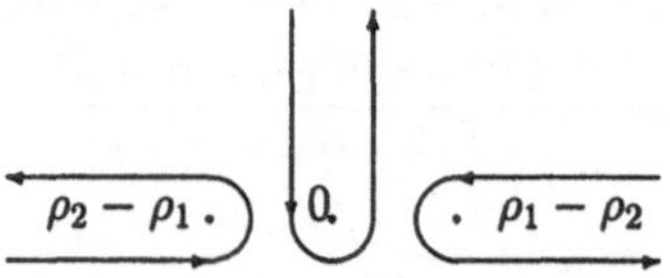

Then, for $j=1$ and

$$\arg(\rho_1-\rho_2)-2\pi<\theta<\arg(\rho_1-\rho_2),$$

$$(\exp(2i\pi\kappa_1)-1)^{-1}\exp(\rho_1 z)\mathcal{L}(C(0;\theta);\Phi(\rho_1;\zeta);z)$$

is a solution to the generalized confluent hypergeometric differential equation and asymptotic to the formal solution

$$\exp(\rho_1 z)\phi(\rho_1;z)$$

in the sector (mod. 2π)

$$\frac{\pi}{2}<\arg(-\zeta z)<\frac{3\pi}{2},$$

namely,

$$-\frac{\pi}{2}-\theta<\arg z<\frac{\pi}{2}-\theta,$$

because $\exp(-z\zeta)$ tends to 0 as ζ tends to the infinity. Therefore, by considering the analytic continuation, we obtain the solution

$$(\exp(2i\pi\kappa_1)-1)^{-1}\exp(\rho_1 z)\tilde{\Phi}(\rho_1;z)$$

asymptotic to

$$\exp(\rho_1 z)\phi(\rho_1;z)$$

in the sector

$$-\frac{\pi}{2}-\arg(\rho_1-\rho_2)<\arg z<\frac{5\pi}{2}-\arg(\rho_1-\rho_2).$$

Similarly, for $j=2$,

$$\arg(\rho_2-\rho_1)-2\pi=\arg(\rho_1-\rho_2)-\pi<\theta<\arg(\rho_1-\rho_2)+\pi=\arg(\rho_2-\rho_1),$$

$$(\exp(2i\pi\kappa_2)-1)^{-1}\exp(\rho_2 z)\mathcal{L}(C(0;\theta);\Phi(\rho_2;\zeta);z)$$

is a solution to the differential equation which is asymptotic to the formal solution

$$\exp(\rho_2 z)\phi(\rho_2;z)$$

in the sector

$$\frac{\pi}{2}<\arg(-\zeta z)<\frac{3\pi}{2},$$

namely,

$$-\frac{\pi}{2}-\theta<\arg z<\frac{\pi}{2}-\theta,$$

because $\exp(-z\zeta)$ tends to 0 as ζ tends to the infinity. Therefore, by considering the analytic continuation, we obtain the solution

$$(\exp(2i\pi\kappa_2)-1)^{-1}\exp(\rho_2 z)\tilde{\Phi}(\rho_2;z)$$

asymptotic to

$$\exp(\rho_2 z)\phi(\rho_2;z)$$

in the sector

$$-\frac{\pi}{2}-\arg(\rho_2-\rho_1)=$$

$$-\frac{3\pi}{2}-\arg(\rho_1-\rho_2)<\arg z<\frac{3\pi}{2}-\arg(\rho_1-\rho_2)$$

$$=\frac{5\pi}{2}-\arg(\rho_2-\rho_1).$$

Therefore, from the relations (R-1) and (R-2), by taking the Laplace transforms with the pathes

$C(\rho_1-\rho_2;\arg(\rho_1-\rho_2))$:

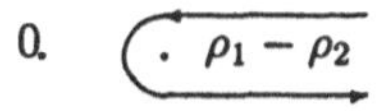

and

$C((\rho_1-\rho_2)\exp(i\pi);\arg(\rho_1-\rho_2)+\pi)$:

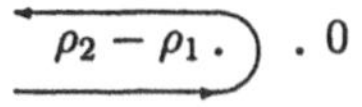

respectively, and after that, by deforming the pathes as in the figures below

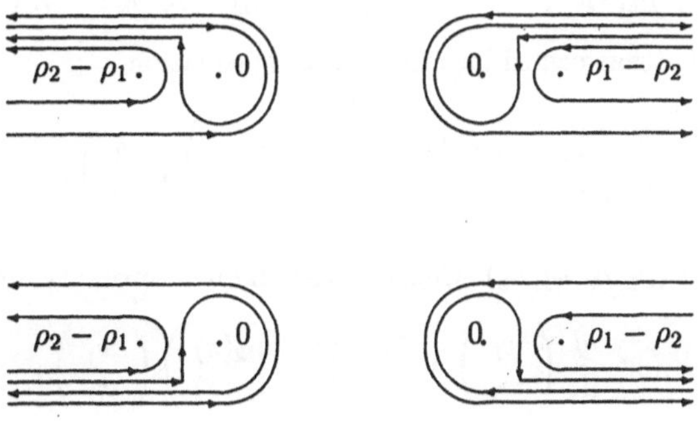

we obtain the relation (S-1)

$$
\begin{aligned}
&-\exp(\rho_1 z)\tilde{\Phi}(\rho_1; z\exp(2i\pi))\exp(-2i\pi\gamma_1) + \exp(\rho_1 z)\tilde{\Phi}(\rho_1; z) \\
&\qquad = c(\exp(2i\pi\kappa_1) - 1)a_{12}\exp(\rho_2 z)\tilde{\Phi}(\rho_2; z),
\end{aligned}
$$

in the sector

$$-\frac{\pi}{2} - \arg(\rho_1 - \rho_2) < \arg z < \frac{\pi}{2} - \arg(\rho_1 - \rho_2),$$

and the relation (S-2)

$$
\begin{aligned}
&-\exp(\rho_2 z)\tilde{\Phi}(\rho_2; z\exp(2i\pi))\exp(-2i\pi\gamma_2) + \exp(\rho_2 z)\tilde{\Phi}(\rho_2; z) \\
&\qquad = c^{-1}(\exp(2i\pi\kappa_2) - 1)a_{21}\exp(\rho_1 z)\tilde{\Phi}(\rho_1; z\exp(2i\pi)),
\end{aligned}
$$

in the sector

$$-\frac{3\pi}{2} - \arg(\rho_1 - \rho_2) < \arg z < -\frac{\pi}{2} - \arg(\rho_1 - \rho_2),$$

respectively.

3.3. Invariants of the Generalized Confluent Hypergeometric Differential Equations of the Second Order.

Thereafter, we fix the constants (A_0, A_1, B_0, B_1), namely, $(\rho_1, \rho_2, \kappa_1, \kappa_2)$ and consider B_2 as a parameter. The generalized confluent hypergeometric differential equation is written in the form of linear system,

$$\frac{d}{dz}\begin{pmatrix} w \\ \frac{d}{dz}w \end{pmatrix} = \begin{pmatrix} 0 & 1 \\ -(B_0 + \frac{B_1}{z} + \frac{B_2}{z^2}) & -(A_0 + \frac{A_1}{z}) \end{pmatrix}\begin{pmatrix} w \\ \frac{d}{dz}w \end{pmatrix}.$$

The existence of formal solutions implies that, by the formal transformation at the infinity,

$$
\begin{aligned}
&\begin{pmatrix} w \\ \frac{d}{dz}w \end{pmatrix} \\
&= \begin{pmatrix} \phi(\rho_1; z)z^{\kappa_1} & \phi(\rho_2; z)z^{\kappa_2} \\ \frac{d}{dz}(\phi(\rho_1; z)e^{\rho_1 z})z^{\kappa_1}e^{-\rho_1 z} & \frac{d}{dz}(\phi(\rho_2; z)e^{\rho_2 z})z^{\kappa_2}e^{-\rho_2 z} \end{pmatrix}\begin{pmatrix} v_1 \\ v_2 \end{pmatrix},
\end{aligned}
$$

the above equation is transformed into the following equation

$$\frac{d}{dz}\begin{pmatrix} v_1 \\ v_2 \end{pmatrix} = \begin{pmatrix} \rho_1 - \frac{\kappa_1}{z} & 0 \\ 0 & \rho_2 - \frac{\kappa_2}{z} \end{pmatrix}\begin{pmatrix} v_1 \\ v_2 \end{pmatrix},$$

which is denoted by $E(\rho_1, \rho_2, \kappa_1, \kappa_2)$. Consider the all linear ordinary differential equation of the form

$$E_A: \quad \frac{d}{dz}\begin{pmatrix} w_1 \\ w_2 \end{pmatrix} = \begin{pmatrix} a_{11}(z) & a_{12}(z) \\ a_{21}(z) & a_{22}(z) \end{pmatrix}\begin{pmatrix} w_1 \\ w_2 \end{pmatrix},$$

where $A = (a_{ij}(z))_{i,j=1,2}$ is meromorphic at the infinity. Denote by

$$\mathcal{E}(\rho_1, \rho_2, \kappa_1, \kappa_2)$$

the set of all linear ordinary differential equations of the above form for each of which there exists a formal transformation which transforms it into the equation

$$E(\rho_1, \rho_2, \kappa_1, \kappa_2).$$

For two linear ordinary differential equations E_A and E_B, we say that they are equivalent, noted by $E_A \sim E_B$, if there exists a holomorphic transformation by which E_A is transformed into E_B. Then, by the classification theory of linear ordinary differential equation at an irregular singular point, which is due to Sibuya[11], Malgrange[10] and Babbitt-Varadarajan[1], we have the isomorphism of sets,

$$\mathcal{E}(\rho_1, \rho_2, \kappa_1, \kappa_2)/\sim \;\simeq\; H^1(S^1, \Lambda)/\sim_H,$$

where Λ is the sheaf on S^1(the set of all directions to the infinity) of germs of holomorphic matricial solutions P to the equation

$$\frac{d}{dz}P = \begin{pmatrix} \rho_1 - \frac{\kappa_1}{z} & 0 \\ 0 & \rho_2 - \frac{\kappa_2}{z} \end{pmatrix} P - P \begin{pmatrix} \rho_1 - \frac{\kappa_1}{z} & 0 \\ 0 & \rho_2 - \frac{\kappa_2}{z} \end{pmatrix},$$

which are asymptotic to the 2-by-2 identity matrix, and two cohomology classes (P_{ij}) and (Q_{ij}) are equivalent, noted by

$$(P_{ij}) \sim_H (Q_{ij}),$$

if there exists a constant matrix

$$G = \begin{pmatrix} \nu & 0 \\ 0 & \mu \end{pmatrix}$$

such that

$$G(P_{ij})G^{-1} = (Q_{ij}).$$

By choosing the open covering $\{U_1, U_2\}$,

$$U_1 = \{\exp(i(\arg z)) : -\frac{\pi}{2} - \arg(\rho_1 - \rho_2) < \arg z < \frac{3\pi}{2} - \arg(\rho_1 - \rho_2)\},$$

$$U_2 = \{\exp(i(\arg z)) : \frac{\pi}{2} - \arg(\rho_1 - \rho_2) < \arg z < \frac{5\pi}{2} - \arg(\rho_1 - \rho_2)\},$$

$H^1(S^1, \Lambda)$ is identified with

$$\{\begin{pmatrix} 1 & c_{12} \\ 0 & 1 \end{pmatrix}, \begin{pmatrix} 1 & 0 \\ c_{21} & 1 \end{pmatrix}\}$$

and so

$$H^1(S^1, \Lambda)/\sim_H$$

is identified with

$$\{c_{12}c_{21}\}.$$

For the generalized confluent hypergeometric differential equation, as basis of solutions asymptotic to

$$(\exp(\rho_1)\phi(\rho_1; z), \exp(\rho_2)\phi(\rho_2; z)),$$

we choose

$$(e_1 \exp(\rho_1 z)\tilde{\Phi}(\rho_1; z), e_2 \exp(\rho_2 z)\tilde{\Phi}(\rho_2; z)),$$

in U_1, and

$$(e_1 \exp(\rho_1 z)\tilde{\Phi}(\rho_1; z), e_2 \exp(\rho_2 z)\tilde{\Phi}(\rho_2; z\exp(-2i\pi)) \exp(2i\pi\gamma_2)),$$

in U_2, where

$$e_1 = (\exp(2i\pi\kappa_1) - 1)^{-1},$$

and

$$e_2 = (\exp(2i\pi\kappa_2) - 1)^{-1}.$$

Then, for z such that

$$\frac{\pi}{2} - \arg(\rho_1 - \rho_2) < \arg z < \frac{3\pi}{2} - \arg(\rho_1 - \rho_2),$$

by the relation (S-2), we see

$$\begin{aligned}&(e_1 \exp(\rho_1 z)\tilde{\Phi}(\rho_1; z), e_2 \exp(\rho_2 z)\tilde{\Phi}(\rho_2; z\exp(-2i\pi)) \exp(2i\pi\gamma_2))\\ &= (e_1 \exp(\rho_1 z)\tilde{\Phi}(\rho_1; z), e_2 \exp(\rho_2 z)\tilde{\Phi}(\rho_2; z)) \begin{pmatrix} 1 & c_{12} \\ 0 & 1 \end{pmatrix},\end{aligned}$$

similarly, by the relation (S-1), we obtain

$$\begin{aligned}&(e_1 \exp(\rho_1 z)\tilde{\Phi}(\rho_1; z\exp(-2i\pi)), e_2 \exp(\rho_2 z)\tilde{\Phi}(\rho_2; z\exp(-2i\pi)))\\ &= (e_1 \exp(\rho_1 z)\tilde{\Phi}(\rho_1; z), e_2 \exp(\rho_2 z)\tilde{\Phi}(\rho_2; z\exp(-2i\pi)) \exp(2i\pi\gamma_2))\\ &\quad\times \begin{pmatrix} 1 & 0 \\ c_{21} & 1 \end{pmatrix} \begin{pmatrix} \exp(-2i\pi\gamma_1) & 0 \\ 0 & \exp(-2i\pi\gamma_2) \end{pmatrix},\end{aligned}$$

for z such that

$$\frac{3\pi}{2} - \arg(\rho_1 - \rho_2) < \arg z < \frac{5\pi}{2} - \arg(\rho_1 - \rho_2),$$

where

$$\begin{aligned}c_{12} &= c^{-1}{e_1}^{-1}e_2(\exp(2i\pi\kappa_2) - 1)a_{21}\exp(2i\pi\gamma_2),\\ c_{21} &= ce_1{e_2}^{-1}(\exp(2i\pi\kappa_1) - 1)a_{12}\exp(2i\pi(\gamma_1 - \gamma_2)).\end{aligned}$$

(See also [4].) Therefore, we have the invariant

$$c_{12}c_{21} = a_{12}a_{21}(\exp(2i\pi\kappa_2) - 1)(\exp(2i\pi\kappa_1) - 1)\exp(2i\pi\gamma_1).$$

By using the formulae of Γ function and the definition of ρ_j, κ_j $(j = 1, 2)$, we have

$$c_{12}c_{21} = -2\exp(i\pi(\kappa_2 - \kappa_1))(\cos(\kappa_1 - \kappa_2)\pi + \cos(\beta - \alpha)\pi).$$

As we have

$$(\beta - \alpha)^2 = (A_1 - 1)^2 - 4B_2,$$

two generalized confluent hypergeometric differential equations with parameters B_2 and B_2' are analytically equivalent if and only if

$$((A_1-1)^2-4B_2)^{\frac{1}{2}} = \pm((A_1-1)^2-4B_2')^{\frac{1}{2}}+2n,$$

namely,

$$(B_2'-B_2+n^2)^2 = n^2((A_1-1)^2-4B_2),$$

for some integer n.

4. Special Cases:Bessel, Kummer, Whittaker, Weber and Airy.

4.1. Bessel Equations.

$$A_0=0,\ A_1=1,\ B_0=1,\ B_1=0,\ B_2=-\nu^2$$
$$\rho_1=i,\ \kappa_1=\frac{1}{2},\ \gamma_1=\frac{3}{2},$$
$$\rho_2=-i,\ \kappa_2=\frac{1}{2},\ \gamma_2=\frac{3}{2},$$
$$\alpha=1-\nu,\ \beta=1+\nu,$$
$$a_{12}=a_{21}=-\cos(\pi\nu),$$
$$c_{12}c_{21}=-4\cos^2(\pi\nu).$$

As $B_2=-\nu^2$, two Bessel equations with parameters ν and ν' are analytically equivalent if and only if $\nu'=\pm\nu+n$ for some integer n. (See also [2].)

4.2. Kummer Equation.

$$A_0=-1,\ A_1=c,\ B_0=0,\ B_1=-a,\ B_2=0,$$
$$\rho_1=0,\ \kappa_1=a,\ \gamma_1=2-a,$$
$$\rho_2=1,\ \kappa_2=c-a,\ \gamma_2=2-c+a,$$
$$\alpha=2-c,\ \beta=1,$$
$$a_{12}=a_{21}=1,$$
$$c_{12}c_{21}=-2\exp(i\pi(c-2a))(\cos((2a-c)\pi)-\cos(c\pi)).$$

4.3. Whittaker Equation.

$$A_0=0,\ A_1=0,\ B_0=-\frac{1}{4},\ B_1=k,\ B_2=\frac{1}{4}-m^2,$$
$$\rho_1=\frac{1}{2},\ \kappa_1=k,\ \gamma_1=2-k,,$$
$$\rho_2=-\frac{1}{2},\ \kappa_2=-k,\ \gamma_2=2+k,,$$
$$\alpha=\frac{3}{2}+m,\ \beta=\frac{3}{2}-m,$$
$$a_{12}=\frac{\Gamma(k)\Gamma(1+k)}{\Gamma(\frac{1}{2}+m+k)\Gamma(\frac{1}{2}-m+k)},\quad a_{21}=\frac{\Gamma(-k)\Gamma(1-k)}{\Gamma(\frac{1}{2}+m-k)\Gamma(\frac{1}{2}-m-k)}$$

$$c_{12}c_{21} = -2(\cos 2k\pi + \cos 2m\pi)\exp(-2ki\pi).$$

4.4. Weber Equation.

The Weber equation is of the form

$$\frac{d^2v}{dz^2} + (2\nu + 1 - z^2)v = 0,$$

which is transformed into the Kummer equation with the parameter $(a, c) = (-\frac{\nu}{2}, \frac{1}{2})$ by the transformation

$$v(z) = \exp(-\frac{z^2}{2})w(z^2).$$

$$\rho_1 = \frac{1}{2}, \ \kappa_1 = -\frac{\nu}{2},, \ \gamma_1 = 2 + \frac{\nu}{2},$$

$$\rho_2 = -\frac{1}{2}, \ \kappa_2 = \frac{1+\nu}{2},, \ \gamma_2 = 2 - \frac{1+\nu}{2},$$

$$\alpha = \frac{3}{2}, \ \beta = 1,$$

$$a_{12} = a_{21} = 1$$

$$c_{12}c_{21} = \exp(2i\pi\nu) - 1.$$

4.5. Airy Equation.

The Airy equation is of the form

$$\frac{d^2v}{dz^2} - zv = 0,$$

which is transformed into the Bessel equation with the parameter $\nu = \frac{1}{3}$ by the transformation

$$v(z) = (z^{\frac{3}{2}})^{\frac{1}{3}}w(\frac{2}{3}iz^{\frac{3}{2}}).$$

$$\rho_1 = i, \ \kappa_1 = \frac{1}{6}, \ \gamma_1 = \frac{11}{6},$$

$$\rho_2 = -i, \ \kappa_2 = \frac{1}{6}, \ \gamma_2 = \frac{11}{6},$$

$$\alpha = \frac{5}{3}, \ \beta = 1,$$

$$a_{12} = a_{21} = 1,$$

$$c_{12}c_{21} = -1.$$

References

[1] Babbitt, D.-G. and Varadarajan, V.S.: Local Moduli for Meromorphic Differential Equations, Bull. Amer. Math. Soc. (New Series), Vol.12, No.1 (1985), p.95-p.98.

[2] Babbitt, D.-G. and Varadarajan, V.S.: Local Moduli for Meromorphic Differential Equations, vol. 169-170, S.M.F. (1989).

[3] Balser, W., Jurkat, W.-B., and Lutz, D.A.: On the Reduction of Connection Problems for Differential Equations with an Irregular Singular Point to Ones with only Regular Singularities,I, SIAM J. Math. Anal. Vol. 12, No. 5 (1981), p.691-p.721.

[4] Balser, W., Jurkat, W.-B. and Lutz, D.A.: Birkhoff Invariants and Stokes' Multipliers for Meromorphic Linear Differential Equations, J. Math. Anal. App. Vol 71, no.1 (1979), p.48-p.94.

[5] Birkhoff, G.-D.: On a Simple Type of Irregular Singular Point, Trans. Amer. Math. Soc., Vol.14 (1913), p.462-p.476.

[6] Ecalle, J.: Les Fonctions Résurgents, Tome III; L'équation du Pont et la Classification Analytique des Objets Locaux, Publications Mathématiques d'Orsay, 85-05.

[7] Erdelyi, A., Magnus, W., Oberhettinger, F., and Tricomi, F. G.: Higher Transendental Functions, I-III, Bateman Manuscript Project, McGraw-Hill (1953).

[8] Inui, T.: Special Functions(Tokushukansuu written in japanese), Iwanami-shoten (1962).

[9] Majima, H: Asymptotic Analysis for Integrable Connections with Irregular Singular Points, Lect. Note in Math. no. 1075, Springer-Verlag(1984).

[10] Malgrange, B.: Remarques sur les Equations Différentielles à Points Singuliers Irreguliers, in Equations Différentielles et Systèmes de Pfaff dans le Champ Complexe edited by R. Gérard and J.-P. Ramis, Lecture Notes in Math., No.712, Springer-Verlag, p.77-p.86.

[11] Sibuya, Y.: Stokes Phenomena, Bull. Amer. Math. Soc, Vol.83(1977), p.1075-p.1077.

[12] Whittaker, E. T., and Watson, G. N.: A Course of Modern Analysis, Cambridge, 1902.

AN INFINITESIMALLY QUASI INVARIANT MEASURE ON THE GROUP OF DIFFEOMORPHISMS OF THE CIRCLE

Marie Paule Malliavin and Paul Malliavin
10 rue Saint Loui en L'Isle, 75004 Paris, France

The group of diffeomorphisms of a riemannian manifold of dimension >1 is very "ramified". The existence of a "reasonnable" quasi invariant measure seems therefore very doubtfull. The case of the group of the diffeomorphisms of the circle S_1 is quite different. We shall in this paper identify it with a loop space which carries a natural Wiener measure.

We shall then prove the infinitesimal quasi invariance using the two following tools :

(i) Itô's stochastic calculus,

(ii) Restriction of the Itô's calculus to a finite codimension manifold by the quasi sure analysis.

This strategy have been already used in [2].

1. Notations.

We shall denote by G the connected component of the identity of the group of C^1-diffeomorphisms of the circle. It can also be defined as the group of diffeomorphisms of index +1.

We shall choose an origin $0\in S_1$ and we will identify S_1 with $[0,2\pi[$. Then we consider the subgroup

$$G_0=\{g\in G \ ;\ g(0)=0\}.$$

2. Identification of G_0 with a loop space.

We denote by $\mathbb{L}_0(\mathbb{R})$ the loop space of continuous maps ℓ of the circle into $\mathbb{R}$ such that $\ell(0)=0$.

2.1. Theorem. *There exist an homeomorphism* Q *between* G_0 *and* $\mathbb{L}_0(\mathbb{R})$.

<u>Proof</u>. Given $g\in G_0$, then we have $\dot{g}(\theta)>0$; we define

2.2 $$\ell_g(\theta)=\log \dot{g}(\theta)-\log \dot{g}(0).$$

<u>Injectivity</u>. If $\ell_{g_1}=\ell_{g_2}$, then $\dot{g}_1(\theta)=c\dot{g}_2(\theta)$. As

$$\int_0^{2\pi}\dot{g}_1(\theta)\,d\theta=2\pi=\int_0^{2\pi}\dot{g}_2(\theta)\,d\theta,$$

we deduce that $\dot{g}_1=\dot{g}_2$. As $g_1(0)=g_2(0)=0$ we get $g_1=g_2$.

<u>Surjectivity</u>. Given $s \in L_0(\mathbb{R})$ we define

$$\dot{g}(\theta) = c e^{s(\theta)}$$

where the constant c is determined by the condition $\int_0^{2\pi} \dot{g}(\theta)d\theta = 2\pi$.

2.2. Corollary. *The Wiener measure* $\mu_{L_0(\mathbb{R})}$ *defines by transport by Q a borelian measure* μ_{G_0} *on* G_0.

<u>Remark</u>. See [2] for the property of the Wiener measure on loops.

3. Infinitesimal action.

We will denote by $\mathcal{G}^\infty$ the Lie algebra of smooth vector fields on S_1. We shall denote by $\mathcal{G}^k$ the Hilbert space of C^{k-1} vector fields on S_1, which have their k^{th}-derivative in L^2. Denote

$$\mathcal{G}_0^k = \{z \in \mathcal{G}^k \; ; \; z(0) = 0\}$$

3.1. Theorem. *The measure* μ_{G_0} *is infinitesimally quasi invariant under the left action of* $\mathcal{G}_0^3$. *More precisely, given* $z \in \mathcal{G}_0^3$, *there exist* $K_z \in \bigcap_p L^p(G_0; \mu_{G_0})$ *such that for every smooth test function* φ *we have :*

$$\left\{ \frac{d}{d\varepsilon} \int_{G_0} \varphi(\exp(\varepsilon z)g)\mu_{G_0}(dg) \right\}_{\varepsilon=0} = \int_{G_0} \varphi(g) K_z(g) \mu_{G_0}(dg).$$

<u>Proof</u>. We shall read the left action of $\mathcal{G}_0^3$ in the chart Q. Denote

$$k(\lambda, \theta) := \exp(\lambda z)(\theta).$$

Then k is determined by the system :

(i)
$$\begin{cases} \frac{\partial k}{\partial \lambda}(\lambda, \theta) = z(k(\lambda, \theta)) \\ k(0, \theta) = \theta \end{cases}$$

We consider the Jacobian (1×1) matrix J defined by

$$\frac{\partial}{\partial \theta} k(\lambda, \theta) = J(\lambda, \theta).$$

Differentiate in θ the first equation of (i), we obtain, by an interversion of the derivations, the following system for the Jacobian matrix

$$\begin{cases} \frac{\partial J}{\partial \lambda} = \dot{z}(k(\lambda, \theta))J \\ J(0, \theta) = \text{Identity} \end{cases}$$

which leads to the following formula :

(ii) $$\frac{\partial k}{\partial \theta}(\lambda_0,\theta)=\exp\left(\int_0^{\lambda_0}\dot{z}(k(\lambda,\theta))d\lambda\right).$$

Denote

$$g_\varepsilon(\theta):= \exp(\varepsilon z)g = k(\varepsilon,g(\theta)) \;;$$

then we get by differentiation :

$$\dot{g}_\varepsilon=\frac{\partial k}{\partial \theta}\dot{g}$$ or

(iii) $$\log(\dot{g}_\varepsilon(\theta))=\int_0^{\varepsilon}\dot{z}(k(\lambda,g(\theta)))d\lambda+\log\dot{g}(\theta).$$

3.2. Lemma.

$$\left\{\frac{d}{d\varepsilon}Q(\exp(\varepsilon z)g)\right\}_{\varepsilon=0}=q$$

where $q(\theta)=\dot{z}(g(\theta))$.

Proof. We have by (iii) $\log(\dot{g}_\varepsilon)=\log(\dot{g})+\varepsilon q+\circ(\varepsilon)$.

In order to be able to use Itô's calculus we want to extend the precedent computation from loops to paths. We denote by X the Wiener space, that is the space of continuous function on $[0,2\pi]$ taking the value 0 at zero

(iv) $$\mathbb{L}_0(R)=\{x\in X \;;\; x(2\pi)=0\}$$

Now we associate to $x\in X$ the function :

3.3 $$\begin{cases} g_x(\theta)=c(x)\int_0^{\theta}e^{x(\xi)}d\xi \\ c(x)=\left(\int_0^{2\pi}e^{x(\xi)}d\xi\right)^{-1}\end{cases}$$ where then

$\theta\longrightarrow g_x(\theta)$ is an homeomorphism of S_1.

We have

$$\lim_{\theta\to 0}\dot{g}_x(\theta)=c(x)$$
$$\lim_{\theta\to 2\pi}\dot{g}_x(\theta)=c(x)e^{x(2\pi)}.$$

We remark that g is a C^1-diffeomorphism if and only if these two limits are equal which means :

$$x(2\pi)=0.$$

Consider now a vector field $Z(x)$ on the Wiener space X. Denote by D_Z the directionnal derivative following the vector Z defined by

$$(D_Z\psi)(x)=\left\{\frac{d}{d\varepsilon}\psi(x+\varepsilon Z(x))\right\}_{\varepsilon=0}.$$

3.4. Lemma. *Denote*

$$Z(x)(\theta)=\dot{z}(g_x(\theta)) \quad \textit{with } z\in\mathcal{G}_0^3 ;$$

then

$$D_Z c=0.$$

Proof.

$$D_Z(\tfrac{1}{c})=\int_\theta^{2\pi} e^{x(\xi)}\dot{z}(g_x(\xi))d\xi ;$$

then by 3.3 this can be written :

$$=c\int_0^{2\pi} \dot{z}(g_x(\xi))\dot{g}_x(\xi)d\xi=c(z(2\pi)-z(0))=0.$$

3.5. Lemma.

$$(D_Z g_x)(\theta)=z(g_x(\theta)).$$

Proof. By 3.4

$$(D_Z g_x)(\theta)=c(x)\int_0^\theta e^{x(\xi)}\dot{z}(g_x(\xi))d\xi$$

$$=\int_0^\theta \dot{z}(g_x(\xi))\dot{g}_x(\xi)d\xi$$

$$=z(g_x(\theta)).$$

which proves the lemma. An alternative proof would have to reverse the reading of 3.2.

Now we make the following observations :

$\dot{z}(g_x(\cdot))$ is an adapted process

$$\frac{d}{d\theta}[\dot{z}(g_x)]=\ddot{z}\dot{g}_x=\ddot{z}e^x\in L^2$$

The following Itô's stochastic integral is well defined

$$H(x)=\int_0^{2\pi} \ddot{z}(g_x(\theta))e^{x(\theta)}dx(\theta).$$

For every smooth test function ψ we have :

(vi) $$\left\{\frac{d}{d\varepsilon}\int_X \psi(x+\varepsilon Z(x))d\mu_x\right\}_{\varepsilon=0}=\int_X \psi(x)H(x)d\mu_x(x).$$

We want to restrict this identity to the hyperplane $x^{-1}(2\pi)$ of X. We shall use two different approaches *quasi sure analysis* and the *pinned* Brownian.

4. Quasi sure analysis.

Being in codimension 1, there exist [1] constants C_p such that :

4.1 $$\int_{G_0} [H^*(\ell_g)]^p \mu_{G_0}(dg) < c_p\Big[\int_X \|\nabla H\|^p(x)\,d\mu_X(x) + \int \|H\|^p(x)\,d\mu(x)\Big]$$

where H^* *denotes a redefinition of* H.

We have, using lemma 3.4,

$$(D_h H)(x) = \left\{\frac{d}{d\varepsilon}\int_0^{2\pi} \ddot{z}(g_{x+\varepsilon h}(\theta))\,e^{x(\theta)+\varepsilon h(\theta)}\,d(x+\varepsilon h)\right\}$$

$$= c(x)\int_0^{2\pi} \dddot{z}(g_x(\theta))\Big[\int_0^{\theta} e^{x}(\xi)h(\xi)\,d\xi\Big]dx(\theta)$$

$$+\int \ddot{z}(g_x(\theta))h(\theta)e^{x(\theta)}dx(\theta) + \int \ddot{z}(g_x(\theta))e^{x(\theta)}\dot{h}(\theta)\,d\theta.$$

We remark that :

$$\int_X [c(x)]^p\ \mu_X(dx) \le E(\exp(p \min_{\theta\in[0,2\pi]} x(\theta))) \le \exp(\frac{2\pi p^2}{2})$$

the last inequality being a consequence of the exponential martingale inequality. In the same way

$$\int_X\Big[\int_0^{2\pi} |e^{x(\theta)}|^p d\theta\Big]\mu_X(dx) \le \exp(\frac{2\pi p^2}{2})$$

$$\int_0^{2\pi} |\ddot{z}(g_x(\theta))|^p d\theta = \int_0^{2\pi} |\ddot{z}(g_x(\theta))|^p \dot{g}_x(\theta)[\dot{g}_x(\theta)]^{-1} d\theta$$

$$\le \Big(\int_0^{2\pi} [\ddot{z}(\theta)]^{2p} d\theta\Big)^{1/2}\Big(\int [\dot{g}_x(\theta)]^{-2}\,d\theta\Big)^{1/2}$$

$$\le c\ \exp(4p^2\pi).$$

Furthermore

$$\|\nabla H\|(x) = \sup_h (D_h H)(x) \qquad \text{with } \|h\|_{H_1} \le 1.$$

Then the first term can be written :

$$\iint_{0<\xi<\theta<2\pi} \dddot{z}(g_x(\theta))e^{x(\xi)}h(\xi)\,d\xi \otimes dx(\theta)$$

which can be written as :

$$I(h,x) = \int_0^{2\pi} h(\xi)\,d\xi\Big[\int_\xi^{2\pi} \dddot{z}(g_x(\theta))e^{x(\xi)}\,dx(\theta)\Big]$$

Therefore

$$\mathrm{Sup}|I(h,x)| \le c\ \sup_\xi(e^{x(\xi)})\sup_\xi\Big|\int_\xi^{2\pi} \dddot{z}(g_x(\theta))\,dx(\theta)\Big|$$

The last expression is, granted L^p-martingale inequalities, dominated in L^p norm by $\exp(4p^2\pi)$. Therefore $\nabla H \in L^p(X,\mu_X)$ for all p, and by 4.2 its redefinition belong to $L^p(\mu_{G_0})$.

We shall now desintegrate μ following the non degenerated map $x \longrightarrow x(2\pi)$. We obtain that for every continuous function v defined on $\mathbb{R}$, we have for any smooth function φ :

$$\int_X v(x(2\pi))\varphi(x)d\mu(x)=\int_{\mathbb{R}} v(\eta)dV(\eta)\int_X \varphi^*(x)\rho_\eta(dx).$$

Now use the fact that

$$(D_Z x)(2\pi)=\dot{z}(2\pi)=0.$$

Therefore denoting $\tilde{v}(x)=v(x(2\pi))$

$$D_Z(\tilde{v}\varphi)=\tilde{v}(D_Z\varphi).$$

We write the definition of the divergence of Z

$$\int D_Z(\tilde{v}\varphi)d\mu_X=\int \tilde{v}\varphi H d\mu_X$$

or

$$\int \tilde{v}(D_Z\varphi)d\mu_X=\int \tilde{v}\varphi H d\mu_X.$$

Now desintegrating μ_X along the map $x \longrightarrow x(2\pi)$ we get

$$\int v(\eta)dV(\eta)\int_X (D_Z\varphi)^* d\rho_\eta=\int v(\eta)dV(\eta)\int H^*\varphi d\rho_\eta$$

or

$$\int_X (D_Z\varphi)^* d\rho_\eta=\int H^*\varphi d\rho_\eta \qquad V(d\eta) \text{ a.e.}$$

The two members of this equality being continuous on the interior O of the support of V, this equality holds true on O. As $\eta=0$ belongs to O we obtain the proof of the theorem 3.1 *under the stronger hypothesis* 4.1.

5. Pinned brownian motion.

We *enlarge* the filtration of $x(\cdot)$ by adding $x(1)$. Then pinned brownian $x(\tau)$ is a semi martingale satisfying the stochastic differential equation :

$$dx_b(\theta)=db(\theta)-\frac{x}{2\pi-\theta}\, d\theta \qquad \theta\in[0,2\pi[$$

where db is a new brownian ;

$$H(x)=I_1(b)+I_2(b)$$

where :

$$I_1(b)=\int_0^{2\pi} \ddot{z}(g_b(\theta))e^{x_b(\theta)}\, db(\theta)$$

$$I_2(b)=-\int_0^{2\pi} \ddot{z}(g_b(\theta))e^{x_b(\theta)}\, \frac{x_b(\theta)}{2\pi-\theta}\, d\theta$$

5.1. Theorem. *Assume that* $z\in\mathcal{G}_0^2$. *Assume furthermore that* $\ddot{z}\in L^{2+\varepsilon}(\varepsilon>0)$. *Then there exist* $\varepsilon'>0$ *and* $K_z\in L^{1+\varepsilon'}$ *such that the conclusion of theorem* 3.1 *holds true.*

<u>Proof</u>.

$$E\left([I_1(b)]^{p'}\right)<c_{p'}\left[E\left(\left[\int[\ddot{z}(g_x(\theta))]^{p'}e^x d\theta\right]^p\right)\right]^{1/p}$$

$$\left[E\left([e^{(p'-1)x}]^q\right)\right]^{1/q}$$

where $p'>1$, p and q are conjugate exponents.

We remark that :

$$\int_0^{2\pi}|\ddot{z}(g_x(\theta)|^{p'}e^x d\theta=\int_0^{2\pi}|\ddot{z}(\lambda)|^{p'}d\lambda$$

$$E\left([I_2(b)]^{1+\varepsilon'}\right)\leq\left[E\left(\int[\ddot{z}(g_b(\theta))]^p\ e^{px}\right)\right]^{1/p}$$

$$\left[E\left(\int_0^{2\pi}\left[\frac{x_b(\theta)}{2\pi-\theta}\right]^q d\theta\right)\right]^{1/q}.$$

We evaluate the last integral by returning the time for the pinned brownian. We get

$$E(|x_b(\theta)|^q)=0([2\pi-\theta]^{q/2})$$

In order to get convergence we have to take $q<2$. We have therefore proved :

5.2. Lemma. *Under the assumption on theorem* 5.1., *there exist* $\varepsilon'>0$ *and* c *such that :*

$$E(|H_z(x_b)|^{1+\varepsilon'})\leq c\|\ddot{z}\|_{L^{2+\varepsilon}}.$$

5.3. Proof of the theorem 5.1.

Given $z\in\mathcal{G}_0^2$, $z\in L^{2+\varepsilon}$, we can find a sequence $z_n\in\mathcal{G}_0^3$ such that

$$\|z-z_n\|_{L^{2+\varepsilon}}\longrightarrow 0.$$

We have by the theorem 3.1

$$\left\{\frac{d}{dt}\int_{G_0}\varphi(\exp(tz_n)g)\mu_{G_0}(dg)\right\}_{t=0}=\int_{G_0}\varphi(g)K_{z_n}(g)\mu_{G_0}(dg).$$

When z_n converge to z the left handside converges to the corresponding expression for z.

By Lemma 5.2 K_{z_n} converges to K_z in $L^{1+\varepsilon}$. As we can limit ourself to *bounded* test functions the right hand side converges and this proves the theorem 5.1.

6. Measure on G.

Denote by r_φ the rotation of angle φ. Define a map

$s: \mathbb{T}\times G_0 \longrightarrow G$ by

$(\varphi, g_0) \longrightarrow r_\varphi \circ g_0.$

6.1. Lemma. s *is an homeomorphism.*

Proof. The relation

$$g(0)=r_\varphi(g_0(0))=r_\varphi(0)$$

determines uniquely φ by the relation

$$\varphi=-g(0).$$

6.2. Definition of μ_G.

We define

$$\mu_G=(s)_*(d\varphi\otimes d\mu_{G_0}).$$

This relation means that

$$\int_G u(g)\mu_G(dg)=\iint_{\mathbb{T}\times G_0} u(r_\varphi g_0)\frac{d\varphi}{2\pi}\,\mu_{G_0}(dg_0).$$

6.3. Lemma. *Denote by* z *a* C^1*-vector field on* S_1*, by* $U_{\cdot,z}$ *the one parameter subgroup of* G *defined by* z*, then we have*

$$(U_{\varepsilon,z}\circ r_\varphi)(\theta)=(r_\varphi\circ U_{\varepsilon,z^\varphi})(\theta)+\circ(\varepsilon)$$

with

$$z^\varphi(\theta)=z(\theta+\varphi).$$

Proof.

$$\left\{\left(\frac{d}{d\varepsilon}U_{\varepsilon,z}\circ r_\varphi\right)(\theta)\right\}_{\varepsilon=0}=z(\theta+\varphi).$$

Therefore

$$(U_{\varepsilon,z}\circ r_\varphi)(\theta)=\theta+\varphi+\varepsilon z(\theta+\varphi)+\circ(\varepsilon)$$

$$U_{\varepsilon,z^\varphi}(\theta)=\theta+\varepsilon z(\theta+\varphi)+\circ(\varepsilon).$$

This two identity proves the lemma.

6.4. Theorem. *The measure* μ_G *is quasi invariant under the left action of* $\mathcal{G}^3$ *and the modulus of quasi invariance belongs to all the* L^p*-space.*

<u>Proof</u>. We fix a smooth test function u and $z\in\mathcal{G}^3$. Then

$$I:=\int_G (D_z u)(g)\mu_G(dg)=\iint_{\mathbb{T}\times G_0} (D_z u)(r_\varphi\circ g)\,d\varphi\otimes\mu_{G_0}(dg_0)=\dot{I}(0)$$

with

$$I(\varepsilon)=\int_{\mathbb{T}} d\varphi\int_{G_0}\frac{d}{d\varepsilon}(u(r_\varphi \exp(\varepsilon z^\varphi)g_0))\mu_{G_0}(dg_0)$$

we can write

$$z^\varphi=\eta_\varphi+z(\varphi)\frac{\partial}{\partial r}$$

where $\eta_\varphi(0)=0$ and with $\frac{\partial}{\partial r}$ is the constant vector field on S_1 which generates the rotation. Then we shall write

$$I(\varepsilon)=I_1(\varepsilon)+I_2(\varepsilon)$$

where

$$I_1(\varepsilon)=\int_{\mathbb{T}} d\varphi\,\frac{d}{d\varepsilon}\int_{G_0} u^\varphi(\exp(\varepsilon\eta_\varphi)g_0)\mu_{G_0}(dg_0)$$

with

$$u^\varphi(g)=u(r_\varphi g).$$

As $\eta_\varphi\in\mathcal{G}_0^3$ we can apply the quasi invariance on G_0 and we get :

$$\dot{I}_1(0)=\int_{\mathbb{T}} d\varphi\int_{G_0} u^\varphi(g_0)\;K_{\eta_\varphi}(g_0)\;\mu_{G_0}(dg_0)$$

$$=\iint u(r_\varphi g_0)K_{\eta_\varphi}(g_0)\,d\varphi\;\mu_{G_0}(dg_0)$$

which proves the quasi invariance as far as I_1 is concerned. The L^p-norm of the modulus of quasi invariance is given by

$$\iint |K_{\eta_\varphi}(g_0)|^p d\varphi\;\mu_{G_0}(dg)<+\infty.$$

We have now to evaluate the contribution on I_2

$$\dot{I}_2(\varepsilon)=\int_{\mathbb{T}} d\varphi\,\frac{d}{d\varepsilon}\int_{G_0} u(r_\varphi\exp(\varepsilon z(\varphi)\frac{\partial}{\partial r})g_0)\mu_{G_0}(dg).$$

We remark now that

$$r_\varphi\exp(\varepsilon\frac{\partial}{\partial r})=\exp(\varepsilon\frac{\partial}{\partial r})r_\varphi.$$

Therefore

$$I_2(0)=\int_{\mathbb{T}} z(\varphi)\,d\varphi\,\frac{\partial}{\partial\varphi}\left[\int_{G_0} u(r_\varphi g_0)\;\mu_{G_0}(d\varphi)\right].$$

We make now an integration by part on $\mathbb{T}$

$$I_2=-\iint_{G_0\times\mathbb{T}} u(r_\varphi g_0)\dot{z}(\varphi)\mu_{G_0}(d\varphi)\,d\varphi$$

which proves the theorem.
The modulus of quasi invariance $\tilde{K}_z$ is given by :

$$\tilde{K}_z(r_\varphi g_0)=K_{\eta_\varphi}(g_0)-\dot{z}(\varphi).$$

7. A unitary representation of the Virasoro algebra.

7.1. A symmetric operator.

The Virasoro algebra V_c is a central extension of $\mathfrak{F}^\infty$. Therefore any representation of $\mathfrak{F}^\infty$ can be lifted to a representation of V_c. Furthermore $\mathfrak{F}^\infty$ can be looked upon as V_0.

Denote by $K_z(g)$ the modulus of quasi-invariance of z operating on μ_G from the left. Now define on smooth functions φ on G the operator

$$A_z\varphi=D_z\varphi-\frac{1}{2}K_z\varphi.$$

Then

$$\int(A_z\varphi)\psi\, d\mu+\int\varphi A_z\psi$$
$$=\int D_z(\varphi\psi)\,d\mu-\int K_z\varphi\psi d\mu=0$$

which means that $\sqrt{-1}\, A_z$ is a *symmetric* operator in $L^2(G,\mu_G)$.

7.2. Theorem. *The product* $A_{z_1}A_{z_2}$ *is well defined. Furthermore*

$$A_{z_1}A_{z_2}-A_{z_2}A_{z_1}=A_{[z_1,z_2]}.$$

<u>Proof</u>. We have shown that $\tilde{K}_{z_1}$ is a function which is differentiable on the Wiener space and therefore differentiable relatively to $\mathfrak{F}^\infty$. Therefore $D_{z_1}\tilde{K}_{z_2}$ exists. Furthermore

$$A_{z_1}A_{z_2}=\left(D_{z_1}-\frac{1}{2}\tilde{K}_{z_1}\right)\left(D_{z_2}-\frac{1}{2}\tilde{K}_{z_2}\right)$$
$$=D_{z_1}D_{z_2}-\frac{1}{2}\tilde{K}_{z_1}D_{z_2}-\frac{1}{2}\left(D_{z_1}\tilde{K}_{z_2}\right).$$
$$-\frac{1}{2}\tilde{K}_{z_2}D_{z_1}+\frac{1}{4}\tilde{K}_{z_1}\tilde{K}_{z_2}$$

exists because $\tilde{K}_{z_1}\tilde{K}_{z_2}$ belongs to L^p by Hölder inequality.

Let us compute

$$I=\left\{\frac{\partial}{\partial\varepsilon_1}\left[\frac{\partial}{\partial\varepsilon_2}\int_G u(\exp(\varepsilon_1 z_1)\exp(\varepsilon_2 z_2)g)\mu_G(dg)\right]\right\}_{\varepsilon_1=\varepsilon_2=0}.$$

Denote

$$u_{\varepsilon_1}(g)=u(\exp(\varepsilon_1 z_1)g).$$

Then

$$I=\left\{\frac{\partial}{\partial\varepsilon_1}\left[\int u(\exp(\varepsilon_1 z_1)g)\ \tilde{K}_{z_2}(\exp(\varepsilon_1 z_1)g)\mu_G(dg)\right]\right\}_{\varepsilon=0}$$

$$=\int u(g)\left[\tilde{K}_{z_1}(g)\tilde{K}_{z_2}(g)+D_{z_1}\tilde{K}_{z_2}(g)\right]dg$$

$$\left\{\frac{\partial}{\partial\eta}\int_G u(\exp(\eta[z_1,z_2])g)\mu_G(dg)\right\}_{\eta=0}=$$

$$\int u(D_{z_1}\tilde{K}_{z_2}-D_{z_2}\tilde{K}_{z_1})\mu_G(dg)$$

which implies

$$\tilde{K}_{[z_1,z_2]}=D_{z_1}\tilde{K}_{z_2}-D_{z_2}\tilde{K}_{z_1}.$$

Finally as :

$$A_{[z_1,z_2]}=D_{z_1}D_{z_2}-D_{z_2}D_{z_1}-\frac{1}{2}\tilde{K}_{[z_1,z_2]},$$

we get the theorem.

Remark. We have get in the preceding construction essentially a representation of V_0.
In order that the central charge c will appear effectively, we will have to construct a *central* extension of G along the fundamental cocycle of $\mathfrak{g}^{\infty}$. This construction does not seem impossible to realize. Some other developments can be found in [3].

BIBLIOGRAPHY

[1] **H. AIRAULT** et **P. MALLIAVIN**. Intégration géométrique sur l'espace de Wiener, Bull. Sc. Math., 112, 1988, 3-52.

[2] **M.P. MALLIAVIN** et **P. MALLIAVIN**. Quasi invariant integration on loop group, J. of Funct. Analysis, 1990, 93, 207-236.

[3] **M.P. MALLIAVIN** et **P. MALLIAVIN**. Mesures quasi invariantes sur certains groupes de dimension infinie, Note aux C.R. Acad. Sc. Paris, octobre 1990.

The Chiral Potts Model: from Physics to Mathematics and back

Barry M. McCoy

Institute for Theoretical Physics
State University of New York at Stony Brook
Stony Brook, NY 11794-3840

Abstract

We review the physics and mathematics of the chiral Potts spin chain with particular emphasis on computation of the energy levels by means of the solution of functional equations. We discuss in detail the relation of the integrable to the superintegrable case and the construction of the ground state in the massless level crossing regime.

1 Introduction

In 1983 Howes, Kadanoff and den Nijs [1] initiated the study of the quantum chiral Potts chain by studying a special case ($N = 3, \phi = \bar{\phi}$) of the following one dimensional spin chain

$$\mathcal{H} = -\sum_{j=1}^{L}\sum_{n=1}^{N-1}\{\bar{\alpha}_n(X_j)^n + \alpha_n(Z_jZ_{j+1}^{\dagger})^n\} \tag{1.1}$$

where

$$X_j = I_N \otimes \ldots X^{j^{th}} \otimes \ldots \otimes I_N \tag{1.2a}$$

$$Z_j = I_N \otimes \ldots \otimes Z^{j^{th}} \otimes \ldots \otimes I_N \tag{1.2b}$$

I_N is the $N \times N$ identity matrix, the $N \times N$ matrices Z and X are

$$Z = \begin{pmatrix} 1 & 0 & & \\ 0 & \omega & & \\ & & \ddots & \\ & & & \omega^{N-1} \end{pmatrix}, X = \begin{pmatrix} 0 & 0 & \cdots & 0 & 1 \\ 1 & 0 & \cdots & 0 & 0 \\ \cdot & \cdot & \cdots & \cdot & \cdot \\ 0 & 0 & & 1 & 0 \end{pmatrix}, \tag{1.3}$$

$\omega = e^{2\pi i/N}$ and

$$\alpha_n = \frac{\exp[i(2n-N)\phi/N]}{\sin(\pi n/N)}, \quad \bar{\alpha}_n = k'\frac{\exp[i(2n-N)\bar{\phi}/N]}{\sin(\pi n/N)}. \tag{1.4}$$

This Hamiltonian has several properties to be noted:
1) $\mathcal{H}$ is hermitian for $\phi, \bar{\phi}$, and k' real;

2) $\mathcal{H}$ commutes with the spin rotation operator

$$R = \prod_{j=1}^{L} X_j \, ;$$

3) $\mathcal{H}$ is translationally invariant in space;
4) Even when $\mathcal{H}$ is hermitian the matrix elements are real only for $\phi = \bar{\phi} = 0$.

Property 1 guarantees that the eigenvalues are real. Thus $\mathcal{H}$ can be interpreted as an energy operator.

Property 2 is called Z_N invariance and guarantees that the eigenvalues of R (which are $e^{2\pi i Q/N}$ for $Q = 0, 1, \cdots N-1$) are good quantum numbers. These Z_N invariant spin chains have come to be called Potts models.

Property 3 guarantees that the eigenvalues of the shift operator (e^{iP} for $P = \frac{2\pi}{L} l, l = 0, 1, \cdots L-1$) are good quantum numbers. P is the momentum of the state.

These 3 properties are generally shared with most systems of physical interest. The special interest in the chiral Potts model lies in property 4, the complexity of the matrix elements. This property of complex matrix elements leads to 3 important distinctions from the real case $\phi = \bar{\phi} = 0$.
1) The system is not time reversal invariant.
2) For $\phi \neq 0$ the system is not parity invariant. This feature is the reason the model is called chiral.
3) The Perron-Frobenius theorem cannot be invoked and thus ground state level crossing may occur.

It might be asked why there should be any interest in thermodynamic systems which violate the discrete symmetries of parity and time reversal since for systems of macroscopic interest the forces which violate P and T are small and can be ignored. The answer is that at least in the real world of 3 dimensions with a finite density of electrons and protons the discrete symmetries of the Hamiltonian can be strongly violated in the ground state. For example organic molecules such as DNA all have left handed helical twists. This produces the phenomena that right and left circularly polarized light propagate at different speeds. In biological systems there is clearly a spontaneous breakdown of parity. One proposed mechanism for high T_c, the anyon mechanism, also breaks P and T. It is thus most useful to study a model when these breaking efforts can be studied in detail.

Howes, Kadanoff and den Nijs studied (1.1) for $N = 3$ and $\phi = \bar{\phi}$ by a variety of series and other approximate techniques and in the course of the study found some very remarkable properties at the point

$$\phi = \bar{\phi} = \pi/2 \, . \tag{1.5}$$

For example they did a series expansion for $< Z_0 >$ and found for $0 < k' < 1$ that to order k'^{13}

$$< Z_0 >= (1 - k'^2)^{1/9} \, . \tag{1.6}$$

They also computed various mass gaps to many orders in k' and found that the gaps seemed to be exactly $1 - k'$.

The special nature of the point $\phi = \bar{\phi} = \pi/2$ was further revealed 2 years later in the paper of von Gehlen and Rittenberg.[2] They wrote (1.1) in the form

$$\mathcal{H} = A_0 + k' A_1 \tag{1.7}$$

and showed for all N that A_0 and A_1 enjoy the remarkable property that

$$[A_0,[A_0,[A_0,A_1]]] = \quad \text{const } [A_0,A_1] \tag{1.8a}$$

$$[A_1,[A_1,[A_1,A_0]]] = \quad \text{const } [A_0,A_1] \, . \tag{1.8b}$$

This condition is sufficient to guarantee that A_0 and A_1 generate a whole sequence of operators A_k and G_k which satisfy the algebra first used by Onsager [3] to solve the two dimensional Ising model [which corresponds here to $N = 2$]

$$[A_\ell, A_m] = 4G_{\ell-m} \tag{1.9a}$$

$$[G_\ell, A_m] = 2A_{m+\ell} - 2A_{m-\ell} \tag{1.9b}$$

$$[G_\ell, G_m] = 0 \, . \tag{1.9c}$$

Indeed, it was the desire to satisfy (1.9) that lead von Gehlen and Rittenberg to discover (1.1)-(1.4) with $\phi = \bar{\phi} = \pi/2$. This was the first example since Onsager's original work of a representation of the algebra (1.9). It is somewhat surprising that it took over 40 for this second example to be found. Systems which obey this algebra we have now designated as superintegrable.

But the full understanding of the integrability properties of (1.1) started with the discovery in 1987 of Au-Yang, McCoy, Perk, Tang, and Yan [4] that for $N = 3$ if

$$\cos\phi = k' \cos\bar{\phi}$$

then $\mathcal{H}$ can be obtained from a two dimensional classical statistical mechanical model whose transfer matrix T depends on a spectral variable u such that

$$[T(u), T(u')] = 0 \tag{1.10}$$

and as $\quad u \to 0$

$$T(u) \to 1(1 + \text{ const } u) + u\mathcal{H} + 0(u^2) \, . \tag{1.11}$$

The complex variable u parameterizes a curve in the space of Boltzman weights and this curve has genus higher than 1. This is the first time such a curve of genus greater than 1 appeared in the theory of solvable models. This work was soon extended [5,6] to $N = 4, 5, k' = 1$ and a very simple product form for the Boltzman weights was found. The case of general N and k' was solved by Baxter, Perk, and Au-Yang [7,8] and the following form obtained

$$\frac{W^h_{p,q}(n)}{W^h_{p,q}(0)} = \prod_{j=1}^{n} \left(\frac{d_p b_q - a_p c_q \omega^j}{b_p d_q - c_p a_q \omega^j} \right) \tag{1.12a}$$

$$\frac{W^v_{p,q}(n)}{W^v_{p,q}(0)} = \prod_{j=1}^{n} \left(\frac{\omega a_p d_q - d_p a_q \omega^j}{c_p b_q - b_p c_q \omega^j} \right) \tag{1.13a}$$

where for both p and q

$$a^N + k' b^N = k d^N, \quad k' a^N + b^N = k c^N \tag{1.14}$$

with

$$k^2 + k'^2 = 1 \,. \tag{1.15}$$

These Boltzman weights satisfy the star-triangle equation

$$\sum_{d=1}^{N} W^v_{q,r}(b-d) W^h_{p,r}(a-d) W^v_{p,q}(d-c) = R_{pqr} W^h_{p,q}(a-b) W^v_{p,r}(b-c) W^h_{q,r}(a-c) \tag{1.16}$$

where

$$R_{pqr} = \frac{f_{pq} f_{qr}}{f_{pr}}$$

with

$$f_{pq} = \left[\frac{\prod_{m=1}^{N} \sum_{n=1}^{N} \omega^{mn} W^v_{p,q}(n)}{\prod_{m=1}^{N} W^h_{pq}(m)} \right]^{1/N} \,. \tag{1.17}$$

The transfer matrix $T_{p,q} \,|_{\{\ell\},\{\ell'\}}$ is

$$T_{p,q} \,|_{\{\ell\},\{\ell'\}} = \prod_{j=1}^{L} W^v_{p,q}(l_j - l'_j) W^h_{p,q}(l_j - l'_{j+1}) \tag{1.18}$$

and the parameters ϕ and $\bar{\phi}$ of (1.1) are given as

$$e^{2i\phi/N} = \omega^{1/2} \frac{a_p c_p}{b_p d_p}, \qquad e^{2i\bar{\phi}/N} = \omega^{1/2} \frac{a_p d_p}{b_p c_p} \,. \tag{1.19}$$

We also note that the eigenvalues P and Q are also obtained from $T_{p,q}$ as

$$e^{-iP} = \lim_{q \to Rp} T_{p,q} \tag{1.20}$$

where R is the automorphism

$$R(a,b,c,d) = (b, \omega a, d, c) \tag{1.21}$$

and

$$e^{-2\pi i Q/N} = \omega^L \lim_{p \to q} \left(\frac{d_p b_q - a_p c_q \omega}{b_p d_q - c_p a_q \omega} \right)^L \left(\frac{a_p d_q - d_p a_q}{c_p b_q - b_p c_q \omega} \right)^L T_{p,Tq} \tag{1.22}$$

where T is the automorphism

$$T(a,b,c,d) = (\omega a, \omega^{-1} b, c, d) \,. \tag{1.23}$$

We note too that there is a manifold where the Boltzmann weights W^h and W^v are real and positive, namely.

$$a_p^* c_p = \omega^{1/2} b_p^* d_p \tag{1.24a}$$

$$| a_p | = | d_p | \tag{1.24b}$$

$$| b_p | = | c_p | \tag{1.24c}$$

On this manifold the Perron-Frobenius theorem guarantees that the maximal eigenvalue of T is non degenerate but on this manifold $\mathcal{H}$ fails to be Hermitian.

2 Functional Equation

Unfortunately, a solution to the star triangle equation by itself is not sufficient to solve for the eigenvalues of $\mathcal{H}$. To do this one needs a functional equation (or equations) for T. For $N = 3$ such a equation was first discovered by Albertini, McCoy and Perk, [9,10]

$$T_{p,q}T_{p,Rq}T_{p,R^2q} = e^{-iP}\{f^L_{p,Rq}f^L_{Rq,p}T_{p,q}$$

$$+f^L_{p,q}f^L_{q,p}T_{p,R^2q} + f^L_{p,R^6q}f^L_{R^2q,p}T_{p,R^4q}\} \tag{2.1}$$

who used it to solve for all the eigenvalues of the superintegral case $\phi = \bar{\phi} = \pi/2$. But to derive it and to generalize it proves much more desirable to derive a more powerful set of relations due to Bazhanov and Stroganov [11] and Baxter, Bazhanov and Perk. [12]

To present these recursion relations we first note that the original result can be put in a more useful form if we define new variables

$$x = a/d,\ y = b/c,\ \mu = d/c \tag{2.2}$$

where from (1.14)

$$k(x^N y^N + 1) = x^N + y^N\ , \tag{2.3}$$

and, defining $\lambda = \mu^N$,

$$kx^N = 1 - k'\lambda^{-1},\ ky^N = 1 - k'\lambda\ . \tag{2.4}$$

Then the weights become

$$\frac{W^h_{p,q}(n)}{W^h_{p,q}(0)} = (\mu_p/\mu_q)^n \prod_{j=1}^{n} \frac{y_q - \omega^j x_p}{y_p - \omega^j x_q} \tag{2.5a}$$

$$\frac{W^v_{p,q}(n)}{W^v_{p,q}(0)} = (\mu_p\mu_q)^n \prod_{k=1}^{n} \frac{\omega x_p - \omega^j x_q}{y_q - \omega^j y_p} \tag{2.5b}$$

Note that the curve (2.3) has genus $(N-1)^2$ for $k \neq 0$.

It also will prove most useful to define a variable

$$t = xy \tag{2.6}$$

and to note that t and λ are related by

$$t^N = [1 - k'(\lambda + \lambda^{-1}) + k'^2]/k^2 . \tag{2.7}$$

This defines a hyperelliptic Riemann surface of genus $N-1$.

We can now state the recursion relations of refs. 11 and 12 and in particular use the recent formulation of Baxter. [13] First we change the normalization of $T(x_q, y_q)$ by defining

$$V(x_q, y_q) = T(x_q, y_q)/(f_{p,q})^L . \tag{2.8}$$

Then, using the facts

$$V(\omega x_q, \omega^{-1} y_q) = R^{-1} V(x_q, y_q) , \tag{2.9}$$

that R commutes with T (and hence V), and that λ and t are invariant under $x, y \rightarrow \omega x, \omega^{-1} y$ we may write $V(x_q, y_q)$ in the factorized form

$$V(x_q, y_q) = y_q^Q V(t_q, \lambda_q) . \tag{2.10}$$

We also define a function of λ_q

$$\alpha(\lambda_q) = \{k'(1 - \lambda_p \lambda_q)^2/[k^2 y_p^{2N} \lambda_q]\}^L \tag{2.11}$$

and a function of t

$$z(t) = [\omega \lambda_p^{2/N} (t_p - t)^2 / y_p^4]^L . \tag{2.12}$$

The eigenvalues of $V(t_q, \lambda_q)$ which are meromorphic functions on the t Riemann surface are now determined in terms of auxillary polynomials $\tau_j(t)$ of degree $(j-1)L$ of the single complex variable t as follows

I. The $\tau_j(t)$ satisfy the set of functional equations

$$\tau_j(t)\tau_2(\omega^{j-1}t) = z(\omega^{j-1}t)\omega^Q \tau_{j-1}(t) + \tau_{j+1}(t) \quad j = 1, \ldots N \tag{2.13a}$$

$$\tau_{N+1} \equiv z(t)\omega^Q \tau_{N-1}(\omega t) + \alpha(\lambda) + \alpha(\lambda^{-1}) . \tag{2.13b}$$

II. From $\tau_j(t)$ we compute a set of meromorphic functions on the t, λ Riemann surface $r_j(t, \lambda)$ as

$$r_j(t,\lambda) = \alpha(\lambda)\omega^{-Qj}\tau_j(t) + \tau_{N-j}(\omega^j t) \prod_{\ell=0}^{j-1} z(\omega^\ell t) \tag{2.14}$$

III. From $\tau_N(t)$ we determine a polynomial of λ $S(\lambda)$ of degree $(N-1)L$ from

$$S(\lambda)S(\lambda^{-1}) = \tau_N(t)\tau_N(\omega t) \cdots \tau_N(\omega^{N-1}t) \tag{2.15}$$

IV. The eigenvalue of $V(t, \lambda)$ is then determined as

$$V(t,\lambda)^N = e^{i\pi L(N-1)(N+4)/12} \lambda^{-(N-1)L/2} \alpha(\lambda)^{-N} r_1(t,\lambda) \cdots r_N(t,\lambda)/S(\lambda^{-1}) \tag{2.16}$$

V. We then can use (1.11) to find the eigenvalues of $\mathcal{H}$ as

$$E = 2Nkt_p^{N/2}\lambda_p \frac{\partial}{\partial\lambda} \ln y^Q V(t,\lambda)\,|_{\lambda=\lambda_p} + \text{const} \tag{2.17}$$

when the const is the same for all eigenvalues and ϕ of (1.1) is related to λ_p by

$$e^{2i\phi} = -\left(\frac{1-k'\lambda_p^{-1}}{1-k'\lambda_p}\right). \tag{2.18}$$

We note that the superintegrable point $\phi = \pi/2$ thus corresponds to

$$\lambda_p = 1, \quad t_p^N = (1-k')/(1+k') \tag{2.19}$$

In this case t_p coincides with a branch point of the curve (2.7). On the hermitian manifold $|\lambda_p| = 1$ and t_p^N is real with

$$(1-k')/(1+k') < t_p^N < (1+k')/(1-k')\,. \tag{2.20}$$

On the real manifold (1.24) $|t_p| = 1$ and the maximum eigenvalue of $V(t,\lambda)$ was explicitly found by Baxter using this formalism. No level crossing can occur on the manifold.

3 Solution of Functional Equations

We first consider the superintegrable case $\phi = \bar{\phi} = \pi/2$. Then if we start from the functional equation (2.1) we find that the eigenvalues have the remarkable property that as meromorphic functions on the Riemann surface (1.14) they all factorize as (in terms of a, b, c, d)

$$\begin{aligned} T_{pq} = {} & \frac{N^L(\eta\frac{a}{d}-1)^L}{[(\eta\frac{a}{d})^N-1]^L}(\eta\frac{a}{d})^{Pa}(\eta\frac{b}{c})^{Pb}(\frac{c^N}{d^N})^{Pc} \\ & \prod_{\ell=1}^{mp}\left(\frac{1-\omega^{-1/2}\bar{v}_\ell\eta^2\frac{ab}{cd}}{1-\omega^{-1/2}\bar{v}_\ell}\right)\prod_{\ell=1}^{m_B}\left(\frac{1+k'}{1-k'}\right)^{1/2}\left\{\frac{a^N+b^N}{2d^N} \pm w_\ell\frac{(a^N-b^N)}{(1+k')d^N}\right\} \end{aligned} \tag{3.1}$$

where

$$\eta = [(1+k')/(1-k')]^{1/2N} \tag{3.2}$$

The w_ℓ, and $\bar{v}_\ell$ are determined by using this factorized form in (2.1). In particular [10] if $m_p = 0$ and $N = 3$

$$E = -(1+k')\left[F\left(-\frac{1}{2},\frac{1}{3};1;\frac{4k'}{(1+k')^2}\right) + F\left(-\frac{1}{2},\frac{2}{3};1;\frac{4k'}{(1+k')^2}\right)\right] \tag{3.3}$$

where $F(a,b;c;z)$ is the hypergeometric function. More generally for $N \geq 3$ the eigenvalue for $m_p = 0$ is found to be [14,10].

$$E = -(1+k')\sum_{\ell=1}^{N} F\left(-\frac{1}{2},\frac{\ell}{N};1;\frac{4k'}{(1+k')^2}\right) \tag{3.4}$$

This is the ground state eigenvalue for small k' and only fails to be analytic at $k' = 1$. However when k' is sufficiently close to one the ground state shifts to a value of $m_p \neq 0$. This phenomenon was extensively studied in refs. 9 and 14.

Our intention here is to present the recent extension of these results to the general case. The ground state in the real manifold was recently treated by Baxter [13]. We here discuss the hermitian manifold and will be most interested in the excitations and level crossing [15]. A most important distinction between the superintegrable and be general case is that in the former the factorization property (3.1) is exact for finite lattices whereas in the general case the computations of refs. 13 and 15 are only in the $L \to \infty$ limit.

In ref. 15 we explicitly solved (2.13) for $N = 3$. Here we generalize that solution to arbitrary N.

We first make a Wiener-Hopf splitting of $\alpha(\lambda^{\pm 1})$ as

$$\prod_{n=0}^{N-1} h^{\pm}(\omega^n t) = \alpha(\lambda^{\pm 1}) \tag{3.5}$$

where

$$\begin{aligned} h^+(t) &= [(t_p - t)/y_p^2]^L \bar{h}^+(t) \\ &= [(t_p - t)/y_p^2]^L \exp \frac{L}{2\pi i} \int_{ct} dt' \frac{\ln[\frac{1-\lambda'\lambda_p}{1-\lambda'\lambda_p^{-1}}]}{t'-t} \end{aligned} \tag{3.6}$$

and

$$\begin{aligned} h^-(t) &= \lambda_p^{2L/3} [\omega(t_p - t)/y_p^2]^L \bar{h}(t) \\ &= \lambda_p^{2L/N} [\omega(t_p - t)/y_p^2]^L \exp \frac{-L}{2\pi i} \int_{ct} dt' \frac{\ln[\frac{1-\lambda'\lambda_p}{1-\lambda'\lambda_p^{-1}}]}{t'-t} \,. \end{aligned} \tag{3.7}$$

The contour c_t is clockwise about the branch cut in $\lambda(t)$ that runs from $[(1-k')/(1+k')]^{1/N}$ to $[(1+k')/(1-k')]^{1/N}$ and the point t lies outside the contour.

We note that

$$h^+(t) h^-(t) = z(t) \tag{3.8}$$

and that if $\quad \lambda_p = 1 \quad$ then $\quad \bar{h}^+(t) = \bar{h}^-(t) = 1\,. \qquad (3.9)$

It is now easy to verify that the following is a solution of (2.13)

$$\tau_2(t) = h^+(t) \frac{f(\omega^{\gamma-1}t)}{f(\omega^\gamma t)} + \omega^Q h^-(\omega t) \frac{f(\omega^{\gamma+1}t)}{f(\omega^\gamma t)} \tag{3.10}$$

where $\gamma = (N+1)/2$ and, more generally

$$\begin{aligned} \tau_j(t) &= f(\omega^{\gamma-1}t) f(\omega^{\gamma-1+j}t) \\ &\sum_{\ell=0}^{j-1} \frac{\omega^{\ell Q} \prod_{n=1}^{\ell} h^-(\omega^{j-n}t) \prod_{n=0}^{j-2-\ell} h^+(\omega^n t)}{f(\omega^{\gamma-1+j-\ell}t) f(\omega^{\gamma-2+j-\ell}t)} \end{aligned} \tag{3.11}$$

where $f(t)$ is an arbitrary function still to be determined. We then find from (2.14) that

$$r_j(t,\lambda) = \omega^{-jQ}\Big(\prod_{n=0}^{j-1} h^+(\omega^n t)\Big)\, \frac{f(\omega^{\gamma-1}t)}{f(\omega^{\gamma+j-1}t)}\, \tau_N(\omega^j t) \tag{3.12}$$

and from (2.15)

$$S(\lambda)S(\lambda^{-1}) = \prod_{k=0}^{N-1} f^2(\omega^{\gamma+k}t)$$
$$\prod_{k=0}^{N-1}\sum_{\ell=0}^{N-1} \frac{\omega^{\ell Q}\prod_{n=1}^{\ell} h^-(\omega^{k-n}t)\,\prod_{n=0}^{N-2-\ell} h^+(\omega^{k+n}t)}{f(\omega^{\gamma-1-\ell+k}t)f(\omega^{\gamma-2-\ell+k}t)} \tag{3.13}$$

We still have to specify $f(t)$. In analogy with the superintegrable case let us write

$$f(t) = \prod_{\ell=1}^{m_p}(tv_\ell + 1) \tag{3.14}$$

Then since $\tau_2(t)$ is to be a polynomial in t the expression (3.10) for $\tau_2(t)$ cannot have any poles and hence we find the restriction on v_ℓ of

$$\prod_{\ell=1}^{m_p}\left(\frac{1-\omega^{-1}v_\ell/v_k}{1-\omega v_\ell/v_k}\right) = -\omega^Q\,\frac{h^-(\omega^{1/2}/v_k)}{h^+(\omega^{-1/2}/v_k)}\,. \tag{3.15a}$$

In the superintegrable case where $\lambda_p = 1$ and $h^{\pm}(t)$ become polynomials in t this reduces to the Bethe's hypothesis equation [16]

$$\prod_{\ell=1}^{m_p}\left(\frac{1-\omega^{-1}v_\ell/v_k}{1-\omega v_\ell/v_k}\right) = -\omega^{Q+L}\left(\frac{t_p-\omega^{1/2}/v_k}{t_p-\omega^{-1/2}/v_k}\right)^L\,. \tag{3.15b}$$

For $N = 3$ this has been previously obtained in ref. 10 and 14.

We must now solve the equation for $S(\lambda)S(\lambda^{-1})$ to find $S(\lambda)$ itself. For $f(t)$ of the form (3.14) we see that $S(\lambda)$ has two sorts of zeroes:

1) $\lambda^{m_p}\prod_{n=0}^{N-1} f(\omega^{\gamma+n}t)$ is a polynomial of order $2m_p$ in λ. It has zeroes at positions $\lambda_\ell^{\pm 1}$ on both sheets of λ.

2) The remaining zeroes may be shared between the sheets of λ in an arbitrary fashion. The minimum eigenvalues of $\mathcal{H}$ occurs when all $|\lambda_\ell| > 1$. These are very conveniently represented as a contour integral. In this form we may let $L \to \infty$ and obtain

$$S(\lambda)/S(0) = \lambda^{m_p}\prod_{n=0}^{N-1} f(\omega^{\gamma+n}t)$$
$$\exp\frac{1}{2\pi i}\int_{c_\lambda}\frac{d\lambda'}{\lambda'-\lambda^{-1}}\ln\left\{\frac{\prod_{\ell=1}^{N-1}(t'_p-\omega^\ell t')^{NL}(\bar{h}^-(\omega t'))^{N-2}\prod_{\ell=2}^{N-1}\bar{h}^+(\omega^\ell t')}{f^N(-\omega^{1/2}t')f^N(-\omega^{-1/2}t')}\right\} \tag{3.16}$$

where c_λ is the contour which is clockwise about the branch cut in $t(\lambda)$ from $\lambda = 0$ to $\lambda = k'$. The point λ is outside the contour.

We thus may use (3.16) and (3.12) in (2.16) and obtain the desired result that as $L \to \infty$

$$V^N(t,\lambda) = \omega^{-\frac{N(N-1)}{2}Q}\, e^{i\pi L(N-1)(N+4)/12} \lambda^{-(N-1)L/2} \prod_{\ell=1}^{N-1} (h^+(\omega^\ell t)^{-\ell}) S(0) f^N(-\omega^{-1/2}t)\lambda^{m_p}$$
$$\exp \frac{1}{2\pi i}\int_{c_\lambda} \frac{d\lambda'}{\lambda' - \lambda^{-1}} \ln\left\{ \frac{\prod_{\ell=1}^{N-1}(t_p - \omega^\ell t')^{NL}(\bar{h}^-(\omega t'))^{N-2} \prod_{\ell=2}^{N-1} \bar{h}^+(\omega^\ell t')}{f^N(-\omega^{1/2}t') f^N(-\omega^{-1/2}t')} \right\} \tag{3.17}$$

where $S(0)$ is determined by the requirement that

$$T(x_p, y_p) = 1 \tag{3.18}$$

for all eigenvalues.

When $m_p = 0$ and hence $f(t)$ is a constant the result (3.17) was obtained by Baxter [13] and can be written in a double integral. When $\lambda_p \to 1$ this double integral reduces to the single integral result (3.4).

We focus our attention here on the case $m_p \geq 1$ and $\Delta Q = 0$. Then we find for m_p finite as $L \to \infty$.

$$E_{mp} - E_0 = \sum_{\ell=1}^{mp} \Delta E(v_\ell; k') \tag{3.19}$$

where

$$\begin{aligned}\Delta E(v;k') = {} & 2(1-k')\bar{t}_p^{N/2} \\ & \pm 2\bar{v}\bar{t}_p^{-\frac{N}{2}+1} \sin\frac{\pi}{N} \frac{\{(\bar{t}_p^N - 1)[(1+k')^2 - (1-k')^2\bar{t}_p^N]\}^{1/2}}{(\omega^{1/2}\bar{t}_p\bar{v} - 1)(\omega^{-1/2}\bar{t}_p\bar{v} - 1)} \\ & + \frac{N\bar{v}\bar{t}_p^{N/2} P}{\pi} \int_1^{(\frac{1+k'}{1-k'})^{2/N}} d\bar{t} \left[\frac{\omega^{1/2}}{\omega^{1/2}\bar{t}\bar{v} - 1} + \frac{\omega^{-1/2}}{\omega^{-1/2}\bar{t}\bar{v} - 1} \right] \\ & \frac{\{(\bar{t}^N - 1)[(1+k')^2 - (1-k')^2\bar{t}^N]\}^{1/2}}{\bar{t}^N - \bar{t}_p^N}\end{aligned} \tag{3.20}$$

with $$t = \left(\frac{1-k'}{1+k'}\right)^{1/N} \bar{t}, \text{ and } \bar{v} = \left(\frac{1-k'}{1+k'}\right)^{1/N} v\,. \tag{3.21}$$

The $\pm$ sign is $+$ if $0 < \varphi \leq \pi/2$, $-$ if $\frac{\pi}{2} < \phi < \pi$ and P indicates a principle value. The integral in (3.20) is a complete Abelian integral of the 3rd kind over the hyperelliptic curve (2.7) which was derived from the Boltzmann weights. In the $N = 3$ superintegrable case ($\bar{t}_p = 1$) (3.20) reduces to (3.14) of ref. 10.

4 Construction of the Ground State

In discussing the energy levels of $\mathcal{H}$ the first property of interest is the ground state energy. As discussed in ref. 13 on the real manifold where the Perron-Frobenius theorem holds the ground state has $m_p = 0$. Similarly for the hermitian manifold if k' is small the ground state is also expected to have $m_p = 0$. Thus, as long as there are no allowed values of v for which $\Delta E(v;k') < 0$ we can regard E_0 as the ground state energy and ΔE as excitations above this ground state.

If k' is sufficiently small we do indeed see from (3.20) that $\Delta E(v;k')$ is positive for all v. However, at $k' = 1$ it is obvious that $\Delta E(v;k') = 0$ for $v = 0$. A numerical study of (3.20) reveals that there is an entire domain in the k', ϕ plane where there is a region on the real v axis in which $\Delta E(v;k')$ is negative. This is a level crossing phenomena and can occur because of property 4 of sec. 1.

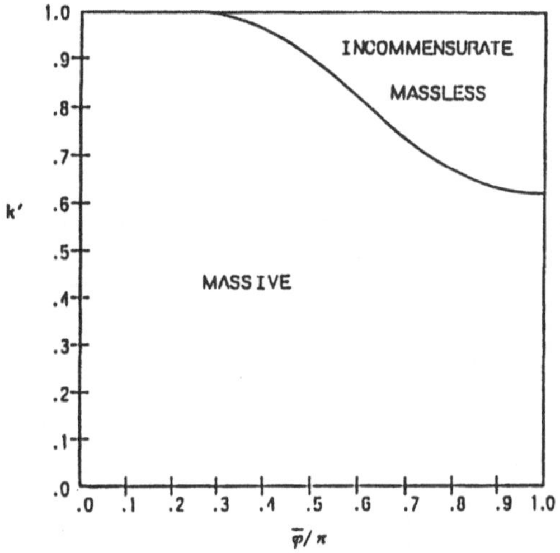

Caption: The phase diagram of the $N = 3$ chiral Potts model in terms of the variables k' and $\bar{\phi}$. The value of k' at the phase boundary is always less than 1 for $\bar{\phi} > 0$.

In the region of negative ΔE the state $m_p = 0$ is surely not the ground state. Indeed m_p can be contemplated to be a finite fraction of L as $L \to \infty$ and thus the true ground state will be a macroscopic amount below $m_p = 0$. In this region more work is needed to compute the true physical ground state and physical excitations. In order to compute the true ground state in the region where $\Delta E < 0$ it is clear that the procedure used above of letting $L \to \infty$ at the very early stage of the computation will not in general be sufficient to study the case where

$$\lim_{L\to\infty} m_p/L > 0 \,. \tag{4.1}$$

Thus it becomes necessary to get better control on the energy levels for L finite.

In the superintegrable case $\phi = \bar{\phi} = \pi/2$ the ground state has been constructed in detail [10,14,17]. The key feature in this construction is the factorization of the eigenvalues for finite L into products of meromorphic functions on the Riemann surface exhibited in (3.1) and (3.3). This factorization is not accidental but is a necessary consequence of the fact that the superintegrable chiral Potts model obeys Onsager's operator algebra (1.8)-(1.9). This factorization was the key tool used by Onsager in his original solution of the Ising model [3] and was found for the superintegrable chiral Potts model by Albertini, McCoy, Perk, and Tang [18]. However, it is more general than either of these models and follows from (1.9) alone as shown by Davies [19].

For the superintegrable chiral Potts case the consequence of the factorization is that the eigenvalues of $\mathcal{H}$ are grouped into a number of sets of the form

$$E = A + Bk' + \sum_{n=1}^{m_E} \pm\sqrt{1 + k'^2 + 2k'a_n}\,. \tag{4.2}$$

Each set has 2^{m_E} eigenvalues which are all connected to each other by analytic continuation in k'. As an example for the chiral Potts model with $N = 3, Q = 0, P = 0$ the sets for $L = 3,4,5,6$, and 7 are given as follows

TABLE I

L	LINEAR TERM	m_E	m_p	NUMBER OF SETS	NUMBER OF EIGENVALUES
3	0	2	0	1	4
		0	3	1	1
4	$-2(1+k')$	2	0	1	4
	$1+k'$	1	2	2	4
5	$-(1+k')$	3	0	1	8
		1	3	3	6
	$2(1+k')$	0	3	3	3
6	0	4	0	1	16
		2	3	5	20
		0	6	10	10
7	$-2(1+k')$	4	0	1	16
		2	3	4	16
		0	6	7	7
	$1+k'$	3	2	3	24
		1	5	21	42

Complete results for all eigenvalues for $N = 3$ and $L = 3, \cdots 7$ are given in ref. 10.

More generally Davies [19] shows from Onsager's algebra above that

$$E = A + Bk' + \sum_{n=1}^{m_E} 2m_n\sqrt{1 + k'^2 + 2k'a_n} \tag{4.3}$$

where

$$m_n = -S_\ell, -S_\ell + 1, \cdots, +S_\ell \tag{4.4}$$

and S_ℓ can be $\frac{1}{2}, 1, \frac{3}{2}, \cdots$. So far there are no known solutions to Onsager's algebra other than chiral Potts so only $S_\ell = \frac{1}{2}$ has yet been found. In this case the spectrum in each set is that of free fermions. If there exist systems with $S_\ell \geq 1$ the additive spectrum would be interpreted as free parafermions.

The procedure of finding the ground state now involves the finding of the value of m_p which gives the lowest eigenvalue. This can be done using the formalism of sec. 3 because when $\phi = \bar{\phi} = \pi/2$ the functions $h^{\pm}(t)$ of (3.7) reduce to polynomials.

$$h^{+}(t) = [(t_p - t)/y_p^2]^L \tag{4.5}$$

$$h^{-}(t) = [\omega(t_p - t)/y_p^2]^L \tag{4.6}$$

and thus τ_2 will indeed be a polynomial in t of order L when $f(t)$ is the polynomial (3.14) of order m_p. The formula for the v_ℓ (3.15b) then involves only polynomials on the right hand side and, indeed, is identical with the Bethe's hypothesis equation for the Heisenberg-Ising chain in the presence of a magnetic field [16]. The process of constructing the ground state in this analagous to the process of filling an interacting Fermi sea and has been done in great detail in ref. 10,14, and 17.

But in the general case $\phi \neq \pi/2$ there is a dramatic change because now the breakup of the eigenvalue into sets no longer happens and thus the number m_E and m_p are no longer good quantum numbers. As an example consider [18]

1) $L = 3, N = 3, Q = 0, P = 0$ where

$$E_\ell = \omega^\ell \Gamma^{+} + \omega^{-\ell}\Gamma^{-} \qquad \ell = 1,2,3 \tag{4.7}$$

$$\Gamma^{\pm} = -2\{6\sqrt{3}\cos\phi \pm [-125 + 108\cos^2\phi]^{1/2}\}^{1/3} \tag{4.8}$$

$$E_{4,5} = \pm 2\sqrt{3}\,. \tag{4.9}$$

2) L=4, N=3, Q=0, P=0

If we call $E = 2x\sqrt{3}$, $U = 2\cos\phi/3$ we find that x satisfies an 8th order equation.

$$\begin{aligned} &x^8 + 4ux^7 + (-54 - 5u^2)x^6 - (162 + 20u^2)x^5 \\ &+(621 + 594u^2 - 134u^4)x^4 \\ &-(-918u - 1296u^3 + 278u^5)x^3 + (-7695u^2 + 1782u^4 + 73u^6)x^2 \\ &-(-5292u^3 + 4644u^5 - 832u^7)x + (2808u^4 - 1728u^6 + 256u^8) = 0 \end{aligned} \tag{4.10}$$

If $\phi = \pi/2$ this equation factorizes over the integers but in general there is no such factorization. This factorization (or if you will the solvability of the group of the equation) corresponds to the existence of the sets of factorized eigenvalues.

Thus we see that for $\phi \neq \pi/2$ for finite L

1) All eigenvalues for fixed L, N, P, and Q form one set. These eigenvalue in general have singularities in both k' and ϕ and continue into each other as k' and /or ϕ encircle the branch points.

2) Correspondingly, there is no factorization of the eigenvalues of T into a t polynomial and a λ polynomial.

From a mathematical point of view it cannot be denied that these finite L properties are at variance with the $L \to \infty$ properties we deduced from the functional equation by ignoring the requirement that $\tau_2(t)$ be a polynomial and instead choose $f(t)$ to be a polynomial.

However, from a physicists point of view, as long as the ground state occurs for $m_p = 0$ both the ground state energy of Baxter [13] and the excited states computed here make a great deal of sense. What seems to be going on is as follows: The polynomial τ_2 is the sum of 2 terms. In one half plane one term is exponentially large as $L \to \infty$ and in the other half plane the other term is exponentially larger. On the boundary between the two half planes $\tau_2(t)$ has zeroes. The largest term has no singularities where it is dominant. Thus there should be a sort of Stokes multiplier phenomenon for the solution of the functional equations. It should be most interesting to study the convergence to the $L \to \infty$ limit.

Moreover in the $L \to \infty$ limit the branch point phenomenon in ϕ and λ is supressed. One would like to obtain in the complex ϕ plane an asymptotic (as $L \to \infty$) distribution of branch points. This seems in the spirit of Bender and Wu's [20] 1969 study of the energy levels of the anharmonic oscillator.

But the construction of the true ground state of the general chiral Potts model in the region when $\Delta E < 0$ will have to go beyond these considerations. If you will, we now have to fill up a Fermi sea where not only is there interaction of particles but also creation of particles (correspondence to the fact the m_E and m_p are no longer good quantum number). The phenomenon of production of domain walls was already seen in the original paper of Howes, Kadanoff, and den Nijs [1] where they argue that in the $L \to \infty$ limit it will be "renormalized away". This may indeed happen but from the present point of view this physical argument most certainly needs a mathematical justification. Indeed, the notion of production itself is at variance with the standard discussion of integrability and yet it seems hard to interpret the breakdown of m_p being a good quantum number in any other way. The very interpretation of the true excitation spectrum in terms of particle like excitations in the region $\Delta E < 0$ is no longer obvious and demands a rigorous treatment.

We thus conclude the mathematical discussion of the chiral Potts model with a physical question. In the physics of the general chiral Potts model the same is the superintegrable case or are there observable effects in the general case not seen in the superintegrable case? A mathematical resolution of this question will shed great light on the physics of the situation.

Acknowledgement

The author is most pleased to acknowledge many useful discussions with Prof. R. Baxter, Prof. V.V. Bazhanov, Prof. M. Jimbo, Prof. T. Miwa, and Prof. S.S. Roan. I also wish to thank Prof. M. Kashiwara and Prof. M. Sato for the opportunity of visiting the Research of Mathematical Science of Kyoto University where this work was done and to thank the Hayashibara Foundation for the opportunity of participating in the symposium. This work was supported in part by the National Science Foundation Grant #DMR-8803678.

References

1. S. Howes, L.P. Kadanoff, and M. den Nijs, *Nucl. Phys.* **B215** [FS7] (1983), 169.

2. G. von Gehlen and V. Rittenberg, *Nucl. Phys.* **B257** [FS14] (1985), 351.

3. L. Onsager, *Phys. Rev.* **65** (1944), 117.

4. H. Au-Yang, B.M. McCoy, J.H.H. Perk, S. Tang and M.L. Yan, *Phys. Letts.* **A123** (1987), 219.

5. B.M. McCoy, J.H.H. Perk, S. Tang and C.H. Sah, *Phys. Letts.* **A125** (1987), 9.

6. H. Au-Yang, B.M. McCoy, J.H.H. Perk and S. Tang in "Algebraic Analysis" ed. M. Kashiwara and T. Kawai, Academic Press 1988, 29.

7. R.J. Baxter, J.H.H. Perk and H. Au-Yang, *Phys. Letts.* **A128** (1988), 138.

8. H. Au-Yang and J.H.H. Perk in Advanced Studies in Pure Mathematics, Vol. 19 (Kinokuniya-Academic 1989), 56.

9. G. Albertini, B.M. McCoy and J.H.H. Perk, *Phys. Letts.* **A135** (1989), 159.

10. G. Albertini, B.M. McCoy and J.H.H. Perk in Advanced Studies in Pure Mathematics, vol. 19 (Kinokuniya-Academic 1989), 1.

11. V.V. Bazhanov and Yu. G. Stroganov, *J. Stat. Phys.* **59** (1990), 799.

12. R.J. Baxter, V.V. Bazhanov and J.H.H. Perk, *Int. J. Mod. Phys. B* **4** (1990), 803.

13. R.J. Baxter, *Phys. Letts.* **A146** (1990), 110.

14. G. Albertini, B.M. McCoy and J.H.H. Perk, *Phys. Letts.* **A139** (1989), 204.

15. B.M. McCoy and S.S. Roan, *Phys. Letts. A* **150 (1990)**, 347.

16. C.N. Yang and C.P. Yang, *Phys. Rev.* **150** (1966), 321.

17. G. Albertini and B.M. McCoy, *Nucl. Phys. B* **350** (1991), 745.

18. G. Albertini, B.M. McCoy, J.H.H. Perk and S. Tang, *Nucl. Phys.* **B314** (1989), 741.

19. B. Davies, *J. Phys. A* **23** (1990), 2247.

20. C.M. Bender and T.T. Wu, *Phys. Rev.* **184** (1969), 1231.

QUANTUM GROUPS AND q-ORTHOGONAL POLYNOMIALS

— Towards a realization of Askey-Wilson polynomials on $SU_q(2)$ —

Masatoshi NOUMI

Department of Mathematics, College of Arts and Sciences
University of Tokyo, Komaba, Meguro-Ku, Tokyo 153, Japan

Many links are already known between quantum groups and q-orthogonal polynomials. In this article, we will give a survey on recent works concerning the realization of q-analogues of the Jacobi polynomials as spherical functions on the quantum group $SU_q(2)$. We would like to emphasize that quantum groups and q-orthogonal polynomials have some characteristics in common and that the interaction between the two fields of mathematics will be important for both of them. The latter half of this article is a review on the works [K3-5] by T.H.Koornwinder and [NM1-5] by K.Mimachi and the author. This article, written from the viewpoint of [NM1-5], may be compared with Koornwinder's survey [K4].

Throughout this article, we will freely use the notations of q-shifted factorials

$$(a;q)_n = (1-a)(1-aq)\dots(1-aq^{n-1}),\quad (a;q)_\infty = \prod_{k=0}^{\infty}(1-aq^k) \quad (|q|<1),$$

and q-hypergeometric series

$${}_{m+1}\varphi_m\left(\begin{matrix} a_0,a_1,\dots,a_m \\ b_1,\dots,b_m\end{matrix};q,x\right) = \sum_{n=0}^{\infty}\frac{(a_0,a_1,\dots,a_m;q)_n}{(q,b_1,\dots,b_m;q)_n}x^n,$$

where

$$(a_0,a_1,\dots,a_m;q)_n = (a_0;q)_n(a_1;q)_n\dots(a_m;q)_n.$$

Note that, if $a_i=q^{-N}$ ($N\in\mathbf{N}=\{0,1,\dots\}$) for some i, then the above series ${}_{m+1}\varphi_m$ reduces to a finite sum. All q-orthogonal polynomials appearing in this article are expressed by such terminating q-hypergeometirc series. For q-hypergeometric series and q-orthogonal polynomials, we refer the reader to the textbook [GR] by Gasper and Rahman and [AW] by Askey and Wilson.

1. Quantum group $SU_q(2)$.

Interpretation of q-orthogonal polynomials by quantum groups began with the discovery of the fact that the matrix elements of (unitary) irreducible representations of the quantum group $SU_q(2)$ are expressed by a q-analogue of the Jacobi polynomials, called the little q-Jacobi polynomials. This result was obtained by Vaksman-Soibelman [VS1], and slightly later by Masuda et al. [MM1] and by Koornwinder [K1], independently. To fix the notation, we start with a review on this result in some details. For the terminology of Hopf algebras, we refer the reader to Abe's book [A]; for the basic facts on the quantum group $SU_q(2)$ and its representations, see [W1, MM2] for example.

Fixing a real number q with $0<q<1$, we denote by $A(SU_q(2))$ the "algebra of functions on $SU_q(2)$", a q-deformation of the algebra of functions on $SU(2)$. Let A be the non-commutative $\mathbb{C}$-algebra generated by the four elements $t_{11}, t_{12}, t_{21}, t_{22}$ with relations

$$(1.1)\qquad qt_{k1}t_{k2} = t_{k2}t_{k1},\ qt_{1k}t_{2k} = t_{2k}t_{1k}\quad (k=1,2)$$

$$t_{12}t_{21} = t_{21}t_{12},\ t_{11}t_{22}- q^{-1}t_{12}t_{21} = t_{22}t_{11}- qt_{12}t_{21} = 1.$$

This algebra A has a structure of Hopf algebra in which the coproduct $\Delta: A \longrightarrow A\otimes A$, the counit $\varepsilon: A \longrightarrow \mathbb{C}$ and the antipode $S: A \longrightarrow A$ are determined by the following values at the generators:

$$(1.2)\qquad \Delta(t_{ij}) = \sum_{k=1}^{2} t_{ik}\otimes t_{kj},\ \varepsilon(t_{ij}) = \delta_{ij}\quad (1\le i,j\le 2),$$

$$S(t_{11}) = t_{22},\ S(t_{12}) = -qt_{12},\ S(t_{21}) = -q^{-1}t_{21},\ S(t_{22}) = t_{11}.$$

Note that the antipode S is an antiautomorphism of $\mathbb{C}$-algebra, while the other two are $\mathbb{C}$-algebra homomorphisms. This Hopf algebra A is regarded as the coordinate ring of the complex quantum group $SL_q(2;\mathbb{C})$ with canonical coordinates t_{ij} $(1\le i,j\le 2)$. To specify the "real form $SU_q(2)$", we fix a *-operation on A so that

$$(1.3)\qquad t_{11}^* = t_{22},\ t_{12}^* = -q^{-1}t_{21},\ t_{21}^* = -qt_{12},\ t_{22}^* = t_{11}.$$

A *-operation on a $\mathbb{C}$-algebra A is a conjugate linear mapping $\varphi\mapsto\varphi^*$ from A to itself such that $1^*= 1$, $(\varphi\psi)^*= \psi^*\varphi^*$ and $\varphi^{**} = \varphi$ for all $\varphi,\psi\in A$. With this *-operation, the algebra A becomes a Hopf *-algebra, which we denote by $A(SU_q(2))$. A Hopf *-algebra is a Hopf algebra A with a *-operation such that the coproduct and the counit are $*$-homomorphisms and that $S(S(\varphi)^*)^* = \varphi$ for all φ. The last condition of compatibility with the antipode is due to Woronowicz, who was the first to recognize its importance in the duality argument (see Woronowicz [W1, W2]).

In the following, we think of a (real) quantum group G as a virtual geometric object represented by a Hopf $*$-algebra $A(G)$, called the algebra of functions on G.

A <u>representation</u> of a quantum group G is a right (or left) $A(G)$-comodule M; its coaction will be denote by $R_G: M \longrightarrow M\otimes A(G)$ (or by $L_G: M \longrightarrow A(G)\otimes M$). A representation M is said to be <u>unitarizable</u> if it has a positive definite Hermitian form $\langle\ ,\ \rangle$, conjugate linear in the first argument, which is <u>G-invariant</u> in the sense

$$\langle R_G(u), R_G(v)\rangle = \langle u, v\rangle \quad \text{for all } u,v\in M. \tag{1.4}$$

On the left hand side, $\langle\ ,\ \rangle$ denotes the Hermitian form naturally extended to $A(G)\otimes M$. Note that right $A(G)$-comodules corresponds to left G-modules of the classical case. For the unitarity of left $A(G)$-comodules, it is appropriate to use Hermitian forms that are conjugate linear in the second argument. In what follows, G stands for $SU_q(2)$ unless otherwise stated.

It is known that all the finite dimensional representations of the quantum group $SU_q(2)$ are completely reducible and unitarizable. The quantum group $SU_q(2)$ has a family V_j ($j\in\frac{1}{2}\mathbb{N}$) of irreducible (unitary) representations of dimension $2j+1$, parametrized by the non-negative half integers $j\in\frac{1}{2}\mathbb{N}$. This family of <u>spin j representations</u> gives a complete list of the isomorphism classes of finite dimensional irreducible representations of $SU_q(2)$.

We now recall the Peter-Weyl Theorem for $SU_q(2)$. Note first that the algebra of functions $A(SU_q(2))$ can be regarded as a two-sided $A(SU_q(2))$-comodule with coactions $R_G=L_G=\Delta$, which we call the <u>regular representation</u> of $SU_q(2)$. For each $j\in\frac{1}{2}\mathbb{N}$, the dual space $V_j^\vee= \mathrm{Hom}_{\mathbb{C}}(V_j,\mathbb{C})$ has a natural structure of left $A(SU_q(2))$-comodule. Define the linear mapping $\Phi_j: V_j^\vee\otimes V_j \longrightarrow A(SU_q(2))$ by

$$\Phi_j(f\otimes v) = (f\otimes \mathrm{id})\circ R_G(v) \quad \text{for all } f\in V_j^\vee,\ v\in V_j. \tag{1.5}$$

Its image, which we denote hereafter by W_j, is the vector subspace spanned by the matrix elements of the representation V_j. Then it turns out that each Φ_j induces an isomorphism of two-sided $A(SU_q(2))$-

comodules $V_j^\vee \otimes V_j \xrightarrow{\sim} W_j$ and that the regular representation is decomposed into the direct sum

$$(1.6) \qquad A(SU_q(2)) = \bigoplus_{j\in\frac{1}{2}\mathbb{N}} W_j, \quad W_j \xleftarrow{\sim} V_j^\vee \otimes V_j.$$

We also recall that there exists a unique right (and left) G-invariant linear functional $h_G: A(SU_q(2)) \longrightarrow \mathbb{C}$ with $h_G(1)=1$, called the Haar measure of $SU_q(2)$. The Haar measure h_G is nothing but the projection $A(SU_q(2)) \longrightarrow W_0=\mathbb{C}$ to the trivial representation in decomposition (1.6). By setting

$$(1.7) \qquad \langle\varphi,\psi\rangle_L = h_G(\varphi^*\psi) \quad \text{for} \quad \varphi,\psi\in A(SU_q(2)),$$

we obtain a right G-invariant, positive definite Hermitian form $\langle\ ,\ \rangle_L$; we use the subscript L to remember that it is conjugate linear in the left argument. A left G-invariant Hermitian form is obtained similarly by setting $\langle\varphi,\psi\rangle_R = h_G(\varphi\,\psi^*)$. The Peter-Weyl decomposition (1.6) is then orthogonal under these Hermitian forms $\langle\ ,\ \rangle_L$ and $\langle\ ,\ \rangle_R$.

2. Little q-Jacobi polynomials as matrix elements.

The spin j representations V_j ($j\in\frac{1}{2}\mathbb{N}$) are realized as right $A(SU_q(2))$-subcomodules of the regular representation $A(SU_q(2))$ as

$$(2.1) \qquad V_j = \mathbb{C}\, t_{11}^{2j} \oplus \mathbb{C}\, t_{11}^{2j-1} t_{12} \oplus \cdots \oplus \mathbb{C}\, t_{12}^{2j},$$

by using the coordinates t_{11}, t_{12} in the first row. For each V_j (realized as above), we define its $\mathbb{C}$-basis $\{ v_m^j;\ m=j,j-1,\dots,-j \}$ by

$$(2.2) \qquad v_m^j = \begin{bmatrix} 2j \\ j+m \end{bmatrix}_{q^2}^{1/2} t_{11}^{j+m}\, t_{12}^{j-m} \qquad (m=j,j-1,\dots,-j),$$

where the symbol $\begin{bmatrix} N \\ k \end{bmatrix}_q$ stands for the Gauss binomial coefficient $(q^N;q^{-1})_k/(q;q)_k$ for $0\le k\le N$. Setting $I_j=\{j,j-1,\dots,-j\}$, we will call an element of I_j a weight for the spin j representation V_j. The matrix elements $w_{mn}^j\in A(SU_q(2))$ ($m,n\in I_j$) of V_j, with respect to our basis $\{ v^j_m \}_{m\in I}$, are defined by the formula

$$(2.3) \qquad R_G(v^j_n) = \sum_{m\in I_j} v^j_m \otimes w^j_{mn} \qquad (n\in I_j).$$

Then it turns out that each w^j_{mn} is expressed as a product of a monomial in t_{rs} (r,s=1,2) and a little q-Jacobi polynomial in a distinguished variable $x = t^*_{12}t_{12} = -q^{-1} t_{12}t_{21}$. The <u>little q-Jacobi polynomials</u> $P^{(\alpha,\beta)}_k(x;q)$ ($k\in\mathbb{N}$) are orthogonal polynomials defined by

$$(2.4) \qquad P^{(\alpha,\beta)}_k(x;q) = {}_2\varphi_1\left(\begin{matrix} q^{-k}, \ q^{\alpha+\beta+k+1} \\ q^{\alpha+1} \end{matrix} ; q, \ qx \right),$$

which are a q-analogue of the Jacobi polynomials on the unit interval [0,1].

Theorem 1 ([VS1, K1, MM1]). The matrix elements w^j_{mn} of the spin j representation V_j, defined as above, are expressed in terms of the little q-Jacobi polynomials in $x = t^*_{12}t_{12}$ of base q^2 as follows, according to the region that the couple of weights (m,n) belongs to:

Case (1) $m+n\geq 0$, $m\geq n$: $\quad w^j_{mn} = c^j_{mn}\, t^{\beta}_{11}\, t^{\alpha}_{12}\, P^{(\alpha,\beta)}_k(x;q^2)$,

Case (2) $m+n\geq 0$, $m\leq n$: $\quad w^j_{mn} = c^j_{mn}\, t^{\beta}_{11}\, t^{\alpha}_{21}\, P^{(\alpha,\beta)}_k(x;q^2)$,

Case (3) $m+n\leq 0$, $m\geq n$: $\quad w^j_{mn} = c^j_{mn}\, P^{(\alpha,\beta)}_k(x;q^2)\, t^{\alpha}_{12}\, t^{\beta}_{22}$,

Case (4) $m+n\leq 0$, $m\leq n$: $\quad w^j_{mn} = c^j_{mn}\, P^{(\alpha,\beta)}_k(x;q^2)\, t^{\alpha}_{21}\, t^{\beta}_{22}$,

where $\alpha=|m-n|$, $\beta=|m+n|$, $k=j-\max\{|m|,|n|\}$ and the constants c^j_{mn} are given by $c^j_{mn} = q^{-k\alpha} \begin{bmatrix} \alpha+k \\ \alpha \end{bmatrix}^{1/2}_{q^2} \begin{bmatrix} \alpha+\beta+k \\ \alpha \end{bmatrix}^{1/2}_{q^2}$. ■

The matrix elements w^j_{mn} $(m,n\in I_j)$ of V_j form a $\mathbb{C}$-basis for the two-sided $A(SU_q(2))$-comodule $W_j = \mathrm{Image}(\Phi_j : V^{\vee}_j \otimes V_j \longrightarrow A(SU_q(2)))$, defined in Section 1: $W_j = \bigoplus_{m,n\in I_j} \mathbb{C}\, w^j_{mn}$. One can show that the above bases $\{ v^j_m \}_{m\in I_j}$ for V_j are orthogonal under the right G-invariant Hermitian form $\langle\ ,\ \rangle_L$. Accordingly, the matrix elements w^j_{mn} are also orthogonal to each other. In fact one has

$$(2.5)\qquad \langle w^j_{mn}, w^k_{rs}\rangle_L = \delta_{jk}\delta_{mr}\delta_{ns}\,\frac{1-q^2}{1-q^{2(2j+1)}}\,q^{2(j-m)}.$$

On the right hand side, one sees the inverse of a q-analogue of the dimension $2j+1$. As we will see later, this orthogonality (2.5) of matrix elements is interpreted as the orthogonality relations for the little q-Jacobi polynomials.

As being the matrix elements of a right $A(SU_q(2))$-comodule, the elements w^j_{mn} $(m,n\in I_j)$ has the property

$$(2.6)\qquad \Delta(w^j_{mn}) = \sum_{r\in I_j} w^j_{mr}\otimes w^j_{rn} \quad\text{and}\quad \varepsilon(w^j_{mn}) = \delta_{mn}.$$

This formula is an abstract addition formula involving non-commuting variables. In fact Koornwinder [K2] derived from (2.6) an addition formula for the little q-Legendre polynomials $P^{(0,0)}_k(x;q)$.

The unitarity of the representation V_j is equivalent to the relation of the matrix elements

$$(2.7)\qquad \sum_{r\in I_j} (w^j_{rm})^* \, w^j_{rn} = \delta_{mn} \quad\text{for}\quad m,n\in I_j;$$

this formula can also be regarded as the orthogonality relation for a q-analogue of the Krawtchouk plynomials (see [K1]).

3. Quantum G-spaces and quantum subgroups.

To understand the meaning of Theorem 1 properly, we need some geometric language for quantum groups. In many cases, it is important to investigate A(G)-comodules with algebra structure; such objects are the key to geometric arguments.

Let A(G) be a Hopf $*$-algebra representing a quantum group G and let A(X) be a $*$-algebra representing a quantum space X. We say that X is a right quantum G-space if A(X) has a right A(G)-comodule structre such that the coaction R_G: $A(X) \longrightarrow A(X)\otimes A(G)$ is a homomorphism of $*$-algebras. Then one can consider the quotient space X/G by taking the $*$-subalgebra of all right G-invariants

(3.1) $$A(X/G) := \{ \varphi \in A(G);\ R_G(\varphi) = \varphi \otimes 1 \} \subset A(G).$$

We will say that the quantum G-space X is homogenous if $A(X/G)=\mathbf{C}$. We do not repeat the similar definitions concerning left quantum G-spaces. Note that the quantum group G is a right (and left) quantum homogeneous space over itself in the above sense, with coaction $R_G = \Delta$ ($L_G = \Delta$).

In considering quantum G-spaces, we have to ask: What is a (quantum) subgroup of a quantum group? At present we do not know what is the right definition of quantum subgroups. In any case, the following one should be examined first: Let $A(H)$ be a Hopf *-algebra and π_H: $A(G) \longrightarrow A(H)$ is a surjective homomorphism of Hopf *-algebras. Then we say that the couple $(A(H),\pi_H)$ represents a (closed) quantum subgroup H of G. If H is a quantum subgroup of G, one can regard the quantum group G as a quantum H-space by taking the coaction

(3.2) $$R_H = (\mathrm{id}\otimes\pi_H)\circ\Delta : A(G) \longrightarrow A(G)\otimes A(H) \quad \text{or}$$
$$L_H = (\pi_H\otimes\mathrm{id})\circ\Delta : A(G) \longrightarrow A(H)\otimes A(G).$$

Then one can easily show that the quotient space G/H is homogenous as a left quantum G-space.

What about the converse? The answer is No. In fact there exist homogenous quantum G-spaces that cannot be expressed as quotient spaces of the form G/H as long as one takes the above definition of quantum subgroups (see Section 7). Such quantum homogeneous spaces might be said to be "non-standard".

4. Interpretation of orthogonality relations.

Returning to the setting of Section 2, we consider the diagonal subgroup of the quantum group $G=SU_q(2)$. We denote by $A(U(1))$ the algebra of functions on the unit circle $U(1)$; $A(U(1))$ is the Hopf *-algebra of Laurent polynomials $\mathbf{C}[t,t^{-1}]$ such that

(4.1) $$\Delta(t)=t\otimes t,\ \varepsilon(t)=1 \quad \text{and} \quad t^*=t^{-1}.$$

We define the diagonal subgroup K of $SU_q(2)$ by the algebra of functions $A(K)=A(U(1))$ and the "restriction mapping" π_K: $A(SU_q(2)) \longrightarrow A(K)$ such that

(4.2) $$\pi_K(t_{11})=t,\ \pi_K(t_{12})=0,\ \pi_K(t_{21})=0,\ \pi_K(t_{22})=t^{-1}.$$

In a sharp contrast to the SU(2) case, one can show that the diagonal is the only place in $SU_q(2)$ where the unit circle $U(1)$ can be embedded. (Where is $SO(2)$ gone?)

We now regard $SU_q(2)$ as the two-sided quantum (K,K)-space by the coactions $R_K = (id\otimes\pi_K)\circ\Delta$ and $L_K=(\pi_K\otimes id)\circ\Delta$. From this viewpoint, the variable $x=-q^{-1}t_{12}t_{21}$ in Theorem 1 is a generator for the $*$-subalgebra of (K,K)-invariants

$$(4.3)\qquad A(K\backslash SU_q(2)/K) := \{\ \varphi\in A(SU_q(2));\ L_K(\varphi) = 1\otimes\varphi,\ R_K(\varphi) = \varphi\otimes 1\ \},$$

which we regard as the algebra of functions on the double coset space $K\backslash SU_q(2)/K$. In fact one can show that $A(K\backslash SU_q(2)/K)$ is a polynomial ring $\mathbf{C}[x]$ with the variable $x = -q^{-1}t_{12}t_{21}$.

For each couple of weights (m,n) $(m,n\in\frac{1}{2}\mathbf{Z})$, we define the vector subspace $\mathcal{A}_{mn}$ of relative (K,K)-invariants in $A(SU_q(2))$ by

$$(4.4)\qquad \mathcal{A}_{mn} = \{\ \varphi\in A(SU_q(2));\ L_K(\varphi) = t^{2m}\otimes\varphi,\ R_K(\varphi) = \varphi\otimes t^{2n}\ \},$$

so that $\mathcal{A}_{00} = A(K\backslash SU_q(2)/K) = \mathbf{C}[x]$. Then the algebra of functions $A(SU_q(2))$ decomposes into

$$(4.5)\qquad A(SU_q(2)) = \bigoplus_{m,n\in\frac{1}{2}\mathbf{Z}} \mathcal{A}_{mn}$$

as a two-sided $\mathbf{C}[x]$-module. Furthermore, one can show that, if $m-n\in\mathbf{Z}$, $\mathcal{A}_{mn}$ is a free left (or right) $\mathbf{C}[x]$-module of rank one having one of the monomials $t_{11}^{\beta}t_{12}^{\alpha}$, $t_{11}^{\beta}t_{21}^{\alpha}$, $t_{12}^{\beta}t_{22}^{\alpha}$, $t_{21}^{\beta}t_{22}^{\alpha}$ with $\alpha=|m-n|$, $\beta=|m+n|$ as its basis.

We remark that the $\mathbf{C}$-basis $\{\ v_m^j\ \}_{m\in I_j}$ of (2.2) for V_j consists of relative K-invariants: $R_K(v_m^j) = v_m^j\otimes t^{2m}$. Accodingly the matrix elements w_{mn}^j have the relative invariance

$$(4.6)\qquad L_K(w_{mn}^j) = t^{2m}\otimes w_{mn}^j \quad\text{and}\quad R_K(w_{mn}^j) = w_{mn}^j\otimes t^{2n};$$

namely, $w_{mn}^j\in\mathcal{A}_{mn}$. In Theorem 1, the expréssion of the matrix elements w_{mn}^j is given based on the above $\mathbf{C}[x]$-module structure of $\mathcal{A}_{mn}$.

If the quantum group $SU_q(2)$ had a subgroup corresponding to $SO(2)$, we could have also used the subgroup instead of the double coset space $K\backslash SU_q(2)/K$. In fact the variable $x=-q^{-1}t_{12}t_{21}$ corresponds to $\sin^2\theta$ in the decomposition

(4.7) $$SU(2) = KAK, \quad A = SO(2) = \left\{ \begin{pmatrix} \cos\theta & -\sin\theta \\ \sin\theta & \cos\theta \end{pmatrix} ; \theta \in \mathbb{R}/2\pi\mathbb{Z} \right\}$$

of $SU(2)$.

Why did the little q-Jacobi polynomials arise in the expression of matrix elements in Theorem 1? One explanation can be given by representing the Haar measure in the variable $x = t_{12}^{*}t_{12}$. The Haar measure $h_G : A(SU_q(2)) \to \mathbb{C}$ is factored through the projection $A(SU_q(2)) \to \mathcal{A}_{00} = \mathbb{C}[x]$ in the above decomposition; this projection represents the averaging by the two-sided action of the unit circle K. On the subalgebra $\mathcal{A}_{00} = \mathbb{C}[x]$ of (K,K)-invariants, it turns out that the Haar measure is represented by the Jackson integral

(4.8) $$h_G(F(x)) = \int_0^1 F(x)\, d_{q^2}x \quad \text{for all} \quad F(x) \in \mathbb{C}[x],$$

on the interval $[0,1]$. The Jackson integral on the interval $[0,c]$ is by definition an infinite sum

(4.9) $$\int_0^c F(x)\, d_q x = \sum_{k=0}^{\infty} F(cq^k)\,(cq^k - cq^{k+1}).$$

Formula (4.8) was already known by Woronowicz [W1] in the form of an infinite sum.

Fix a couple of weights (m,n) such that $m-n \in \mathbb{Z}$, $m+n \geq 0$, $m \geq n$ (Case (1) of Theorem 1). We now try to rewrite the orthogonality of matrix elements $\langle w^j_{mn}, w^{j'}_{mn} \rangle_L = 0$ ($j \neq j'$) in terms of the Jackson integral. By using

(4.10) $$(t_{11}^{\beta}\, t_{12}^{\alpha})^{*}\, t_{11}^{\beta}\, t_{12}^{\alpha} = x^{\alpha}\,(q^2x;q^2)_{\beta},$$

the orthogonality of w^j_{mn} and $w^{j'}_{mn}$ $(j \neq j')$ is translated into

(4.11) $$\int_0^1 P_k^{(\alpha,\beta)}(x;q^2)\, P_{k'}^{(\alpha,\beta)}(x;q^2)\, x^{\alpha}(q^2x;q^2)_{\beta}\, d_{q^2}x = 0 \quad \text{for} \quad k \neq k'.$$

This formula is exactly the orthogonality relation for the little q-Jacobi polynomials $P_k^{(\alpha,\beta)}(x;q^2)$ in [AA].

5. Differential representations and q-difference equations.

Next we consider the differential representations of V_j. For this purpose one can make use of the quantized universal enveloping algebra $U_q(s\ell(2;\mathbb{C}))$. Following Jimbo [J1], we denote by $U_q(s\ell(2;\mathbb{C}))$ the $\mathbb{C}$-algebra generated by the four elements e, k, k^{-1}, f with relations

$$(5.1)\quad kk^{-1} = k^{-1}k = 1,\ kek^{-1} = qe,\ kfk^{-1} = q^{-1}f,\ ef-fe = \frac{k^2 - k^{-2}}{q - q^{-1}}.$$

We take the Hopf algebra structure of $U_q(s\ell(2;\mathbb{C}))$ such that

$$(5.2)\quad \Delta(k) = k\otimes k,\ \Delta(e) = e\otimes k + k^{-1}\otimes e,\ \Delta(f) = f\otimes k + k^{-1}\otimes f.$$

We also fix a $*$-operation on $U_q(s\ell(2;\mathbb{C}))$ so that $k^*=k$, $e^*=f$, $f^*=e$; the resulting Hopf $*$-algebra will be denoted by $U_q(su(2))$.

Between the two Hopf $*$-algebras $A(SU_q(2))$ and $U_q(su(2))$, one has a pairing $(\ ,\ )$: $U_q(su(2))\times A(SU_q(2)) \longrightarrow \mathbb{C}$ satisfying the following compatibility conditions:

$$(5.3)\quad 1)\ (ab,\varphi) = (a\otimes b,\Delta(\varphi)),\ (1,\varphi)=\varepsilon(\varphi),$$

$$2)\ (a,\varphi\psi) = (\Delta(a),\varphi\otimes\psi),\ (a,1)=\varepsilon(a),$$

$$3)\ (S(a),\varphi) = (a,S(\varphi)),\ (a^*,\varphi) = \overline{(a,S(\varphi)^*)}$$

for all $a,b\in U_q(su(2))$ and $\varphi,\psi\in A(SU_q(2))$. We fix such a pairing by taking the following values at the generators:

$$(5.4)\quad (k,T) = \mathrm{diag}(q^{1/2},q^{-1/2}),\ (e,T) = E_{12},\ (f,T) = E_{21},$$

where $T=\begin{pmatrix} t_{11} & t_{12} \\ t_{21} & t_{22}\end{pmatrix}$ and E_{rs} are the matrix units. In what follows, we regard the elements of $U_q(su(2))$ as linear functionals on $A(SU_q(2))$ through the pairing defined as above.

If M is a right (resp. left) $A(SU_q(2))$-comodule, then M has a natural structure of left (resp. right) $U_q(su(2))$-module by setting

$$(5.5)\quad a.v = (id\otimes a)\circ R_G(v)\quad (\text{ resp. } v.a = (a\otimes id)\circ L_G(v)\ )$$

for $a\in U_q(su(2))$ and $v\in M$, where R_G: $M \longrightarrow M\otimes A(SU_q(2))$ (resp. L_G: $M \longrightarrow A(SU_q(2))\otimes M$) is the coaction defining the comodule structure of M.

If X is a right quantum G-space over $G=SU(2)$, then the left $U_q(su(2))$-module structure of the algebra of functions $A(X)$ has the following properties:

(5.6) 1) If $a \in U_q(su(2))$ and $\Delta(a) = \sum_i a_i^1 \otimes a_i^2$, then $a.1 = 1.\varepsilon(a)$ and $a.(\varphi\psi) = \sum_i (a_i^1.\varphi)(a_i^2.\psi)$ for any $\varphi, \psi \in A(X)$.

2) $a.(\varphi^*) = (S(a)^*.\varphi)^*$ for any $a \in U_q(su(2))$ and $\varphi \in A(SU_q(X))$.

This statement is true in the general setting of quantum G-spaces if G has the corresponding quantized universal enveloping algebra. These properties of (5.6) can be used to define a $U_q(\mathfrak{g})$ version of quantum G-spaces; in a recent preprint [Kor], Korogodsky calls such objects QGS-algebras.

Note that the algebra of functions $A(SU_q(2))$ itself has a structure of two-sided $U_q(su(2))$-module with properties (5.6). With the notation of (5.4), this structure is characterized by

(5.7) $$k.T = T\,\mathrm{diag}(q^{1/2}, q^{-1/2}),\quad e.T = T\,E_{12},\quad f.T = T\,E_{21}$$

$$T.k = \mathrm{diag}(q^{1/2}, q^{-1/2})\,T,\quad T.e = E_{12}\,T,\quad T.f = E_{21}\,T.$$

By using (5.6) and (5.7), the differential representations of the right $A(SU_q(2))$-comodules V_j can be determined as follows:

(5.8) $$k.v_n^j = v_n^j\,q^n \quad \left(\text{or } \frac{k - k^{-1}}{q - q^{-1}}.v_n^j = v_n^j\,[n]\ \right),$$

$$e.v_n^j = v_{n+1}^j([j-n][j+n+1])^{1/2},\quad f.v_n^j = v_{n-1}^j\,([j+n][j-n+1])^{1/2},$$

where $[n]=(q^n-q^{-n})/(q-q^{-1})$. Since the matrix elements w_{mn}^j ($m \in I$) for a fixed m form a right $A(SU_q(2))$-comodule isomorphic to V_j, they satisfy the same equation as (5.8); those formulas written in terms of the variable $x=-q^{-1}t_{12}t_{21}$ give rise to some contiguity relations for little q-Jacobi polynomials.

Recall that the center of $U_q(su(2))$ is generated by the self-adjoint element

(5.9) $$C = \frac{(k-k^{-1})(qk-q^{-1}k^{-1})}{(q-q^{-1})^2} + fe,$$

called the <u>Casimir element</u>. Evaluating the Casimir element at the highest weight vector v_j^j, one sees that C acts on the irreducible representation V_j as the scalar operator with eigenvalue $[j][j+1]$. Hence one has

$$(5.10) \qquad (C - [j][j+1]).w^j_{mn} = 0 \quad \text{for all } j \in \tfrac{1}{2}\mathbb{N} \text{ and } m,n \in I_j.$$

Namely the Peter-Weyl decomposition (1.6) of $A(SU_q(2))$ is reformulated as the eigenspace decomposition of the Casimir operator $C : A(SU_q(2)) \longrightarrow A(SU_q(2))$:

$$(5.11) \quad A(SU_q(2)) = \bigoplus_{j \in \frac{1}{2}\mathbb{N}} W_j, \quad W_j = \{ \varphi \in A(SU_q(2));\ (C-[j][j+1]).\varphi = 0 \}.$$

Noting that the Casimir operator preserves the vector subspaces $\mathcal{A}_{mn}$ of relative (K,K)-invariants, one can express its action as an operator acting on the polynomial ring $\mathbb{C}[x]$ for each $\mathcal{A}_{mn}$. Then equation (5.10) can be read as the q-difference equation of order 2 for the corresponding little q-Jacobi polynomial:

$$(5.12) \quad \{ q^{\alpha+\beta}(q^{-2\beta} - q^2 x) T_{q^2} + q(q^{\alpha+\beta+2k+1} + q^{-\alpha-\beta-2k-1}) x - q^{-\alpha-\beta}(1+q^{2\alpha}) + q^{-\alpha-\beta}(1-x) T^{-1}_{q^2} \} \, P^{(\alpha,\beta)}_k(x;q^2) = 0,$$

where T_q is the q-shift operator defined by $T_q F(x) = F(xq)$.

In [MM2], the Rodrigues formula for the little q-Jacobi polynomials is also derived by using the differential representation.

6. Quantum 2-sphere $SU_q(2)/K$.

As a typical example of quantum G-spaces, let us examine the quotient space $SU_q(2)/K$ by the diagonal subgroup $K=U(1)$:

$$(6.1) \qquad A(SU_q(2)/K) = \{ \varphi \in A(SU_q(2));\ R_K(\varphi) = \varphi \otimes 1 \}.$$

This quantum G-space is regarded as a quantization of the 2-sphere. With the notations of Section 4, we can decompose the $*$-subalgebra in two ways:

$$(6.2) \qquad A(SU_q(2)/K) = \bigoplus_{m \in \mathbb{N}} \mathcal{A}_{m0} = \bigoplus_{j \in \mathbb{N}} W_j \cap A(SU_q(2)/K),$$

where $W_j \cap A(SU_q(2)/K) = \bigoplus_{m \in I_j} \mathbb{C} w^j_{m0}$ is a left $A(SU_q(2))$-subcomodule of $A(SU_q(2)/K)$ isomorphic to the spin j representation $V^\vee_j$. By (6.2), one sees that the algebra of functions on this quantum 2-sphere is generated by the three elements

$$(6.3) \qquad \xi = t_{11}t_{12}, \quad x = -q^{-1}t_{12}t_{21}, \quad \eta = t_{21}t_{22};$$

the algebraic relations among these generators and the $*$-structure are given by

$$(6.4)\quad q^2\xi x = x\xi,\ q^2 x\eta = \eta x,\ -q\xi\eta = x(1-x),\ -q^{-1}\eta\xi = x(1-q^2x)$$

$$\text{and}\quad \xi^* = -q^{-1}\eta,\ x^* = x,\ \eta^* = -q\xi,$$

respectively. For the moment, let us define the "real coordinates" x_0, x_1, x_2 by

$$(6.5)\quad x_0 = x,\quad x_1 = \mathrm{Re}(\eta) = (\eta+\eta^*)/2,\ x_2 = \mathrm{Im}(\eta) = (\eta-\eta^*)/2i;$$

Then one sees that (6.4) implies the following quadratic equation

$$(6.6)\qquad x_1^2 + x_2^2 = \frac{1+q^2}{2}x_0 - \frac{1+q^4}{2}x_0^2,$$

which corresponds to the equation of the 2-sphere with a diameter on the unit interval $[0,1]$ of the x_0-axis.

By restricting the Haar mesure $h_G: A(SU_q(2)) \longrightarrow \mathbb{C}$ to the $*$-subalgebra $A(SU_q(2)/K)$, one obtains a left G-invariant functional $A(SU_q(2)/K) \longrightarrow \mathbb{C}$; we readily know that it is the composition of the averaging by the unit circle K acting from the left and the Jackson integral on the unit interval $[0,1]$ in the variable x. The equation (6.6) and this description of the left G-invariant measure remind us of a picture of the quantum sphere like a sliced onion.

By (6.2) and Theorem 1, we know that the relative right K-invariants in $W_j \cap A(SU_q(2)/K)$ are given by

$$(6.7)\quad w^j_{m0} = \text{const.}\xi^m P^{(m,m)}_{j-m}(x;q^2)\quad\text{and}\quad w^j_{-m0} = \text{const.}\ P^{(m,m)}_{j-m}(x;q^2)\eta^m$$

for each $m\in\mathbb{N}$ with $0\le m\le j$. In this sense, the (associated) spherical functions on the quantum 2-sphere, relatively invariant under the action of the diagonal subgroup K, are expressed by little q-Jacobi polynomials. Note also that w^j_{m0} are solutions to the equation $\varphi.(C-[j][j+1])=0$ for the Casimir operator on $A(SU_q(2)/K)$.

Hereafter, we say that an element φ of the algebra of functions on a quantum G-space X is a <u>spherical function</u> (resp. <u>zonal spherical function</u>) if φ belongs to an irreducible component of $A(X)$ and has some property corresponding to the relative invariance (resp. invariance) with respect to a subgroup of G.

7. Podles' quantum 2-spheres and big q-Jacobi polynomials.

As can be found in [AA], the little q-Jacobi polynomials are not the only q-analogue of the Jacobi polynomials. Besides these, the Jacobi polynomials have a number of q-analogues that have orthogonality relations for different kind of measures. We repeat the question: Why did the little q-Jacobi polynomials appear in the expression of the matrix elements w^j_{mn} ? Are they the only q-analogue arising from quantum G-spaces? These naive questions were a motive to the recent works [NM1-5] and Koornwinder's [K3-5]. In this section, we review the result of [NM1] concerning the spherical functions on the quantum 2-spheres of Podles.

The quotient space $SU_q(2)/K$ is not the only quantization of the 2-sphere, either. In fact, Podles [P] found a family of homogeneous quantum G-spaces that can be regarded as quantization of the 2-sphere. Following the parametrization of [NM1], we denote Podles' quantum 2-spheres by $S^2_q(c,d)$ $(c,d\in\mathbb{R})$. The algebra of functions on the quantum 2-sphere $S^2_q(c,d)$ is defined as the $\mathbb{C}$-algebra generated by the three letters ξ, x, η with relations

$$(7.1)\quad q^2\xi x = x\xi,\ q^2x\eta = \eta x,\ -q\xi\eta = (c-x)(d+x),\ -q\eta\xi = (c-q^2x)(d+q^2x);$$

its $*$-operation is given similarly to (6.4). This quantum 2-sphere $S^2_q(c,d)$ becomes a left quantum G-space over $G=SU_q(2)$ by the coaction L_G: $A(S^2_q(c,d))\longrightarrow A(SU_q(2)) \otimes A(S^2_q(c,d))$ such that

$$(7.2)\qquad L_G(\varphi_m) = \sum_{n=-1}^{1} w^1_{mn} \otimes \varphi_n \quad \text{for} \quad m=1,0,-1,$$

where $\varphi_1= (1+q^2)^{1/2}\,\xi$, $\varphi_0= c-d-(1+q^2)x$, $\varphi_{-1}= (1+q^2)^{1/2}\eta$. This time the quadratic equation corresponding to (6.6) represents a 2-sphere with a diameter on the interval $[-d,c]$ of the x-axis. In considering quantum 2-spheres as right G-spaces, one needs to modify the $*$-operation so that $\xi^*=-q\eta$, $x^*=x$, $\eta^*=-q^{-1}\xi$.

We remark that the quantum sphere $SU_q(2)/K$ is isomorphic to Podles' quantum 2-sphere $S^2_q(1,0)$ as a left G-space over $G=SU_q(2)$. Furthermore one can show that $S^2_q(c,d)$, for general (c,d), are not isomorphic to each other and cannot be written in the form of quotient spaces of $SU_q(2)$ by quantum subgroups in the sense of Section 3.

The algebra of functions $A(S^2_q(c,d))$ is decomposed into the sum

(7.3) $$A(S_q^2(c,d)) = \bigoplus_{j\in\mathbb{N}} U_j, \quad U_j \simeq V_j^{\vee}$$

of irreducible $A(SU_q(2))$-comodules of integral spin, in the same way as $A(SU_q(2)/K)$. Hence the quantum sphere $S_q^2(c,d)$ has a unique left G-invariant functional $h_S: A(S_q^2(c,d)) \to \mathbb{C}$ with $h_S(1)=1$; the subscript S stands for $S_q^2(c,d)$. One can also introduce a left G-invariant Hermitian form on $A(S_q^2(c,d))$ by $\langle\varphi,\psi\rangle_R = h_S(\varphi\, \psi^*)$.

We now consider the left action of the diagonal subgroup K of $SU_q(2)$ on $S_q^2(c,d)$. Then the $*$-subalgebra $A(K\backslash S_q^2(c,d))$ is the polynomial ring generated by x. This time the left G-invariant functional h_S turns out to be the composition of the averaging by K and the Jackson integral on the interval $[-d,c]$: For any $F(x)\in A(K\backslash S_q^2(c,d))$, one has

(7.4) $$h_S(F(x)) = \frac{1}{c+d} \int_{-d}^{c} F(x)\, d_{q^2}x,$$

where

(7.5) $$\int_{-d}^{c} F(x)\, d_q x := \int_0^c F(x)\, d_q x - \int_0^{-d} F(x)\, d_q x.$$

In the case of Podles' quantum 2-sphere $S_q^2(c,d)$, the spherical functions that diagonalize the action of $K=U(1)$ are expressed by the big q-Jacobi polynomials $P_k^{(\alpha,\alpha)}(x;c,d:q^2)$. The big q-Jacobi polynomials are a family of orthogonal polynomials

(7.6) $$P_k^{(\alpha,\beta)}(x;c,d:q) = {}_3\varphi_2\left[\begin{matrix} q^{-k},\ q^{\alpha+\beta+k+1},\ q^{\alpha+1}x/c \\ q^{\alpha+1},\ -q^{\alpha+1}d/c \end{matrix} ;\ q,\ q\right],$$

containing the parameters (c,d). When $(c,d)=(1,0)$, the polynomial $P_k^{(\alpha,\beta)}(x;1,0:q)$ is a constant multiple of the little q-Jacobi polynomial $P_k^{(\beta,\alpha)}(x:q)$ with α and β interchanged.

Theorem 2 ([NM1]). Let $\{\varphi_m^j\}_{m\in I_j}$ be a $\mathbb{C}$-basis for the irreducible component U_j $(j\in\mathbb{N})$ of $A(S_q^2(c,d))$ and assume that $L_K(\varphi_m^j) = t^{2m}\otimes\varphi_m^j$ for all $m\in I_j$. Then one has

$$\varphi_m^j = \text{const. } \xi^{\alpha} c^{j-\alpha} (-q^{2(\alpha+1)}d/c;q^2)_{j-\alpha} \; P_{j-\alpha}^{(\alpha,\alpha)}(x;c,d:q^2) \qquad \text{if} \quad 0 \le m \le j,$$

$$\varphi_m^j = \text{const. } c^{j-\alpha} (-q^{2(\alpha+1)}d/c;q^2)_{j-\alpha} \; P_{j-\alpha}^{(\alpha,\alpha)}(x;c,d:q^2) \; \eta^{\alpha} \qquad \text{if} \quad -j \le m \le 0,$$

where $\alpha = |m|$. ■

The big q-Jacobi polynomials $P_k^{(\alpha,\beta)}(x;c,d:q)$ have the orthogonality relation of type

$$(7.7) \qquad \int_{-d}^{c} p_k(x)\, p_{k'}(x)\, (qx/c;q)_{\alpha} (-qx/d;q)_{\beta} \, d_q x \;=\; 0 \qquad (\, k \neq k' \,)$$

for the Jackson integral on the interval $[-d,c]$ if $c>0$ and $d>0$. This type of orthogonality for $\alpha=\beta$ is interprerted by the orthogonality of the decomposition (7.3) under the left G-invariant Hermitian form $\langle\ ,\ \rangle_R$. The q-difference equation for the big q-Jacobi polynomials with $\alpha=\beta$ is also obtained from the action of the Casimir element of $A(S_q^2(c,d))$.

So far we have not referred to the nondegeneracy of the Hermitian form $\langle\ ,\ \rangle_R$ on $A(S_q^2(c,d))$. If $c+q^{2r}d \neq 0$ for any $r \in \mathbf{Z}\setminus\{0\}$, the Hermitian form $\langle\ ,\ \rangle_R$ is nondegenerate. If $c+q^{2r}d=0$ for some $r \in \mathbf{Z}\setminus\{0\}$, however, it is nondegenerate only on the components U_j with $j<|r|$. One should be careful about this kind of arguments since remarkable differences may occur between the algebraic approach and the C^*-algebra approach. According to Podles [P], the C^*-algebra corresponding to $A(S_q^2(c,d))$ collapses to a finite dimensional vector spaces for the special values of (c,d) with $c+q^{2r}d=0$ for some $r \in \mathbf{Z}\setminus\{0\}$, where "bad" components are killed.

It is interesting to examine the special cases when $c+q^{2r}d=0$ for some $r \in \mathbf{Z}\setminus\{0\}$. In such a case, the representation of the G-invariant functional (7.4) reduces to a finite sum. The expression of spherical functions in Theorem 2 also falls into that by <u>q-Hahn polynomials</u>

$$(7.8) \qquad Q_k(x;a,b,N:q) = {}_3\varphi_2\left[\begin{matrix} q^{-k},\ abq^{k+1},\ x \\ aq,\ q^{-N} \end{matrix}\ ;\ q,\ x\right],$$

where $k,N \in \mathbf{N}$ and $0 \le k \le N$. Then the orthogonality of spherical functions φ_m^j $(j<|r|)$ corresponds to the orthogonality of q-Hahn polynomials $Q_k(x;a,a,N:q^2)$ $(0 \le k \le N)$ for a finite sum. This fact should be compared with the result, by Kirillov-Reshetikhin [KR] and by Koelink-

Koornwinder [KK], that the q-Hahn polynomials appear as a natural q-analogue, in the description of the Clebsch-Gordan coefficients for $SU_q(2)$.

8. A family of quantum 3-spheres.

In the previous section, we gave an interpretation of the big q-Jacobi polynomials $P_k^{(\alpha,\beta)}(x;c,d:q)$ with $\alpha=\beta$ as spherical functions on Podles' quantum 2-spheres $S_q^2(c,d)$. What about the other big q-Jacobi polynomials with $\alpha\neq\beta$? Regard the quantum group $SU_q(2)$ itself as a quantum 3-sphere S_q^3; then the quantum 2-sphere $S_q^2=S_q^2(1,0)$ is a quotient space of S_q^3 by the action of $K=U(1)$. Does there exist a family of quantum 3-spheres "$S_q^3(c,d)$" $(c,d\in\mathbb{R})$ from which Podles' quantum 2-spheres are obtained as the orbit space $S_q^3(c,d)/K$? If one can find such a quantum 3-sphere, it will be the place for the big q-Jacobi polynomials to live. One answer to this question is given in [NM2-3]. In this section we will give a review on this work.

We will define a new quantum space denoted by $\tilde{S}_q^3$ below. It gives a quantum deformation family $\tilde{S}_q^3 \longrightarrow \mathbb{R}^2$ whose fiber over the point $(c,d)=(1,0)$ is the quantum 3-sphere $S_q^3=SU_q(2)$. The total space $\tilde{S}_q^3$ has the structure of a two-sided quantum (G,K)-space over the quantum group $G=SU_q(2)$ and its diagonal subgroup $K=U(1)$. Then Podles' family of quantum 2-spheres will be recovered as the quotient $\tilde{S}_q^3/K \longrightarrow \mathbb{R}^2$. A remarkable point about this "quantum family" is the fact that the algebra of functions $A(\mathbb{R}^2)$ on the base space is <u>not</u> contained in the center of $A(\tilde{S}_q^3)$.

We define the algebra of functions $A(\tilde{S}_q^3)$ on $\tilde{S}_q^3$ to be the $\mathbb{C}$-algebra generated by the six elements $\tilde{t}_{11},\tilde{t}_{12},\tilde{t}_{21},\tilde{t}_{22}$ and c,d with relations

$$(8.1)\qquad q\tilde{t}_{k1}\tilde{t}_{k2} = \tilde{t}_{k2}\tilde{t}_{k1},\ q\tilde{t}_{1k}\tilde{t}_{2k} = \tilde{t}_{2k}\tilde{t}_{1k}\quad (k=1,2)$$

$$\tilde{t}_{21}\tilde{t}_{12}-\tilde{t}_{12}\tilde{t}_{21} = (q-q^{-1})d,$$

$$\tilde{t}_{11}\tilde{t}_{22}-q^{-1}\tilde{t}_{12}\tilde{t}_{21} = \tilde{t}_{22}\tilde{t}_{11}-q\tilde{t}_{21}\tilde{t}_{12} = c+d,$$

$$q^2\tilde{t}_{k1}d = d\tilde{t}_{k1},\ q^2d\tilde{t}_{k2}= \tilde{t}_{k2}d\quad (k=1,2).$$

This algebra has a $*$-operation such that

(8.2) $\tilde{t}_{11}^{*} = \tilde{t}_{22}$, $\tilde{t}_{12}^{*} = -q^{-1}\tilde{t}_{21}$, $\tilde{t}_{21}^{*} = -q\tilde{t}_{12}$, $\tilde{t}_{22}^{*} = \tilde{t}_{11}$, $c^{*}=c$, $d^{*}=d$.

First observe that the $*$-algebra $A(\tilde{S}_q^3)$ reduced to the algebra of functions $A(SU_q(2))$ by the specialization $(c,d)=(1,0)$. From (8.1), one can show that the element c belongs to the center of $A(\tilde{S}_q^3)$, while d does not. Taking the commutative subalgebra $A(\mathbb{R}^2)=\mathbb{C}[c,d]$ of $A(\tilde{S}_q^3)$, we regard $\tilde{S}_q^3$ as the total space of a deformation family $\tilde{S}_q^3 \longrightarrow \mathbb{R}^2$, although the "parameter" d is not in the center of $A(\tilde{S}_q^3)$. If one uses "real coordinates" as in (6.5), then the relation $\tilde{t}_{11}\tilde{t}_{22}-q^{-1}\tilde{t}_{12}\tilde{t}_{21} = \tilde{t}_{22}\tilde{t}_{11}-q\tilde{t}_{21}\tilde{t}_{12} = c+d$ corresponds to the defining equation of a family of 3-spheres. It should be noted, however, that our quantum family does not imply the existence of individual fibers "$S_q^3(c,d)$", except for the case $d=0$, since the "parameter" d is not in the center of $A(\tilde{S}_q^3)$.

One can also check by a direct calculation that the quantum space $\tilde{S}_q^3$ has a structure of two-sided (G,K)-space whose coactions $L_G: A(\tilde{S}_q^3) \longrightarrow A(SU_q(2))\otimes A(\tilde{S}_q^3)$ and $R_K: A(\tilde{S}_q^3) \longrightarrow A(\tilde{S}_q^3)\otimes A(K)$ are characterized by the formulas

(8.3) $L_G(\tilde{t}_{ij}) = \sum_{k=1}^{2} t_{ik}\otimes \tilde{t}_{kj}$ $(1\le i,j\le 2)$, $L_G(c)=1\otimes c$, $L_G(d)=1\otimes d$,

$$R_K(\tilde{t}_{k1})=\tilde{t}_{k1}\otimes t,\ R_K(\tilde{t}_{k2})=\tilde{t}_{k2}\otimes t^{-1}\ (1\le k\le 2),\ R_K(c)=c\otimes 1,\ R_K(d)=d\otimes 1.$$

Note that the elements c, d are two-sided (G,K)-invariants. In fact $\tilde{S}_q^3$ gives a family of quantum homogeneous G-spaces in the sense that $A(G\backslash\tilde{S}_q^3)=A(\mathbb{R}^2)$.

As a left $A(SU_q(2))$-comodule, the algebra of functions $A(\tilde{S}_q^3)$ is decomposed into irreducible components as follows:

(8.4) $$A(\tilde{S}_q^3) = \bigoplus_{j\in\frac{1}{2}\mathbb{N}} \tilde{W}_j\otimes A(\mathbb{R}^2), \quad \text{with } \tilde{W}_j \simeq (2j+1)V_j^{\vee},$$

where $\tilde{W}_j$ is a $(2j+1)$-tuple copy of the left $A(SU_q(2))$-comodule $V_j^{\vee}$ and $SU_q(2)$ acts on $A(\mathbb{R}^2)$ trivially as we mentioned above. From this decomposition, one sees that there exists a unique left G-invariant, $A(\mathbb{R}^2)$-homomorphism $h: A(\tilde{S}_q^3) \longrightarrow A(\mathbb{R}^2)$ with $h(1)=1$. As to the two-sided action of $K=U(1)$, it turns out that the algebra of functions on the double coset space

$$(8.5) \qquad A(K\backslash\tilde{S}^3_q/K) = A(\mathbb{R}^2)[x] = \mathbb{C}[c,d,x]$$

is a commutative $*$-subalgebra isomorphic to the polynomial ring in three indeterminates. Here the "zonal" variable x is given by $x = c-\tilde{t}_{11}\tilde{t}_{22} = -d-q^{-1}t_{12}t_{21}$. On this subalgebra, the G-invariant $A(\mathbb{R}^2)$-valued functional h is represented by the Jackson integral on the interval $[-d,c]$ just as in (7.4).

What are the spherical functions on this (G,K)-space $\tilde{S}^3_q$? The answer is similar to Theorem 1 except for the point that the little q-Jacobi polynomials $P_k^{(\alpha,\beta)}(x:q^2)$ are replaced by the big q-Jacobi polynomials $c^k(-q^{2(\beta+1)}d/c;q^2)_k P_k^{(\beta,\alpha)}(x;c,d:q^2)$ as in Theorem 2.

The quotient space $\tilde{S}^3_q/K$ recovers Podles' family of quantum 3-spheres. In fact, the $*$-subalgebra $A(\tilde{S}^3_q/K)$ of $A(\tilde{S}^3_q)$ is generated by the five elements

$$(8.6) \qquad \xi=\tilde{t}_{11}\tilde{t}_{12},\ x=c-\tilde{t}_{11}\tilde{t}_{22} = -d-q^{-1}t_{12}t_{21},\ \eta=\tilde{t}_{21}\tilde{t}_{22},\ c,\ d;$$

these elements satisfy the same relations as those of Podles' quantum 2-spheres (7.1), with c, d regarded as central elements.

9. Where is SO(2) gone? —— Koornwinder's infinitesimal approach.

After the realization of the little q-Jacobi polynomials as matrix elements, Koornwinder proceeded in a different way. In [K3], he investigated a quantum analogue of the double coset space $SO(2)\backslash SU(2)/SO(2)$. As we mentioned in Section 4, the diagonal is the only place in $SU_q(2)$ where the unit circle $U(1)$ can be embedded. Probably this means that one cannot obtain such a quantum analogue as long as one sticks to the subgroups recognized by quotient Hopf $*$-algebras of $A(SU_q(2))$. In fact Koornwinder [K3] succeeded in passing the action of the Lie algebra $so(2)$ to the algebra of functions $A(SU_q(2))$ by using a twisted primitive element in $U_q(su(2))$. The resulting zonal spherical functions were the continuous q-Legendre polynomials. This fact is remarkable: They are orthogonal polynomials for an absolutely continuous measure!

For a couple of integers (m,n), we say that an element D of $U_q(su(2))$ is a twisted primitive element of type (k^m,k^n) if $\Delta(D) = D\otimes k^m + k^n\otimes D$ and $\varepsilon(D)=0$. This means that, if X is a right quantum G-space over $G=SU_q(2)$, then D acts on $A(X)$ as a twisted derivation in the sense that

(9.1) $D.(\varphi\psi) = (D.\varphi)(k^m.\psi) + (k^n.\varphi)(D.\psi)$ for all $\varphi,\psi\in A(X)$.

Note also that k^m acts as a $\mathbb{C}$-algebra automorphism. One can show that, for any integer m, an element D of $U_q(su(2))$ is a twisted primitive element of type (k^m,k^{m-2}) if and only if

(9.2) $$D = k^{m-1}\left(a\, e + b\,\frac{k - k^{-1}}{q - q^{-1}} + c\, f \right) \quad \text{for some } a,b,c\in\mathbb{C}.$$

One can also show that, if $m-n\neq 2$, then the twisted primitive elements of type (k^m,k^n) must be a constant multiple of k^m-k^n.

Here we think of the twisted primitive elements of type (k,k^{-1}) as correspond to the elements of the Lie algebra $s\ell(2;\mathbb{C})$. Following Koornwinder, let us take the twisted primitive element

(9.3) $$\theta = q^{-1/2}\, e - q^{1/2} f$$

as an analogue of the generator for the Lie algebra so(2). Note that the coeffients are so chosen that $S(\theta)^* = \theta$. Then we define the quantum analogue of SO(2)\SU(2)/SO(2) by taking the following subalgebra of θ-invariants:

(9.4) $$A_q(SO(2)\backslash SU(2)/SO(2)) := \{ \varphi\in A(SU_q(2));\ \theta.\varphi=0,\ \varphi.\theta=0 \}.$$

(The left hand side is an <u>ad hoc</u> notation.) By $S(\theta)^* = \theta$ and the property (5.6.2), one sees that (9.4) becomes a *-subalgebra of $A(SU_q(2))$. By investigating the action of θ on each spin j representation V_j, one can show that the *-algebra of (9.4) has a basis $\{ \varphi_j \}_{j\in\mathbb{N}}$ such that $\varphi_j\in W_j$, or equivalently, $(C-[j][j+1]).\varphi = 0$. These elements are thought of as the zonal spherical functions with respect to the action of θ.

Following Rahman's parametrization, we denote by $p_k^{(\alpha,\beta)}(x;q)$ $(k\in\mathbb{N})$ the <u>continuous q-Jacobi polynomials</u>:

(9.5) $$p_k^{(\alpha,\beta)}(x;q) = \frac{(q^{\alpha+1},-q^{\beta+1};q)_k}{(q,-q;q)_k}\, {}_4\varphi_3\left(\begin{matrix} q^{-k},\ q^{\alpha+\beta+k+1},\ q^{1/2}z,\ q^{1/2}z^{-1} \\ q^{\alpha+1},\ -q^{\beta+1},\ -q \end{matrix} ;\ q,\ q \right),$$

where $x=\frac{1}{2}(z+z^{-1})$. The right hand side of (9.5) is a polynomial in $x=\frac{1}{2}(z+z^{-1})$ of degree k. In the case $\alpha=\beta=0$, the polynomials

$p_k^{(0,0)}(x;q)$ are called the <u>continuous q-Legendre polynomials</u>.

Theorem 3 ([K3]). The $*$-subalgebra (9.6) of $A(SU_q(2))$ is a polynomials ring generated by the element

$$x = \frac{1}{2}(t_{11}^2 + qt_{12}^2 + q^{-1}t_{21}^2 + t_{22}^2).$$

Moreover, for each $j\in\mathbb{N}$, the zonal spherical function φ_j is a constant multiple of the continuous q-Legendre polynomial $p_j^{(0,0)}(x;q^2)$ in the above x of base q^2. ■

The continuous q-Jacobi polynomials $p_k(x)=p_k^{(\alpha,\beta)}(x;q)$ have the orthogonality relation

$$(9.6)\qquad \int_{-1}^{1} p_k(x)\, p_{k'}(x)\, \frac{h(x;1)h(x;-1)(1-x^2)^{-1/2}}{h(x;q^{\alpha+1/2})h(x;-q^{\beta+1/2})}\, dx = 0 \quad \text{for} \quad k\neq k',$$

where $h(x;a)=(az,az^{-1};q)_\infty = \prod_{k=0}^{\infty}(1-2xaq^k+a^2q^{2k})$. The orthogonality relation for the continuous q-Legendre polynomials $p_k^{(0,0)}(x;q)$ also corresponds to the orthogonality of the Peter-Weyl decomposition (1.6).

Generalizing the above consideration, Koornwinder [K4,5] studied the quantum analogue of a 1-parameter family of circles K_σ in SU(2) that connects the diagonal subgroup and SO(2). For any positive real number σ, set

$$(9.7)\qquad \theta_\sigma = -iq^{-1/2}e + (q^\sigma - q^{-\sigma})\,\frac{k-k^{-1}}{q-q^{-1}} + iq^{1/2} f;$$

this element satisfies $S(\theta_\sigma)^* = -\theta_\sigma$. Note that θ_1 is a constant multiple of θ in (9.3) and that, when $\sigma\longrightarrow+\infty$, $q^\sigma\theta_\sigma$ tends to $-(k-k^{-1})/(q-q^{-1})$, which corresponds to the diagonal subgroup $K=U(1)$. He investigates the zonal spherical functions φ in the sense

$$(9.8)\qquad (C-[j][j+1]).\varphi = 0,\quad \theta_\sigma.\varphi=0,\quad \varphi.\theta_\tau=0,$$

taking two twisted primitive elements, one for the left and the other for the right action of $U_q(su(2))$. Then it turns out that the $*$-subalgebra

$$(9.9)\qquad \{\ \varphi\in A(SU_q(2));\ \theta_\sigma.\varphi = 0,\ \varphi.\theta_\tau = 0\ \}$$

is a polynomial ring generated by a quadratic form x in $t_{11}, t_{12}, t_{21}, t_{22}$. Then the zonal spherical functions belonging to W_j give rise to a 2-parameter subfamily of Askey-Wilson polynomials.

These are regarded as a 2-parameter extension of continuous q-Legendre polynomials. The two parameters σ, τ in this family now symbolize the position of two circles "moving around in $SU_q(2)$". Theorem 3 is the special case when $\sigma=\tau=0$. The original setting of $K \backslash SU_q(2)/K$ with the diagonal subgroup K is also recovered as the limiting case when $\sigma, \tau \longrightarrow +\infty$. The explicit description of this family will be given later in (11.6).

10. Semisimple twisted elements in $U_q(\mathfrak{sl}(2;\mathbf{C}))$.

In [K3], Koornwinder also pointed out that the twisted primitive element $k\theta_\sigma$ is diagonalizable on every spin j representation V_j. In relation to this fact, we should remark that, in the SU(2) case, θ_σ corresponds to a constant multiple of the tangent vector of the circle gKg^{-1} for some $g \in SU(2)$. Let us reverse the problem: How many twisted primitive elements does the quantized universal enveloping algebra $U_q(\mathfrak{sl}(2;\mathbf{C}))$ has? As is stated in [NM5], there exists a family of twisted primitive elements that corresponds to the semisimple orbit $\{ ghg^{-1};\ g \in SL(2;\mathbf{C}) \}$ of $h \in \mathfrak{sl}(2;\mathbf{C})$.

For each element $g = \begin{pmatrix} a_{11} & a_{12} \\ a_{21} & a_{22} \end{pmatrix}$ of $GL(2;\mathbf{C})$ (the usual one!), we define the twisted primitive element $\theta(g)$ by

$$(10.1)\quad \theta(g) := -a_{11}a_{12}q^{-1/2}e + (a_{11}a_{22}+a_{12}a_{21})\frac{k-k^{-1}}{q-q^{-1}} + a_{21}a_{22}q^{1/2}f.$$

This element corresponds to $\mathrm{Ad}(g)\frac{h}{2}$ in the Lie algebra $\mathfrak{sl}(2;\mathbf{C})$. Then one can show that the twisted primitive element $k\theta(g)$ of type $(k^2,1)$ is <u>semisimple</u> in the sense that it is diagonalizable on every finite dimensional representations, if $a_{11}a_{22}-q^{2r}a_{12}a_{21} \neq 0$ for all $r \in \mathbf{Z}$ (and if q is not a root of unity). On the spin j representation V_j ($j \in \frac{1}{2}\mathbf{N}$), $k\theta(g)$ has mutually distinct eigenvalues $\lambda_m(g)$ ($m \in I_j$), depending on g, where

$$(10.2)\qquad \lambda_m(g) = \frac{q^m - q^{-m}}{q - q^{-1}} \cdot (q^m a_{11}a_{22} - q^{-m} a_{12}a_{21}).$$

Now we assume that $\overline{a_{11}}=a_{22}$ and $\overline{a_{21}}=-a_{12}$ so that $g\in SU(2)\times\mathbb{R}_{>0}$ and $S(\theta(g))^*=-\theta(g)$. Note that Koornwinder's θ_σ of (9.3) is the case when $a_{11}=q^{\sigma/2}$ and $a_{21}=iq^{-\sigma/2}$. Let us consider the $*$-subalgebra of $\theta(g)$-invariants

$$(10.3) \quad A_q(SU(2)/K(g)) := \{ \varphi\in A(SU_q(2));\ \theta(g).\varphi = 0\},$$

which can be regarded as a quantization of the 2-sphere $SU(2)/K(g)$, where K is the diagonal subgroup $U(1)$ of $SU(2)$ and $K(g)=gKg^{-1}$. In fact the $*$-algebra (10.3) represents the function algebra on a left quantum $SU_q(2)$-space in the sense of Section 3. Furthermore, one can show that it is isomorphic to Podles' quantum 2-sphere $S_q^2(c,d)$ with $c=|a_{11}|^2$ and $d=|a_{21}|^2$. An explicit correspondence of generators is given as follows:

$$\xi = t_{11}^2 a_{11}a_{12}q^{-1} + t_{11}t_{12}(a_{11}a_{22}+a_{12}a_{21}) + t_{12}^2 a_{21}a_{22}$$

$$(10.4)\quad -x = t_{11}t_{21}a_{11}a_{12}q^{-1} + q^{-1}t_{12}t_{21}(a_{11}a_{22}+a_{12}a_{21}) + t_{12}t_{22}a_{21}a_{22}$$

$$\eta = t_{21}^2 a_{11}a_{12}q^{-1} + t_{21}t_{22}(a_{11}a_{22}+a_{12}a_{21}) + t_{22}^2 a_{21}a_{22}$$

This means that Podles' quantum 2-sphere $S_q^2(c,d)$ with positive c,d is constructed as the "quotient space" of $SU_q(2)$ by a semisimple twisted primitive element, even though it has no corresponding quantum subgroup in the sense of quotient Hopf $*$-algebras.

11. Askey-Wilson polynomials as spherical functions.

At the end of Section 9, we referred to Koornwinder's realization [K4,5] of a 2-parameter subfamily of Askey-Wilson polynomials as zonal spherical functions on $SU_q(2)$. This work is generalized to the non-zonal cases by [NM5] and Koelink [Koe]. Keeping the notation of the previous section, we now examine the analogy of spherical functions on $SU(2)$ relatively invariant with respect to $K(g_1)$ acting from the left and $K(g_2)$ from the right.

Taking a couple of matrices

$$(11.1)\quad g_1=\begin{pmatrix} b_{11} & b_{12} \\ b_{21} & b_{22}\end{pmatrix},\ g_2=\begin{pmatrix} a_{11} & a_{12} \\ a_{21} & a_{22}\end{pmatrix} \in SU(2)\times\mathbb{R}_{>0},$$

we assume that b_{11}, b_{21}, a_{11}, a_{21} are all nonzero. Then one can show that the *-subalgebra of $A(SU_q(2))$ consisting of two-sided invariants

$$(11.2)\quad A_q(K(g_1)\backslash SU(2)/K(g_2)):=\{\ \varphi\in A(SU_q(2));\ \theta(g_2).\varphi = \varphi.\theta(g_1) = 0\ \}$$

is a polynomial ring generated by a quadratic form in $t_{11}, t_{12}, t_{21}, t_{22}$. We take a specified generator, denoted by $x(g_1,g_2)$, although we do not give here the explicit formula for $x(g_1,g_2)$.

We now investigates the (associated) spherical functions φ on $SU_q(2)$ with respect to the couple $(\theta(g_1),\theta(g_2))$ of twisted primitive elements. For any $j\in\frac{1}{2}\mathbf{N}$ and $m,n\in I_j$, we say that an element φ of $A(SU_q(2))$ is a (doubly associated) spherical functions of type $(j;m,n)$ if

$$(11.3)\quad (C-[j][j+1]).\varphi = 0 \quad (\text{ or } \ \varphi\in W_j\ ) \quad \text{and}$$

$$(k\theta(g_2)-\lambda_n(g_2)).\varphi = 0, \quad \varphi.(\theta(g_1)k-\lambda_m(g_1)) = 0.$$

Note that this equation makes sense since the two elements $k\theta(g_2)$ acting from the left and $\theta(g_2)k$ from the right are simultaneously diagonalizable on $A(SU_q(2))$. One can also show that there is a nonzero spherical function ψ^j_{mn} of type $(j;m,n)$ unique up to a constant multiple. Fixing a couple of weights (m,n) $(m,n\in\frac{1}{2}\mathbf{Z})$ with $m-n\in\mathbf{Z}$, we take a nonzero spherical function $e_{mn}(g_1,g_2)$ of minimal spin $\max\{|m|,|n|\}$. Then the spherical function ψ^j_{mn} of spin j is written in the form $\psi^j_{mn}= p^j_{mn}(x(g_1,g_2))e_{mn}(g_1,g_2)$ for some polynomial $p^j_{mn}(x)$ of degree $k=j-\max\{|m|,|n|\}$. We call $p^j_{mn}(x)$ the zonal part of ψ^j_{mn}. Then it turns out that the zonal parts of our spherical functions realize a 4-parameter family of the Askey-Wilson polynomials.

The Askey-Wilson polynomials $p_k(x;a,b,c,d|q)$ $(k\in\mathbf{N})$ are a 4-parameter family of orthogonal polynomials expressed by terminating q-hypergeometric series ${}_4\varphi_3$:

$$(11.4)\quad p_k(x;a,b,c,d|q)$$

$$= a^{-k}\ (ab,ac,ad;q)_k\ {}_4\varphi_3\left(\begin{matrix} q^{-k},\ abcdq^{k-1},\ az,\ az^{-1} \\ ab,\ ac,\ ad \end{matrix}\ ;\ q,\ q\right),$$

where $x=\frac{1}{2}(z+z^{-1})$. It is known that these polynomials $p_k(x;a,b,c,d|q)$ are symmetric in the parameters (a,b,c,d), although one cannot directly see from expression (11.4). Note also that, for general (a,b,c,d), they have the orthogonality relation with respect to a measure consisting of both continuous and discrete parts. The Askey-Wilson polynomials, introduced by [AW], can be regarded as a "master family" of q-orthogonal polynomials. In fact, they supply a large number of q-orthogonal polynomials including the little, the big and the continuous q-Jacobi polynomials as special (limiting) cases.

Theorem 4 ([NM5], see also [Koe]). For any $j\in\frac{1}{2}\mathbf{N}$ and $m,n\in I_j$, the zonal part of the doubly associated spherical function ψ^j_{mn} with respect to $(\theta(g_1),\theta(g_2))$ is a constant multiple of the Askey-Wilson polynomial in $x=x(g_1,g_2)$ of base q^2

$$(11.5)\qquad p_k(\ x\ ;\ \frac{t}{s}q,\ \frac{s}{t}q^{2\alpha+1},\ -\frac{1}{st}q,\ -stq^{2\beta+1}\ \mid q^2),$$

where $\alpha=|m-n|$, $\beta=|m+n|$, $k=j-\max\{|m|,|n|\}$. The parameters s,t are determined as follows, according to the region which the couple (m,n) belongs to:

Case (1) $m+n\geq 0$, $m\geq n$: $(s,t)=(|b_{11}/b_{21}|,|a_{11}/a_{21}|)$,

Case (2) $m+n\geq 0$, $m\leq n$: $(s,t)=(|a_{11}/a_{21}|,|b_{11}/b_{21}|)$,

Case (3) $m+n\leq 0$, $m\geq n$: $(s,t)=(|a_{21}/a_{11}|,|b_{21}/b_{11}|)$,

Case (4) $m+n\leq 0$, $m\leq n$: $(s,t)=(|b_{21}/b_{11}|,|a_{21}/a_{11}|)$. ■

Koornwinder's zonal spherical functions with respect to $(\theta_\sigma,\theta_\tau)$, explained in Section 9, are the cases when $\alpha=\beta=0$:

$$(11.6)\qquad p_k(\ x\ ;\ \frac{t}{s}q,\ \frac{s}{t}q,\ -\frac{1}{st}q,\ -stq\ \mid q^2)\quad (\text{with}\ \ s=q^\sigma,\ t=q^\tau);$$

these are the "Legendre part" of (11.5).

In the special case when $(s,t)=(1,1)$ and $K(g_1)=K(g_2)=SO(2)$, the above expression (11.5) represents the continuous q-Jacobi polynomials (9.5). In this sense, expression (11.5) suggests a way to reparametrize the Askey-Wilson polynomials as a 2-parameter extension of the continuous q-Jacobi polynomials:

$$(11.7)\ p_k^{(\alpha,\beta)}(x;s,t:q)=p_k(x;\frac{t}{s}q^{1/2},\frac{s}{t}q^{\alpha+1/2},-\frac{1}{st}q^{1/2},-stq^{\beta+1/2}|q).$$

Our realization of the Askey-Wilson polynomials also passes naturally to Podles' quantum 2-spheres. In fact, under the identification (10.4) one has

$$(11.8)\qquad A\,(S_q^2(c,d)) = \bigoplus_{j\in\mathbb{N}} U_j, \quad U_j = \bigoplus_{m\in I_j} \mathbb{C}\,\psi^j_{m0},$$

if $c=|a_{11}|^2$, $d=|a_{21}|^2$. Then Theorem 4 implies that the spherical functions on this Podles' quantum 2-sphere, relatively invariant with respect to the right action of $\theta(g_1)k$, are expressed by Askey-Wilson polynomials. Especially, Rogers' q-ultraspherical polynomials $p_k^{(\alpha,\alpha)}(x;q)$ (continuous q-Jacobi polynomials with $\alpha=\beta$) appear as spherical functions on Podles' quantum sphere $S_q^2(1,1)$ (see also [NM4]).

The big and the little q-Jacobi polynomials are also obtained as special limiting cases of the Askey-Wilson polynomials. In the framework of Theorem 4, we obtain spherical functions expressed by the big q-Jacobi polynomials as special limiting cases when $K(g_1)$ approaches to the diagonal (after a suitable rescaling of the variable x). Theorem 1, the little q-Jacobi case, can also be understood as a special case of Theorem 4 when $K(g_1)$ and $K(g_2)$ are both on the diagonal.

We remark that the Askey-Wilson polynomials, parametrized as in (11.7), all reduce to the Jacobi polynomials on some intervals. In fact one has

$$(11.9)\qquad \lim_{q\to 1} \frac{1}{(1-q)^k}\, p_k^{(\alpha,\beta)}(x;s,t:q)$$

$$= (\alpha)_k (s+\tfrac{1}{s})^k (t+\tfrac{1}{t})^k\ {}_2F_1\left[\begin{matrix} -k,\ \alpha+\beta+k+1 \\ \alpha+1 \end{matrix}\ ;\ \frac{s^2+t^2-2stx}{(1+s^2)(1+t^2)}\right]$$

In the classical case, the Jacobi polyonimals and the linear transformation of the variable x describe all these types of spherical functions on SU(2). Comparing these two cases, one can say that the part of linear transformations, in the $SU_q(2)$ case, is played by parameters of the Askey-Wilson polynomials.

The Jacobi polynomials have a number of q-analogues that are unified as the family of Askey-Wilson polynomials. The above realization of the Askey-Wilson polynomials on $SU_q(2)$ gives an interpretation of this variety of q-Jacobi polynomials in the "geometric" terms of quantum groups. This point is related to the existence of non-standard quantum homogeneous spaces and also to the problem of quantum subgroups.

To end this survey, I would like to give some additional remarks on the realization of q-orthogonal polynomials as spherical functions on quantum G-spaces. The (standard) odd dimensional quantum sphere S_q^{2n+1} can be defined as the quotient space $SU_q(n+1)/SU_q(n)$. In this case, the zonal spherical functions are also described by the little q-Jacobi polynomials $P_k^{(n-1,\beta)}(x;q)$ as a natural q-analogue (see [NYM1,2] and [VS2]). The non-standard quantum $SU_q(2)$-spaces $S_q^2(c,d)$ and $\tilde{S}_q^3$ can also be generalized to higher dimensional ones, non-standard $\mathbb{CP}^n$'s and $\tilde{S}_q^{2n+1}$ (see [Kor]); they also have zonal spherical functions expressed by big q-Jacobi polynomials. Still I do not know if Koornwinder's infinitesimal approach can be carried out in these higher dimensional cases.

References

[A] E.Abe: Hopf algebras, Cambridge tracts in mathematics 74, Cambridge University Press, 1980.

[AA] G.E.Andrews and R.Askey: Classical orthogonal polynomials, Lecture Notes in Math., **1171**, Springer, 1985, 36-62.

[AW] R.Askey and J.Wilson: Some basic hypergeometric orthogonal polynomials that generalize Jacobi polynomials, Memoirs Amer. Math. Soc. **54**(1985), No. 319.

[D] V.G.Drinfeld: Quantum groups, Proceedings of the International Congress of Mathematicians, Berkely, California, U.S.A., 1986, 798-820.

[GR] G.Gasper and M.Rahman: Basic hypergeometric series, Encylclopedia of Mathematics and its Applications Vol.35, Cambridge University Press, 1990.

[J1] M.Jimbo: A q-difference analogue of $U(\mathfrak{g})$ and the Yang-Baxter equation, Lett. Math. Phys. **10**(1985), 63-69.

[J2] M.Jimbo: A q-analogue of $U(\mathfrak{gl}(N+1))$, Heck algebra and the Yang-Baxter equation, Lett. Math. Phys. **11**(1986), 247-252.

[KR] A.N.Kirillov and N.Yu.Reshetikhin: Representations of the algebra $U_q(s\ell(2))$, q-orthogonal polynomials and invariants of links, in Infinite-dimensional Lie algebras and groups, edited by

V.G.Kac, World Scientific, 1989, 285-339.

[Koe] H.T.Koelink: The addition formula for continuous q-Legendre polynomials and associated spherical elements on the SU(2) quantum group related to Askey-Wilson polynomials, preprint 1990.

[KK] H.T.Koelink and T.H.Koornwinder: The Clebsh-Gordan coefficients for the quantum group $S_{\mu}U(2)$ and q-Hahn polynomials, to appear in Nederl. Acad. Wetensch. Proc..

[K1] T.H.Koornwinder: Representations of the twisted SU(2) quantum group and some q-hypergeometric orthogonal polynomials, Nederl. Akad. Wetensch. Proc. Ser.A **92**(1989), 97-117.

[K2] T.H.Koornwinder: The addition formula for little q-Legendre polynomials and the SU(2) quantum group, CWI Rep. AM-R8906, preprint.

[K3] T.H.Koornwinder: Continuous q-Legendre polynomials as spherical matrix elements of irreducible representations of the quantum SU(2) group, CWI Quaterly, 2(1989), 171-173.

[K4] T.H.Koornwinder: Orthogonal polynomials in connections with quantum groups, in Orthogonal Polynomials, Theory and Practice, edited by P.Nevai, NATO ASI Series, Kluwer Academic Publishers, 257-292, 1990.

[K5] T.H.Koornwinder: Askey-Wilson polynomials as zonal spherical functions on the SU(2) quantum group, preprint 1990.

[Kor] L.I.Korogodsky: Quantum projective spaces, spheres and hyperboloids, preprint 1990.

[MM1] T.Masuda, K.Mimachi, Y.Nakagami, M.Noumi and K.Ueno: Representations of quantum groups and a q-analogue of orthogonal polynomials, C. R. Acad. Sci. Paris **307**(1988), 559-564.

[MM2] T.Masuda, K.Mimachi, Y.Nakagami, M.Noumi and K.Ueno: Representations of the quantum group $SU_q(2)$ and the little q-Jacobi polynomials, to appear in J. Functional Analysis.

[NM1] M.Noumi and K.Mimachi: Quantum 2-spheres and big q-Jacobi polynomials, Commun. Math. Phys. **128**(1990), 521-531.

[NM2] M.Noumi and K.Mimachi: Big q-Jacobi polynomials, q-Hahn polynomials and a family of quantum 3-spheres, Lett. Math. Phys. **19**(1990), 299-305.

[NM3] M.Noumi and K.Mimachi: Spherical functions on a family of quantum 3-spheres, to appear in Compositio Mathematica.

[NM4] M.Noumi and K.Mimachi: Rogers' q-ultraspherical polynomials on a quantum 2-sphere, to appear in Duke Mathematical Journal.

[NM5] M.Noumi and K.Mimachi: Askey-Wilson polynomials and the quantum group $SU_q(2)$, Proc. Japan Acad. Ser.A **66**(1990), 146-149.

[NYM1] M.Noumi, H.Yamada and K.Mimachi: Zonal spherical functions on the quantum homogenous space $SU_q(n+1)/SU_q(n)$, Proc. Japan Acad. Ser.A **65**(1989), 169-171.

[NYM2] M.Noumi, H.Yamada and K.Mamachi: Finite dimensional representations of the quantum group $GL_q(n;\mathbf{C})$ and the zonal spherical functions on $U_q(n-1)\backslash U_q(n)$, preprint 1990.

[P] P.Podleś: Quantum spheres, Lett. Math. Phys. **14**(1987), 193-202.

[VS1] L.L.Vaksman and Ya.S.Soibelman: Algebra of functions on the quantum SU(2) group, Funct. Anal. i-ego Pril. **22**(1988), 1036 -1040 (in Russian).

[VS2] L.L.Vaksman and Ya.S.Soibelman: Algebra of functions on quatum group SU(n+1) and odd dimensional quantum spheres, to appear in Algebra and Analysis (in Russian).

[W1] S.L.Woronowicz: Twisted SU(2) group. An example of non-commutative differential calculus, Publ. RIMS, Kyoto Univ., **23**(1987), 117-181.

[W2] S.L.Woronowicz: Compact matrix pseudogroups, Comm. Math. Phys., **111**(1987), 613-665.

R-matrices with gauge parameters and multi-parameter quantized enveloping algebras

By Masato Okado*) and Hiroyuki Yamane**)

1. Introduction. In this note, we introduce a multi-parameter quantized enveloping algebra $U_{q,Q}(\mathfrak{g})$ associated with an arbitrary symmetrizable Kac–Moody Lie algebra $\mathfrak{g}$. Here $Q = (q_{ij})$ denotes a square matrix of degree rank $\mathfrak{g}$ whose entries are parameters satisfying $q_{ii} = 1, q_{ij}q_{ji} = 1$. If we put all q_{ij} equal to 1, a quotient of $U_{q,Q}(\mathfrak{g})$ turns out to be the Drinfeld–Jimbo quantized enveloping algebra $U_q(\mathfrak{g})$ [1,2]. Recently, Takeuchi [6] introduced two-parameter quantized enveloping algebras $U_{\alpha,\beta}$ and $U'_{\alpha,\beta}$ associated with $\mathfrak{gl}(n)$ and $\mathfrak{sl}(n)$, respectively. The Hopf algebra $U'_{\alpha,\beta}$ can be obtained as a special case of $U_{q,Q}(\mathfrak{sl}(n))$ by setting $q = \sqrt{\alpha\beta}, q_{i\,i+1} = \sqrt{\alpha\beta^{-1}}$, $q_{ij} = 1$ for $|j - i| > 1$.

Next we follow the argument of Jimbo [3] to obtain R–matrices. Let us take the n dimensional representation π_0 of $U_{q,Q}(\mathfrak{sl}(n))$. Let π, π' be two representations of $U_{q,Q}(\widehat{\mathfrak{sl}}(n))$ obtained as lifts of π_0. Consider the following equation for $\check{R}$:

$$\check{R}((\pi \otimes \pi') \circ \Delta(X)) = ((\pi' \otimes \pi) \circ \Delta(X))\check{R} \qquad (X \in U_{q,Q}(\widehat{\mathfrak{sl}}(n))). \tag{1.1}$$

We can show that the solution $\check{R}_{\pi\pi'}$ satisfies the Yang–Baxter equation:

$$(\check{R}_{\pi'\pi''} \otimes 1)(1 \otimes \check{R}_{\pi\pi''})(\check{R}_{\pi\pi'} \otimes 1) = (1 \otimes \check{R}_{\pi\pi'})(\check{R}_{\pi\pi''} \otimes 1)(1 \otimes \check{R}_{\pi'\pi''}). \tag{1.2}$$

This $\check{R}_{\pi\pi'}$ coincides with the solution discovered by Perk and Schultz [5] without using the Hopf algebra structure. After having finished this work, at Hayashibara Conference in 1990, Reshetikhin remarked that our $U_{q,Q}(\mathfrak{g})$ can be interpreted as his multi–parametrization of the Hopf algebra structure of $U_q(\mathfrak{g})$ (see the Appendix).

2. Multi–parameter quantized enveloping algebra. Let I denote a finite index set. Let $A = (a_{ij})_{i,j\in I}$ be a symmetrizable generalized Cartan matrix [4], and let $D = (d_i\delta_{ij})_{i,j\in I}$ be a diagonal matrix such that $d_i \in \mathbf{Z} \setminus \{0\}$, ${}^t(DA) = DA$. With each A we can associate a Kac–Moody Lie algebra $\mathfrak{g}$. Let K be a field and let q, q_{ij} be elements in $K \setminus \{0\}$ such that $q^{d_i} - q^{-d_i} \neq 0, q_{ii} = 1, q_{ij}q_{ji} = 1$ for $i, j \in I$. For simplicity, we define the matrix Q by $(q_{ij})_{i,j\in I}$. We define an associative algebra $U_{q,Q}(\mathfrak{g})$, which is generated

*) Department of Mathematical Science, Faculty of Engineering Science, Osaka University, Toyonaka, Osaka 560, Japan.

**) Department of Mathematics, Faculty of science, Osaka University, Toyonaka, Osaka 560, Japan.

by $\{e_i, f_i, L_i, L_i^{-1}, M_i, M_i^{-1}\}_{i\in I}$ satisfying the following relations:

$$
\begin{aligned}
&L_iL_i^{-1} = L_i^{-1}L_i = 1, \quad M_iM_i^{-1} = M_i^{-1}M_i = 1,\\
&L_iL_j = L_jL_i, \quad M_iM_j = M_jM_i, \quad L_iM_j = M_jL_i,\\
&L_ie_jL_i^{-1} = q^{d_ia_{ij}}q_{ij}e_j, \quad M_ie_jM_i^{-1} = q^{d_ia_{ij}}q_{ij}^{-1}e_j,\\
&L_if_jL_i^{-1} = q^{-d_ia_{ij}}q_{ij}^{-1}f_j, \quad M_if_jM_i^{-1} = q^{-d_ia_{ij}}q_{ij}f_j,\\
&[e_i, f_j] = \delta_{ij}\frac{L_i - M_i^{-1}}{q^{d_i} - q^{-d_i}},\\
&\sum_{\nu=0}^{1-a_{ij}} (-q_{ij})^{\nu}\begin{bmatrix}1-a_{ij}\\ \nu\end{bmatrix}_{q^{d_i}} e_i^{1-a_{ij}-\nu}e_je_i^{\nu} = 0 \quad (i\neq j),\\
&\sum_{\nu=0}^{1-a_{ij}} (-q_{ij})^{-\nu}\begin{bmatrix}1-a_{ij}\\ \nu\end{bmatrix}_{q^{d_i}} f_i^{1-a_{ij}-\nu}f_jf_i^{\nu} = 0 \quad (i\neq j).
\end{aligned}
\tag{2.1}
$$

Here the symbol $\begin{bmatrix}m\\ n\end{bmatrix}_t \in \mathbf{C}[t, t^{-1}]$ is defined by

$$
\begin{bmatrix}m\\ n\end{bmatrix}_t = \frac{(t^m - t^{-m})(t^{m-1} - t^{-m+1})\cdots(t^{m-n+1} - t^{-m+n-1})}{(t - t^{-1})(t^2 - t^{-2})\cdots(t^n - t^{-n})}.
$$

Furthermore $U_{q,Q}(\mathfrak{g})$ has the Hopf algebra structure. The coproduct Δ, counit ε and antipode S are defined by

$$
\begin{aligned}
&\Delta(L_i) = L_i\otimes L_i, \quad \Delta(M_i) = M_i\otimes M_i,\\
&\Delta(e_i) = 1\otimes e_i + e_i\otimes L_i, \quad \Delta(f_i) = f_i\otimes 1 + M_i^{-1}\otimes f_i,\\
&\varepsilon(L_i) = \varepsilon(M_i) = 1, \quad \varepsilon(e_i) = \varepsilon(f_i) = 0,\\
&S(L_i) = L_i^{-1}, \quad S(M_i) = M_i^{-1}, \quad S(e_i) = -e_iL_i^{-1}, \quad S(f_i) = -M_i^{-1}f_i.
\end{aligned}
\tag{2.2}
$$

3. Main result. Let us restrict ourselves to the case $\widehat{\mathfrak{g}} = \widehat{\mathfrak{sl}}(n)$. The generalized Cartan matrix $A = (a_{ij})_{0\le i,j\le n-1}$ is given by $a_{ij} = 2\delta_{ij}^{(n)} - \delta_{i\,j+1}^{(n)} - \delta_{i\,j-1}^{(n)}$, where $\delta_{ij}^{(n)} = 1$ ($i\equiv j \bmod n$), $= 0$ (otherwise). We can set D to be the identity matrix.

Set

$$
K = \mathbf{C}\bigl(q, q_{ij}(0\le i,j\le n-1) \mid q_{ii} = 1,\ q_{ij}q_{ji} = 1,\ \prod_{0\le k\le n-1} q_{ik} = 1\,(0\le i,j\le n-1)\bigr),
$$

$$
K' = K\bigl(\lambda_i, \lambda_i', t_i, t_i'\,(0\le i\le n-1) \mid \prod_{0\le i\le n-1} t_i = \prod_{0\le i\le n-1} t_i' = \prod_{0<i<j\le n-1} q_{ij}\bigr).
$$

Put

$$
r_{ij} = \prod_{i\le k\le j} q_{ik}^{-1} \quad (i\le j), \quad = \prod_{j<k\le i} q_{ik} \quad (i>j).
$$

Let us consider the following K-algebra homomorphism $\pi : U_{q,Q}(\widehat{\mathfrak{sl}}(n)) \longrightarrow \mathrm{End}((K')^n)$:

$$\begin{aligned}(3.1)\qquad &\pi(L_i) = t_i \sum_{0\le j\le n-1} q^{\delta^{(n)}_{ij+1}-\delta^{(n)}_{ij}} r_{ij} E_{jj}, \quad \pi(M_i) = t_i^{-1} \sum_{0\le j\le n-1} q^{\delta^{(n)}_{ij+1}-\delta^{(n)}_{ij}} r_{ij}^{-1} E_{jj},\\ &\pi(e_i) = \lambda_i t_i E_{i-1\, i}, \quad \pi(f_i) = \lambda_i^{-1} E_{i\, i-1}.\end{aligned}$$

Here E_{ij} denotes the matrix unit and $E_{-1\,0}, E_{0\,-1}$ should be understood as $E_{n-1\,0}, E_{0\,n-1}$, respectively. We also define the homomorphism $\pi' : U_{q,Q}(\widehat{\mathfrak{sl}}(n)) \longrightarrow \mathrm{End}((K')^n)$ by replacing $\{\lambda_i, t_i\}$ with $\{\lambda'_i, t'_i\}$.

For later use, let us define

$$c_{ij} = \frac{t'_1 t'_2 \cdots t'_j}{t_1 t_2 \cdots t_i} \quad (0 \le i, j \le n-1),$$

$$s_{ij} = r_{i+1\,i} r_{i+2\,i} \cdots r_{ji}, \quad \gamma_{ij} = \frac{\lambda'_{i+1}\lambda'_{i+2}\cdots\lambda'_j}{\lambda_{i+1}\lambda_{i+2}\cdots\lambda_j} \quad (0 \le i < j \le n-1).$$

We extend the definition of γ_{ij}, s_{ij} by $\gamma_{ij} = \gamma_{ji}^{-1}, s_{ij} = s_{ji}^{-1}$ for $i > j$. Let us consider the following K'-linear equation:

$$(3.2)\qquad \check{R}((\pi \otimes \pi') \circ \Delta(e_i)) = ((\pi' \otimes \pi) \circ \Delta(e_i))\check{R} \quad (0 \le i \le n-1).$$

Using the properties of the coproduct Δ and the argument in [3], we have

Theorem. (1) *The equation* (3.2) *has a unique solution given as follows (up to a constant multiple)* :

$$\begin{aligned}(3.3)\qquad \check{R}_{\pi\pi'} = &\sum_i c_{ii}(q^2 x - 1) E_{ii} \otimes E_{ii}\\ &+ \sum_{i\ne j} \{c_{jj}\gamma_{ij}(q^2-1) x^{\theta(i-j)} E_{ii}\otimes E_{jj} + c_{ij} s_{ij} q (x-1) E_{ij}\otimes E_{ji}\}.\end{aligned}$$

Here $x = (\lambda'_0\lambda'_1\cdots\lambda'_{n-1})/(\lambda_0\lambda_1\cdots\lambda_{n-1})$ *and* $\theta(i) = 1$ $(i \ge 0)$, $= 0$ *(otherwise).*

(2) *Let* π *(resp.* π', π''*) be the homomorphism* (3.1) *with parameters* $\{\lambda_i, t_i\}$ *(resp.* $\{\lambda'_i, t'_i\}$, $\{\lambda''_i, t''_i\}$*). Then the Yang-Baxter equation holds:*

$$(\check{R}_{\pi'\pi''} \otimes 1)(1 \otimes \check{R}_{\pi\pi''})(\check{R}_{\pi\pi'} \otimes 1) = (1 \otimes \check{R}_{\pi\pi'})(\check{R}_{\pi\pi''} \otimes 1)(1 \otimes \check{R}_{\pi'\pi''}).$$

Remark 1. For the existence of the solution, we need the condition $\prod_{0\le i\le n-1} t_i = \prod_{0\le i\le n-1} t'_i = \prod_{0<i<j\le n-1} q_{ij}$ in the definition of K'.

Remark 2. The usual R-matrix for $U_q(\widehat{\mathfrak{sl}}(n))$ is obtained by specializing $q_{ij}(0 \le i, j \le n-1)$, $t_i(0 \le i \le n-1)$ and $\lambda_i(1 \le i \le n-1)$ to be 1 in (3.3) (see [3]).

Remark 3. Our $\check{R}_{\pi\pi'}$ was already discovered by Perk–Schultz [5]. More precisely, their solution of the Yang-Baxter equation is as follows (see (26) in [5]):

$$\begin{aligned}&\sum_i c_{ii}(q^2 x^{(1+\epsilon_i)/2} - x^{(1-\epsilon_i)/2}) E_{ii} \otimes E_{ii}\\ &+ \sum_{i\ne j} \{c_{jj}\gamma_{ij}(q^2-1) x^{\theta(i-j)} E_{ii}\otimes E_{jj} + c_{ij} s_{ij} q (x-1) E_{ij}\otimes E_{ji}\}.\end{aligned}$$

Here $\epsilon_i = \pm 1$ are discrete parameters.

Appendix. Suppose that $\mathfrak{g}$ is finite dimensional. Let us recall the definition of the h-adic topological quantized enveloping algebra $U_h(\mathfrak{g})$ over $\mathbf{C}[[h]]$. $U_h(\mathfrak{g})$ is generated by the elements $X_i, Y_i, H_i (i \in I)$ satisfying the following relations:

$$[H_i, H_j] = 0,$$
$$[H_i, X_j] = a_{ij} X_j, \quad [H_i, Y_j] = -a_{ij} Y_j,$$
$$[X_i, Y_j] = \delta_{ij} \frac{\sinh hd_i H_i}{\sinh hd_i},$$
$$\sum_{\nu=0}^{1-a_{ij}} (-)^{\nu} \begin{bmatrix} 1-a_{ij} \\ \nu \end{bmatrix}_{e^{hd_i}} X_i^{1-a_{ij}-\nu} X_j X_i^{\nu} = 0 \quad (i \neq j),$$
$$\sum_{\nu=0}^{1-a_{ij}} (-)^{\nu} \begin{bmatrix} 1-a_{ij} \\ \nu \end{bmatrix}_{e^{hd_i}} Y_i^{1-a_{ij}-\nu} Y_j Y_i^{\nu} = 0 \quad (i \neq j),$$

The coproduct Δ_0 of $U_h(\mathfrak{g})$ is defined by

$$\Delta_0(H_i) = H_i \otimes 1 + 1 \otimes H_i,$$
$$\Delta_0(X_i) = 1 \otimes X_i + X_i \otimes \exp hd_i H_i,$$
$$\Delta_0(Y_i) = Y_i \otimes 1 + \exp(-hd_i H_i) \otimes Y_i.$$

Reshetikhin's remark is as follows: Define the homomorphism $\overline{\Delta}_0 : U_h(\mathfrak{g}) \longrightarrow U_h(\mathfrak{g}) \hat{\otimes} U_h(\mathfrak{g})$ by

$$\overline{\Delta}_0(x) = \exp\big(h \sum_{i,j} \varphi_{ij} H_i \otimes H_j\big) \Delta_0(x) \exp\big(-h \sum_{i,j} \varphi_{ij} H_i \otimes H_j\big).$$

Here $\varphi_{ij} (i,j \in I)$ are parameters satisfying $\varphi_{ij} + \varphi_{ji} = 0$. Then $U_h(\mathfrak{g})$ has another Hopf algebra structure with the coproduct $\overline{\Delta}_0$.

Now put

$$Z_i = \exp\big(h \sum_{j,k} \varphi_{jk} a_{ji} H_k\big),$$
$$e_i = Z_i X_i, \quad f_i = Z_i Y_i,$$
$$L_i = Z_i^2 \exp hd_i H_i, \quad M_i = Z_i^{-2} \exp hd_i H_i.$$

Then setting $q = e^h$, $q_{ij} = \exp\big(2h \sum_{k,l} \varphi_{kl} a_{ki} a_{lj}\big)$, the elements e_i, f_i, L_i, M_i in $\big(U_h(\mathfrak{g}), \overline{\Delta}_0\big)$ satisfy the same formulas (2.1-2) as those of $\big(U_{q,Q}(\mathfrak{g}), \Delta\big)$.

Acknowledgements. The authors would like to express their hearty thanks to Professors E. Date, M. Jimbo, M. Kashiwara, T. Miwa and J. Murakami for their valuable comments. The second author also thanks Professor M. Takeuchi for sending his preprint. The authors also thank Professor N. Yu. Reshetikhin for valuable discussions at Hayashibara Conference.

References

[1] V.G. Drinfeld: Quantum groups, Proc. ICM Berkeley, 798–820 (1987).

[2] M. Jimbo: A q–difference analogue of $U(\mathfrak{g})$ and the Yang–Baxter equation, Lett. Math. Phys. **10**, 63–69 (1985).

[3] M. Jimbo: Quantum R matrix for the generalized Toda system, Commun. Math. Phys. **102**, 537–547 (1986).

[4] V.G. Kac: Infinite–dimensional Lie algebras, 2nd edition, Cambridge Univ. Press, Cambridge (1985).

[5] J.H.H. Perk and C.L. Schultz: 'Families of commuting transfer matrices in q–state vertex models' in Non–linear Integrable systems—Classical theory and quantum theory (Proceedings of RIMS symposium, Kyoto, Japan, 13–16 May 1981) ed. by M. Jimbo and T. Miwa, World Scientific, Singapore (1983).

[6] M. Takeuchi: A two–parameter quantization of $GL(n)$ (Summary), Proc. Japan Acad. **66**, Ser.A, No.5, 112–114 (1990).

ANALYTIC EXPRESSION OF VOROS COEFFICIENTS AND ITS APPLICATION TO WKB CONNECTION PROBLEM

Kanehisa Takasaki

RIMS, Kyoto University, Kitashirakawa, Sakyo-ku, Kyoto-shi 606, Japan

1. Introduction

Usually, the WKB method starts from formal solutions (WKB or Liouville-Green solutions) expanded in powers of the Planck constant, and connects these solutions by asymptotic matching at turning points. Voros [V] proposed a resummation prescription of these formal calculations, and argued that his results should be deeply related with Ecalle's theory of "resurgent functions." Further progress along that line has been made by F. Pham and his coworkers [DDP]. We report another approach based upon an idea of Olver [O].

Olver's method for the WKB connection problem, unlike the asymptotic matching at turning points, is based upon analysis at points at infinity. Naturally, one needs semi-global information on a set of solutions for which to consider the connection problem. Olver's idea is very intriguing, because it directly gives an exact connection formula without using any approximation. His connection formulas, however, contain strange quantities whose analytic properties were fairly obscure at that moment; Olver gave only qualitative results on these quantities. From our present standpoint, it it not hard to notice that these quantities (which should be called the "Olver coefficients" or something like that) are nothing else than the "Voros coefficients" in the terminology of Ecalle and Pham. This inspires us with the hope to find some new analytical expressions to the Voros coefficients.

Our basic strategy is just to combine Olver's idea with basic techniques in scattering theory [AKNS], [ZS]. We first convert the problem, as Olver does, to another linear problem by the so called Liouville transformation. This is a quite standard technique [H], and also used in the derivation of Liouville-Green formal solutions. From an analytical point of view, the transformed equation resembles a linear problem in scattering theory; the potential now decays at the rate of inverse squared of the distance from the origin. This fact also lies in the heart of Olver's method, because the definition of his coefficients relies upon that property. Olver's book stops at this stage, but we attempt to go forward further. From the point of view of scattering theory, Voros coefficients may be identified with the "a"-coefficient (inverse transmission coefficients) that is defined along with the "b"-coefficients. (The ratio b/a is called the reflection coefficients.) Further, it is well known in scattering theory that "a" and "b" are connected with Jost solutions by a simple integral relation. From these observations, we are naturally led to an iterated integral series expansion of Voros coefficients.

Under some additional condition, one can further rewrite this iterated integral series into a Laplace integral; the integrand has again an iterated integral series expansion. This part is largely inspired by work of Grigis and Gérard [GG]. Since such a Laplace integral representation is a basic object in the theory of resurgent functions as well as in the work of Voros, we expect to deduce from Olver's connection formulas some new insight into the Ecalle-Voros theory. We shall show only a few examples anticipating further progress in that direction.

The last section is devoted to issues beyond the scope of the second order Sturm-Liouville problem. These are still speculative, but appear to offer various interesting material.

I wish to express my sincere gratitude to Professor Toshihiko Nishimoto for drawing my attention to Olver's work, and to Professor Nobuyuki Tose for encouragement as well as his help for collecting material unavailable in Japan. I am also indebted to Professors Alain Grigis, Kazuhiko Aomoto, Frederic Pham, Koichi Uchiyama and Masafumi Yoshino for discussions and useful information. A more detailed report on the results announced below will be published elsewhere.

2. Second order linear problem and Liouville transformation

We consider the linear equation

$$\psi_{qq} = \lambda^2 f(q)\psi, \tag{1}$$

in the complex plane, where $f(q)$ is a complex analytic function with some nice analytical properties (see below), and λ is a nonzero complex parameter. The subscript "qq" stands for the second derivative, $\psi_{qq} = d^2\psi/dq^2$. We mostly assume that λ is a real positive number; however, we shall see that extending it to complex values becomes crucial later on.

The above linear problem can be converted into a new equation of the form

$$\phi_{ss} = (\lambda^2 + h)\phi, \tag{2}$$

$$h \underset{\text{def}}{=} -f^{-3/4}(f^{-1/4})_{qq}, \tag{3}$$

by the transformation of variables

$$\phi = f^{1/4}\psi, \quad s = \int_{q_0}^{q} dq' f(q')^{1/2}. \tag{4}$$

The determination of $f^{1/4}$ and the point q_0 are suitably chosen subject to the situation in consideration. This transformation is known for years (since, probably, the days of Liouville). We call it the "Liouville transformation," and the s-plane the "Liouville plane." Actually, this can be slightly generalized to the linear equation

$$\psi_{qq} = (\lambda^2 f(q) + g(q))\psi, \tag{1'}$$

and leads to the same equation as (2) except that h is then given by

$$h \underset{\text{def}}{=} g/f - f^{-3/4}(f^{-1/4})_{qq}. \tag{3'}$$

Equation (2) may be viewed as a "perturbation" of the equation with $h = 0$, the latter being readily solved by the exponential functions $e^{\pm\lambda s}$. This point of view lies in the heart of the so called "Liouville-Green approximation." For further analysis, we assume that

$$h = O(|s|^{-2}) \quad \text{as } |s| \to \infty. \tag{5}$$

A typical case is the following.

Proposition. *Condition (5) is satisfied if $f(q)$ is a polynomial and $g(q) = 0$.*

Roughly, (5) means that the linear problem on the Liouville plane is of "scattering type" as opposed to the original problem on the q-plane. For our actual analysis of connection problems, condition (5) may be further relaxed; it is frequently sufficient to require the decay property only in a subdomain of the Liouville plane or simply along several curves with both ends at infinity.

Of course, even if (5) is satisfied, the Liouville transformation is a somewhat subtle thing, because $h = h(s)$ is in general a multi-valued function on the s-plane with branch point singularities at the image of zero's of $f(q)$, i.e., at "turning points." It should be noted that the multi-valuedness comes only from that of the inverse map $q = q(s)$ of the Liouville transformation $s = s(q)$. On the q-plane, h in (3) is given by

$$h = \frac{4ff_{qq} - 5f_q{}^2}{16f^3}, \tag{6}$$

hence it has poles at the zeros of $f(q)$, but no branch point singularities. It is the multi-valuedness of $q = q(s)$ that makes $h(s)$ a multi-valued function on the Liouville plane.

The sheet structure of $h = h(s)$ is, in general, very complicated. The following general notions are convenient to understand the geometric situation.

Definition.
- Zeros of $f(q)$ are called "turning points."
- A "Stokes curve" is a curve that starts from a turning point and whose image on the Liouville plane is a half-line or a segment parallel to the real axis.
- A "Principal (or anti-Stokes) curve" is a curve that starts from a turning point and whose image on the Liouville plane is a half-line or a segment parallel to the imaginary axis.

More precisely, these are the definitions for the case of $\lambda > 0$; if $\arg\lambda \neq 0$, the part "parallel to the real (imaginary) axis" in the above definition should be modified as "rotated from the real (imaginary) axis by an angle of $-\arg\lambda$."

For illustration, we now consider the case of a harmonic oscillator (see Fig. 1):

$$f(q) = q^2 - E, \quad E > 0 \tag{7}$$

with $\lambda > 0$. In this case, there are two turning points at $q = \pm E^{1/2}$, six Stokes curves and five principal curves. The four points

$$\begin{aligned} \infty_1 &: \ q = -\infty \\ \infty_2 &: \ q = +i\infty \\ \infty_3 &: \ q = -i\infty \\ \infty_4 &: \ q = +\infty \end{aligned}$$

at infinity will play an important role in the formulation of our connection problem. We choose paths for reaching these points as indicated in Fig. 1.

3. Solutions of Liouville-Green form

If one compares the linear problem on the Liouville plane with the situation of potential scattering theory, it would be natural to seek for solutions of the form

$$\phi_\pm = w_\pm e^{\pm s\lambda}. \tag{8}$$

The amplitude part $w_\pm$ should be, in some sense, close to the unity. This can achieved in two different ways.

3.1. Formal solutions

One way is to convert the linear equation by the well known transformation

$$v \underset{\text{def}}{=} (\log \phi)_s \tag{9}$$

to the Riccati equation

$$v_s + v^2 = \lambda^2 + h. \tag{10}$$

Substitution of a formal expansion

$$v_\pm(s, \lambda) = \pm\lambda + \sum_{n=1}^{\infty} v_{\pm,n}(s)\lambda^{1-n} \tag{11}$$

give rise to a set of relations that determine the coefficients recursively as:

$$v_{\pm,1} = 0, \quad v_{\pm,2} = \pm h/2, \quad v_{\pm,3} = h_s/4, \ldots.$$

Solving (9) as

$$\phi_\pm = \exp \int^s ds' v_\pm(s', \lambda), \tag{12}$$

one obtains a pair of solutions as desired, and this is essentially the same as the so called Liouville-Green (or WKB) solutions. (The usual construction is done on the q-plane.)

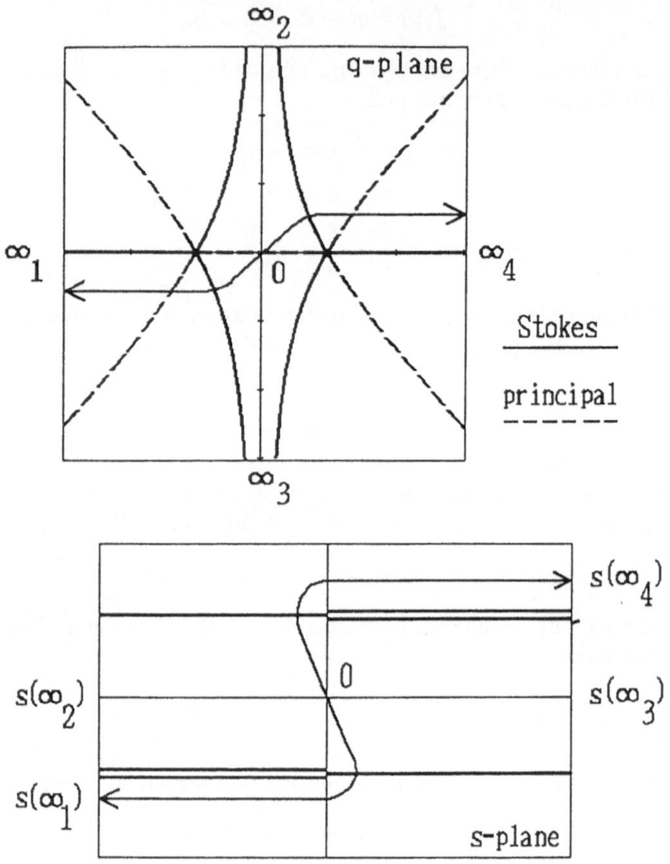

Fig. 1. q-plane and s-plane for $f(q) = q^2 - E,\ E > 0$

These solutions are, however, only *formal* power series of λ, and do not converge in general.

3.2. Analytic solutions

Another way is similar to the usual method in scattering theory. In that approach, one converts the differential equations

$$\left(\frac{d^2}{ds^2} \pm 2\lambda \frac{d}{ds}\right) w_\pm = h w_\pm \tag{13}$$

for $w_\pm$ into the integral equations

$$w_+(s,\lambda) = 1 + \int_{s_0^+}^{s} dt \frac{1-e^{2(t-s)\lambda}}{2\lambda} h(t) w_+,$$

$$w_-(s,\lambda) = 1 + \int_{s}^{s_0^-} dt \frac{1-e^{2(s-t)\lambda}}{2\lambda} h(t) w_+, \tag{14}$$

where $s_0^{\pm}$ are some fixed points. These integral equations can be solved by successive substitution (i.e., Neumann series):

$$w_+(s,\lambda) = 1 + \sum_{n=1}^{\infty} w_+^{(n)}(s,\lambda),$$

$$w_+^{(n)}(s,\lambda) \underset{\text{def}}{=} \int_{s_0^+}^{s=t_{n+1}} dt_n \int_{s_0^+}^{t_n} dt_{n-1} \cdots \int_{s_0^+}^{t_2} dt_1 \prod_{i=1}^{n} \frac{h(t_i)(1-e^{2(t_i-t_{i+1})\lambda})}{2\lambda},$$

$$w_-(s,\lambda) = 1 + \sum_{n=1}^{\infty} w_-^{(n)}(s,\lambda),$$

$$w_-^{(n)}(s,\lambda) \underset{\text{def}}{=} \int_{s=t_0}^{s_0^-} dt_1 \int_{t_1}^{s_0^-} dt_2 \cdots \int_{t_{n-1}}^{s_0^-} dt_n \prod_{i=1}^{n} \frac{(1-e^{2(t_{i-1}-t_i)\lambda})h(t_i)}{2\lambda}. \tag{15}$$

Note that the integral equations implement, as well, some initial condition at the point $s_0^{\pm}$. In our treatment of global connection problems, we choose $s_0^{\pm}$ to be points at infinity (just as in the construction of Jost solutions in scattering theory). It is exactly at this stage that decay property (5) play a crucial role. Further, we have to select a suitable path of integration so that the exponential functions in (15) have a uniform bound; the above Neumann series will then converge. In view of these requirements, we now assume that

- the points $s_0^{\pm}$ are put at infinity with Re $s_0^{\pm} = \mp\infty$;
- the integrals in (14) and (15) are along such paths $\Gamma_{\pm}(s)$ that starts from $s_0^{\pm}$ and ends at q;
- along $\Gamma_{\pm}(s)$, $\pm$Re s is monotonously increasing. (Such a path is said to be "progressive.")

Under that situation, one has indeed the uniform bound

$$|1-e^{2(t_i-t_{i+1})\lambda}| \le 2,$$

and one can prove the following basic result.

Proposition. *$w_{\pm}^{(n)}$ satisfy the inequality*

$$|w_{\pm}^{(n)}(s,\lambda)| \le \frac{V_{\pm}(s)^n}{n!|\lambda|}, \quad V_{\pm}(s) \underset{\text{def}}{=} \int_{\Gamma_{\pm}(s)} dt|h(t)|. \tag{16}$$

Therefore the Neumann series converge and obey the inequality

$$|w_{\pm}(s,\lambda) - 1| \le \exp(V_{\pm}(s)/|\lambda|) - 1. \tag{17}$$

Solutions of the original linear equation on the q-plane are now given by

$$\psi_{\pm}(q,\lambda) = w_{\pm}(s(q),\lambda)(ds(q)/dq)^{-1/2}e^{\pm s(q)\lambda}, \tag{18}$$

which we call exact solutions of the Liouville-Green form. If $s_0^{\pm}$ are the images of points $\infty_{\pm}$ at infinity of the q-plane, these solution are exponentially small ("recessive" in the terminology of the WKB method) in a neighborhood of $\infty_{\pm}$. Such solutions play a basic role in our treatment of the connection problem; a more precise situation is presented in the beginning of the next section. If $\arg\lambda \neq 0$, all conditions on Re $s_0^{\pm}$ and Re s should be replaced to Re $s_0^{\pm}\lambda$ and Re $s\lambda$, and everything goes in a quite parallel way.

3.3. Asymptotic expansion linking formal and analytic solutions

If the above analytic solutions have asymptotic expansion as $\lambda \to \infty$ (in a sector, for example), the asymptotic series should agree, up to a factor independent of q, with a formal solution described in Subsection 3.1 (because asymptotic expansion, if exists, is unique). The Laplace integral representation discussed later on indeed yields such asymptotic expansion.

4. Formulation of connection problem

4.1. General setting

In the following, we consider the case of $\lambda > 0$ alone. This is just for simplicity of presentation; everything carries over to the case of $\arg\lambda \neq 0$.

To consider the WKB connection problem, we first fix a cut-sheet of $f(q)^{1/2}$ and a determination of $s(q)$ and $(ds(q)/dq)^{-1/2}(= f(q)^{-1/4})$, though this is more or less for convenience of calculations. We then select a set of points ∞_I ($I = 1, 2, \ldots$) and attach to them an exact solution of the Liouville-Green form,

$$\psi_I(q,\lambda) = w_I(s(q),\lambda)(ds/dq)^{-1/2}\exp\epsilon_I s(q)\lambda, \tag{19}$$

where ϵ_I is a sign factor, taking values in ± 1, and subject to the condition that

$$\text{Re}\ \epsilon_I s(q) = -\infty \quad \text{at} \quad q = \infty_I. \tag{20}$$

For each solution, we take a domain D_I comprised of points q with a path $C_I(q)$ that links ∞_I with q (avoiding turning points, of course) and deforms continuously as q moves. Further, its image $\Gamma_I(s(q)) =_{\text{def}} s(C_I(q))$ on the Liouville plane is assumed to be progressive with respect to Re $\epsilon_I s$. The amplitude part $w_I(s,\lambda)$ is given, for each point s of $\Delta_I =_{\text{def}} s(D_I)$, by the Neumann series in the previous section. Such a domain D_I can be chosen farly large, as illustrated below; this fact is crucial in Olver's method.

4.2. Example: harmonic oscillator

For illustration, we now consider again the case of the harmonic oscillator with $E > 0$ and $\lambda > 0$. To fix a determination of s and $(ds/dq)^{-1/2}$, we cut the q-plane along two curves that start, respectively, from the two turning points and tend to infinity in the second and fourth quadrant of the q-plane (see Fig. 2).

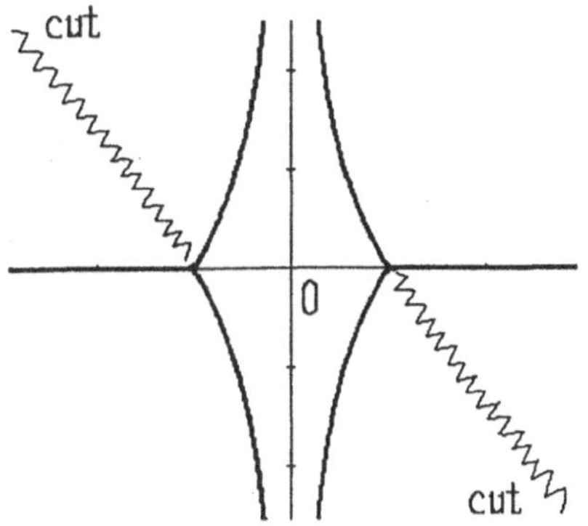

Fig. 2. Cut sheet on the q-plane for the harmonic oscillator

As a set of reference points at infinity, we take the four points $\infty_1, \ldots, \infty_4$ as mentioned in Section 2, and construct four solutions of the Liouville-Green form:

$$\psi_I = w_I(ds/dq)^{-1/2}e^{s\lambda} \quad (I = 1,2),$$
$$\psi_I = w_I(ds/dq)^{-1/2}e^{-s\lambda} \quad (I = 3,4). \tag{21}$$

For each ψ_I, one can determine a maximal domain D_I of points q that can be reached by a progressive path $C_I(q)$ from ∞_I. Let us call P_{IJ} the principal curve between ∞_I and ∞_J.

- D_1 is the complement of the union of the interval $[-E^{1/2}, +E^{1/2}]$ and the domain on the right side of $P_{24} \cup P_{34}$
- D_2 is the whole plane cut along P_{13} and P_{34}.
- D_3 is the whole plane cut along P_{12} and P_{24}.
- D_4 is the complement of the union of the interval $[-E^{1/2}, +E^{1/2}]$ and the domain on the left side of $P_{12} \cup P_{13}$.

4.3. Olver's basic idea

In Olver's method to the WKB connection problem, connection formulas are given as a collection of linear relations

$$\psi_I = c_{IJ}\psi_J + c_{IK}\psi_K. \tag{22}$$

among suitable triples $\{\psi_I, \psi_J, \psi_K\}$ that we call "fundamental triplets." We say a triplet is fundamental if each of the three reference points $\infty_I, \infty_J, \infty_K$ at infinity can be reached from the other two through the corresponding two domains in D_I, D_J, D_K, in other words, if there are three progressive paths

$$C_{IJ} \subset D_I \cap D_J, \quad C_{JK} \subset D_J \cap D_K, \quad C_{KI} \subset D_K \cap D_I$$

and C_{IJ} links ∞_I and ∞_J, etc. In the case of the harmonic oscillator above, fundamental triplets are $\{\psi_1, \psi_2, \psi_3\}$ and $\{\psi_2, \psi_3, \psi_4\}$. Olver gives explicit formulas of the coefficients of (22) for such an fundamental triplet. In fact, Olver's consideration in his book is limited to a *generic case*, i.e., the case where the triplet is associated with three Stokes curves that start from a turning point and tend to the three infinite points without meeting any other turning points. (Note that this also implicitly assume that the turning point at the center is of order one, i.e., a zero of $f(q)$ of order one.) We shall show later that Olver's method can be extended to more general cases.

A key to Olver's observation is the following general result.

Proposition. *Let ψ_I and ψ_J be two solutions of the Liouville-Green form with $\epsilon_I = +1$ and $\epsilon_J = -1$, and suppose that the reference points ∞_I and ∞_J are linked by a curve C_{IJ} whose image Γ_{IJ} on the Liouville plane is progressive (i.e.,* Re *$s(q)$ is increasing as q tends from ∞_I to ∞_J). Then w_I and w_J both have finite boundary values $\lim_{q\to\infty_J} w_I$ and $\lim_{q\to\infty_I} w_J$, and these boundary values actually coincide.*

The boundary values

$$a_{IJ}(\lambda) \underset{\text{def}}{=} \lim_{q\to\infty_J} w_I = \lim_{q\to\infty_I} w_J \tag{23}$$

are exactly the Voros (or Olver) coefficient in the sense of Section 1. For a generic fundamental triplet $\{\psi_I, \psi_J, \psi_K\}$ (Fig. 3), one has three such quantities

$$a_{IJ} = a_{JI}, \quad a_{JK} = a_{KJ}, \quad a_{KI} = a_{IK}. \tag{24}$$

Olver's basic idea is to determine the coefficients c_{IJ} and c_{IK} by comparing the behavior of the tree terms in (22) as q tends to ∞_J and ∞_K. Note that each term in (22) carries a factor (Liouville-Green factor, so to speak) of the form $(ds/dq)^{-1/2}e^{\pm s\lambda}$, the other part being a quantity with a definite boundary value at each reference point at infinity. The two Liouville-Green factors have opposite behavior, one being exponentially large ("dominant") and the other exponentially small ("recessive"). As q tends to ∞_J or to ∞_K, one can select a dominant one, divide both hand sides of (22) by that factor, and consider the limit. This yields two linear equations that determine c_{IJ} and c_{IK} in terms of the boundary values of w's (i.e., a's). In this calculation, however, one should also take into account the multi-valuedness of $(ds/dq)^{-1/2}$ and $e^{\pm s\lambda}$; to this end, we have

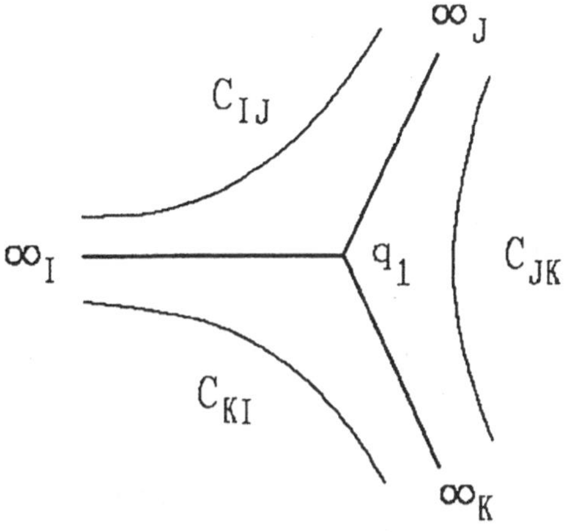

Fig. 3. A generic fundamental triplet

fixed a cut-sheet on the q-plane, and selected a determination of these functions. Across a cut line, they have discontinuity:

$$\begin{aligned} (ds/dq)^{-/2} &\longrightarrow \sigma i (ds/dq)^{-1/2}, \\ e^{\pm s\lambda} &\longrightarrow e^{\pm(-s+2\omega)\lambda}, \end{aligned} \tag{25}$$

where σ takes values in $\{+1, -1\}$ depending on the way to go across the cut line; ω is determined by the turning point to which the cut line is hooked. Doing these calculations, one can explicitly write the coefficients c_{IJ} and $_{IK}$ in terms of the following three quantities.

- powers of $i = \sqrt{-1}$,
- exponential functions of the form $e^{-2\omega\lambda}$, $\omega \in \mathbb{C}$,
- the $a_{IJ}(\lambda)$'s introduce above.

4.4 Connection formulas for harmonic oscillator

In the case of the harmonic oscillator with $E > 0$ and $\lambda > 0$, one can thus obtain the following connection formulas for the fundamental triplets $\{\psi_1, \psi_2, \psi_3\}$ and $\{\psi_2, \psi_3, \psi_4\}$.

$$\begin{aligned} \psi_1 &= \frac{1}{a_{23}(\lambda)}\psi_2 + i\frac{e^{-\pi i E\lambda/2}}{a_{23}(\lambda)}\psi_3, \\ \psi_4 &= i\frac{e^{-\pi i E\lambda/2}}{a_{23}(\lambda)}\psi_2 + \frac{1}{a_{23}(\lambda)}\psi_3. \end{aligned} \tag{26}$$

At first sight, a_{IJ}'s other than a_{23} appear to have no contribution, but this is not the case; they are simply reduced to the unity:

$$a_{12}(\lambda) = a_{13}(\lambda) = a_{42}(\lambda) = a_{43}(\lambda) = 1. \tag{27}$$

As a progressive path C_{23} for a_{23}, one may take the imaginary axis of the q-plane, which is mapped to the real axis of the Liouville plane (see Fig. 1).

The case of quartic oscillators as Voros considered can be analyzed in much the same way. A class of equations with nonpolynomial potentials, such as the Mathieu equation, can also be dealt with along the same line.

5. Iterated integrals and Laplace integrals

Olver's method thus gives exact connection formulas without any approximation. An unsatisfactory feature is that analytic structure of the basic quantities a_{IJ} is obscure from the construction. What Olver did is to show some qualitative estimates of a_{IJ}'s as well as conditions under which a_{IJ}'s become trivial as in (25). As Olver discussed, a_{IJ} always behaves as

$$a_{IJ} = 1 + O(\lambda^{-1}) \quad (\lambda \to \infty). \tag{28}$$

The lowest order WKB approximation is simply to replace a_{IJ} by the unity. What, for example, about higher order corrections? This is the place where we now attempt at a more detailed analysis.

We shall also deal with non-generic cases. If Stokes curves and turning points are in some "degenerate" configuration, Olver's method outlined above should be applied very carefully or, in extremely degenerate case, suitably modified.

5.1. General results

Let us recall the fact that the potential h on the Liouville plane decays at infinity as shown in (5). This means that the new linear problem is of "scattering type."

In scattering theory, two basic quantities, usually written "a" and "b," are introduced as connection coefficients of Jost solutions. A remarkable fact is that these coefficients also have an integral representation in terms of Jost solutions. From the definition of a_{IJ} (as well as Olver's proof of the existence of boundary values), one can see that a_{IJ} is essentially the same as the a-coefficient, whereas w_I and w_J play the role of Jost solutions in scattering theory. With this analogy, one can derive an integral representation of a_{IJ}.

Proposition. *$a_{IJ}(\lambda)$ have two integral expressions as:*

$$\begin{aligned} a_{IJ}(\lambda) &= 1 + (2\lambda)^{-1} \int_{\Gamma_{IJ}} ds h(s) w_I(s, \lambda) \\ &= 1 + (2\lambda)^{-1} \int_{\Gamma_{IJ}} ds h(s) w_J(s, \lambda), \end{aligned} \tag{29}$$

where the setting is the same as in the previous section.

Remark. In fact, the analogy with scattering theory is somewhat subtle. In the usual situation of scattering theory, the term λ^2 is rather $-\lambda^2$, and Jost solutions are thereby oscillatory at infinity. In our setting, we are viewing the region where ϕ_I's behave exponentially small or large; the situation of scattering theory takes place on a line parallel to the imaginary axis of the Liouville plane. Nevertheless, the analogy with the a-coefficient turns out to be valid in the present situation.

From this integral representation, we can derive remarkable conclusions:

• First, substituting the iterated integral series for w_I and w_J, one can obtain the following expression of a_{IJ}.

Proposition. *$a_{IJ}(\lambda)$ has the iterated integral expansion*

$$\begin{aligned} a_{IJ}(\lambda) =& 1 + \frac{1}{2\lambda}\int_{\Gamma_{IJ}} ds\, h(s) \\ &+ \sum_{n=1}^{\infty} \frac{1}{2\lambda}\int_{(*)} dt_0 \cdots dt_n h(t_0) \prod_{i=1}^{n} \frac{(1-e^{2(t_{i-1}-t_i)\lambda})h(t_i)}{2\lambda}. \\ (*):&\quad t_0, \ldots, t_n \in \Gamma_{IJ}, \quad t_0 \preceq \ldots \preceq t_n, \end{aligned} \tag{30}$$

where $\preceq$ means that the points are ordered in that way along Γ_{IJ} from $s(\infty_I)$ to $s(\infty_J)$.

• Further, if Γ_{IJ} can be chosen to be a *straight line* on the Liouville plane, one can rewrite the above iterated integral into a Laplace integral:

$$a_{IJ}(\lambda) = 1 + (2\lambda)^{-1}\int_{\Gamma_{IJ}} dsh(s) + (2\lambda)^{-1}\int_0^{e^{i\theta}\infty} dte^{-2t\lambda}A_{IJ}(t), \tag{31}$$

where θ is the angle between Γ_{IJ} and the real axis of the Liouville plane. This is due to the obvious identity

$$\frac{1-e^{2(t_{i-1}-t_i)\lambda}}{2\lambda} = \int_{t_{i-1}}^{t_i} ds_i\, e^{2(s_i-t_i)\lambda}. \tag{32}$$

Changing the integration variables from (s_i, t_i) to (x_i, y_i) as

$$x_i = s_i - t_{i-1}, \quad y_i = t_i - s_{i-1},$$

one can indeed derive from (28) an expression like (31).

Proposition. *$A_{IJ}(t)$ has the iterated integral expansion*

$$\begin{aligned} A_{IJ}(t) =& \sum_{n=1}^{\infty}\int_{(**)} dx_1 \cdots dx_n\, dy_1 \cdots dy_n \prod_{i=1}^{n} h(s + \sum_{j=1}^{i}(x_j + y_j)) \\ (**):&\quad x_1, \ldots, x_n, y_1, \ldots, y_n \in [0, e^{i\theta}), \quad \sum_{j=1}^{n} y_j = t \end{aligned} \tag{33}$$

A somewhat careful consideration shows that $A_{IJ}(t)$ gives a holomorphic function in a neighborhood of the path of integration above.

• One can show, in the same way, that w_I's themselves have a similar Laplace integral representation. The above Laplace integral representation is, in fact, derived from such a Laplace integral of w_I's and basic relation (29).

These (iterated) integral representations provide detailed information on the analytic structure of a_{IJ} as well as ψ_I's themselves. An immediate consequence of the Laplace integral representation is that $a_{IJ}(\lambda)$ has an asymptotic expansion

$$a_{IJ}(\lambda) \sim 1 + \sum_{n=1}^{\infty} a_{IJ,n}/(2\lambda)^n \quad (\lambda \to \infty, |\arg\lambda + \theta| < \pi/2). \tag{34}$$

The coefficients can be read out from the Taylor coefficients of $A_{IJ}(t)$ at $t = 0$. As Voros observed (in a different formulation), the coefficients $a_{IJ,n}$ can also be evaluated, independently, by the WKB formal solutions mentioned in Section 3. The above result implies that such a formal series expansion is Borel summable in the sector arising in (34).

For the harmonic oscillator discussed above, only a_{23} is non-trivial. For a_{23}, one can choose Γ_{23} to be the real axis of the Liouville plane. In that case, $\theta = 0$. In fact, one can further move Γ_{23} as far as it does not meet the images of turning points. This, in particular, shows that the path of the Laplace integral can be rotated within the range $|\theta| < \pi/2$. A similar analysis can be done in more general cases.

5.2. A degenerate configuration of Stokes curves

Let us now consider the case that allows Stokes curves linking two turning points. This is a kind of "denereracy," which can be resolved by slightly changing the potential or the phase $\arg\lambda$. Voros indeed relies upon the latter trick in actual calculations. Olver, in his book, also avoids to deal with such degeneracy directly; however, Olver's method can be extended to degenerate cases, and this reveals an interesting phenomena. Let us consider a simple case where two turning points, say q_1 and q_2 are connected by a finite Stokes curve (see Fig. 4). This type of configuration frequently occurs in applications. For example, the previous harmonic oscillator with $\arg\lambda = \pm\pi/2$ has such has such Stokes curves.

For calculations, we put two cut lines, one between the Stokes curves S_1 and S_3 and the other between S_2 and S_4. We also assume that

$$\begin{aligned} \operatorname{Re} s(q) &= -\infty \quad \text{at } q = \infty_1, \infty_3, \\ \operatorname{Re} s(q) &= +\infty \quad \text{at } q = \infty_2, \infty_4, \end{aligned} \tag{35}$$

so that the associated exact solutions of the Liouville-Green form are written

$$\begin{aligned} \psi_I &= w_I (ds/dq)^{-1/2} e^{s\lambda} \quad \text{for } I = 1,3, \\ \psi_I &= w_I (ds/dq)^{-1/2} e^{-s\lambda} \quad \text{for } I = 2,4. \end{aligned} \tag{36}$$

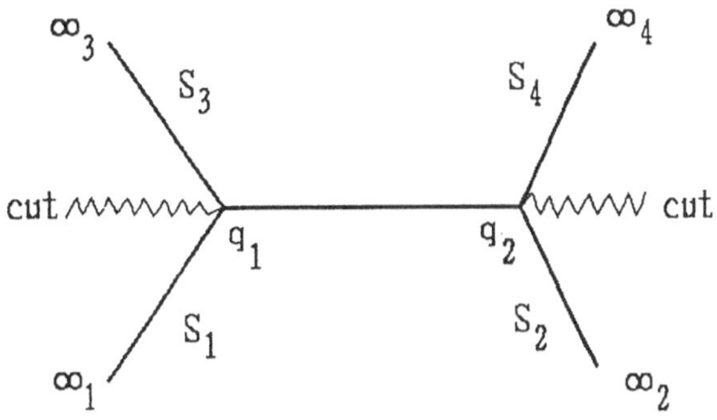

Fig. 4. A degenerate configuration of Stokes curves

This configuration may be part of a larger set of Stokes curves; other part does not affect the treatment of fundamental triplets in $\psi_1, \ldots, \psi_4$. Any triplet of the four ψ_I's is "fundamental," and one can derive a set of connection formulas just as in the nondegenerate case discussed above (with slightest modification due to the difference of the configuration of Stokes curves). Changing triplets and their combinations in (22), one obtains seemingly different formulas for the same triplet. Since such two formulas must have the same contents, the coefficients should satisfy some algebraic relations (consistency conditions). In the present situation, one can thus obtain the relation

$$a_{14}a_{23} = a_{12}a_{34} + e^{-2\omega\lambda}a_{13}a_{24}, \tag{37}$$

where

$$\omega \underset{\text{def}}{=} s(q_2) - s(q_1). \tag{38}$$

In fact, this kind of relations are not specific to "degenerate" configurations. They do exist even in "nondegenerate" configurations, but are simply hidden.

5.3. Algebraic relation of a_{IJ}'s and resurgence

From the geometric situation on the Liouville plane, one can see that linear paths Γ_{IJ} for the definition of $A_{IJ}(t)$ can be selected as indicated in Fig. 5. Accordingly, each of $a_{IJ}(\lambda)$ has the following Laplace integral representation.

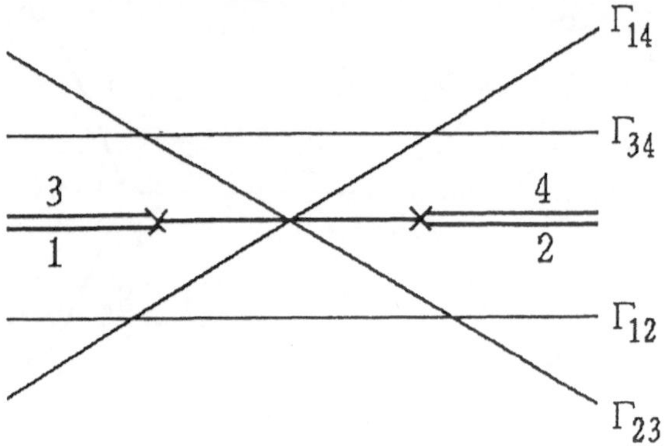

Fig. 5. Linear paths on the Liouville plane

$$a_{14}(\lambda) = 1 + \frac{1}{2\lambda}\int_{\Gamma_{14}} ds\, h(s) + \frac{1}{2\lambda}\int_0^{e^{i\theta}\infty} dt\, e^{2t\lambda} A_{14}(t) \quad (-\theta_0 < \theta < \theta_0), \tag{39}$$

$$a_{23}(\lambda) = 1 + \frac{1}{2\lambda}\int_{\Gamma_{23}} ds\, h(s) + \frac{1}{2\lambda}\int_0^{e^{i\theta}\infty} dt\, e^{2t\lambda} A_{23}(t) \quad (-\theta_0 < \theta < \theta_0). \tag{40}$$

$$a_{14}(\lambda) = 1 + \frac{1}{2\lambda}\int_{\Gamma_{14}} ds\, h(s) + \frac{1}{2\lambda}\int_0^{e^{i\theta}\infty} dt\, e^{2t\lambda} A_{14}(t) \quad (0 < \theta < \theta_0), \tag{41}$$

$$a_{23}(\lambda) = 1 + \frac{1}{2\lambda}\int_{\Gamma_{23}} ds\, h(s) + \frac{1}{2\lambda}\int_0^{e^{i\theta}\infty} dt\, e^{2t\lambda} A_{23}(t) \quad (-\theta_0 < \theta < 0). \tag{42}$$

(θ_0 is a positive constant to be determined by the configuration of Stokes curves and turning points.) In (41) and (42), one will not be able to put $\theta = 0$; A_{14} and A_{23} are expected to have singularities on the positive real axis.

Remarkably, (37) carries precise information on these singularities. To see this, we first rewrite (37) as:

$$a_{14} = \frac{a_{12}a_{34}}{a_{23}} + e^{-2\omega\lambda}\frac{a_{13}a_{24}}{a_{23}}. \tag{37'}$$

All the ingredients of the right hand side, except the exponential function, have a Laplace integral representation along the half line $[0, e^{i\theta}\infty)$ with $-\theta_0 < \theta < 0$. On the other hand, a_{14} has a Laplace integral representation as in (41) along the half line $[0, e^{i\theta}\infty)$ for $0 < \theta < \theta_0$. What will occur if one rotates the path in (41) downward across the positive real axis? That should result in a Laplace integral along $[0, e^{i\theta}\infty)$ with $\theta_0 < \theta < 0$ (the main part) plus some contributions from paths running around singularities of $A_{14}(t)$ (see Fig. 6). The right hand side of (37′) takes exactly such a form; the first term is the main part to be obtained as an analytic continuation of $A_{14}(t)$ through a neighborhood of $t = 0$, whereas the second term is a contribution from a singularity at $t = \omega$. A similar interpretation can be found for a_{23} if one writes (37) as:

$$a_{23} = \frac{a_{12}a_{34}}{a_{14}} + e^{-2\omega\lambda}\frac{a_{13}a_{24}}{a_{14}}. \tag{37''}$$

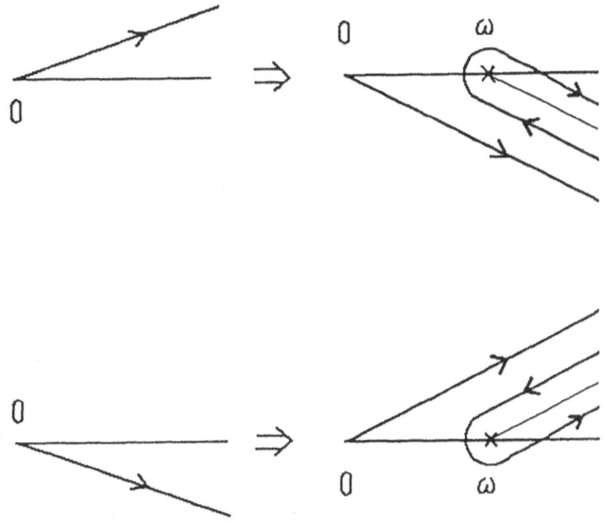

Fig. 6. Interpretation of (37) by deformations of the path

Of particular importance is the fact that the discontinuity along a path on the Liouville plane (now issuing from ω) is again written in a closed form in terms of a_{IJ}'s. This is obviously the same phenomena as originally called "resurgence" by Ecalle, and rediscovered by Voros in the context of the WKB connection problem. The Liouville plane now plays the role of the "Borel plane" in the Ecalle theory. Note, further, that ω is a "half-period" of $f(q)^{1/2}$; this is a general phenomena, as already observed by Voros, Ecalle and Pham.

One can interpret our connection formulas in the same way; now w_I's as well as a_{IJ}'s join the game. The connection formulas, too, thus turn out to be a concise representation of the "resurgence structure" of w_I's on the Borel plane. This also agrees with observations of Voros, Ecalle and Pham. It should be noted that in our treatment based upon Olver's idea, connection formulas are derived without any analysis on singularities on the Borel(=Liouville) plane. Singularity structures can be read out from the connection formulas as a consequence. In this respect, our approach is opposite, or complementary, to the method of Voros; Voros rather starts from a hypothesis on such singularities, and derives his connection formulas from that postulate.

5.4. More degenerate case

A further degenerate case takes place if the two turning points in the above example coalesce to a turning point of second order. In that case, the calculation of connection coefficients cannot be reduced to that of triplets. One should therefore consider the four solutions altogether and seek for a 2×2-connection relation of the form

$$\begin{aligned} \psi_1 &= c_{13}\psi_3 + c_{14}\psi_4, \\ \psi_2 &= c_{23}\psi_3 + c_{24}\psi_4. \end{aligned}$$

From the boundary behavior of each term as $q \to \infty_1, \ldots, \infty_4$, one can derive four linear relations. These relations, however, are not independent and cannot determine the four coefficients. Olver's method thus breaks down in this case.

6. Beyond second order Sturm-Liouville linear problem

6.1. Dirac equations

There are obviously a number of possible directions that deserve further studies. An immediate idea is to develop a similar approach to "Dirac equations" of the form

$$\phi_s = \begin{pmatrix} \lambda & q(s) \\ r(s) & -\lambda \end{pmatrix}, \quad \phi = \begin{pmatrix} \phi_0 \\ \phi_1 \end{pmatrix}, \tag{43}$$

where potentials $q(s)$ and $r(s)$ are assumed to have some analytic properties parallel to the potential $h(s)$ in the "Schrödinger equation" on the Liouville plane. In this case, we directly start from the s-plane rather than seeking for an analogue of the q-plane. Actually, even in the study of the second order Sturm-Liouville problem, Dirac equations have been rather a standard framework; Ecalle and Grigis deal with this problem by converting the original second order equation into a first order Dirac-type system. Further, scattering theory has been also developed for the Dirac equations (for example, by Zakharov and Shabat [ZS], and Ablowitz et al. [AKNS] with applications to soliton theory); our iterated integral series and Laplace integral representation can be naturally extended to that direction.

From our point of view, the analysis on the s-plane is more fundamental than the q-plane, and has enough contents in itself as well as further extensions. One may even forget of the origin of $h(s)$ (the q-plane, the Liouville transformation, etc.); the only thing that plays a crucial role seems to be the "resurgence" of the potential $h(s)$ (here s is identified with the Borel variable). Ecalle indeed reports intriguing results along that line for a more general class of linear systems [E, vol. III].

6.2. String equations

Nonlinear equations should be the next, and even more important subject. Of course, Ecalle has made, in that direction as well, a number of suggestive observations, dealing with quite general situations. Let us rather seek for more concrete examples; a special equation should have its own interesting properties.

In that respect, recent progress in theoretical physics seems to offer a family of very interesting nonlinear differential equations, the "string equations," which are expected to describe the physics of low dimensional string and gravity theory [BK], [DS], [GM]. In the simplest case, the string equation becomes the first Painlevé equation:

$$u_{xx} = u^2 - x, \quad u = u(x). \tag{44}$$

From a physical reason, it is better to put a small parameter G on the left hand side as:

$$Gu_{xx} = u^2 - x, \quad u = u(x), \tag{44'}$$

but for the moment, let us discuss the case of $G = 1$. Physicists seek for solutions with the boundary condition

$$u \sim x^{1/2} \quad \text{as } x \to +\infty, \tag{45}$$

and further ask, for example, if there is any natural condition that determines a remaining arbitrary constant. Several candidates for such a condition have been discussed: – some requirements on the location of possible poles on the real axis (as conjectured by the classical theory of Boutroux [B]), a boundary condition as $x \to -\infty$ (as suggested by F. David [D]), etc. Along with the first Painlevé equation, physicists have introduced an infinite sequence of higher string equations that describe different physical contents:

$$P_{2k+1}[u] = x \quad (k = 1, 2, \ldots), \tag{46}$$

where $P_{2k+1}[u]$ is the generator (the variational derivative of a Hamiltonian $H_{2k+1}[u]$) of the higher (k-th) Korteweg-de Vries (KdV) equation,

$$u_{t_{2k+1}} = P_{2k+1}[u]_x. \tag{47}$$

It is well known in soliton theory [GD] that $P_{2k+1}[u]$'s can be recursively determined as coefficients of a formal solution of the Riccati equation associated with the linear equation

$$\phi_{xx} + u\phi = \lambda^2\phi. \tag{48}$$

(Recall the construction of formal WKB solutions in Section 3!) For example,

$$\begin{aligned} &P_3[u] = (3u^2 - u_{xx})/16, \\ &P_5[u] = -(10u^3 - 10uu_{xx} - 5(u_x)^2 + u_{xxxx})/64, \\ &\text{etc}\ldots \end{aligned} \tag{49}$$

After suitable rescaling of u and x, equation (47) with $k = 1$ reproduce the first Riccati equation. In general, $P_{2k+1}[u]$ is a differential polynomial of u including at most $2k$ order derivatives of u, the highest order term being linear in u. For the higher string equations, physicists require the boundary condition

$$u \sim \text{const.}x^{1/(k+1)} \quad \text{as } x \to +\infty. \tag{50}$$

The constant on the right hand side is to be determined by the equation itself. As in the $k = 1$ case, this boundary condition is still too weak to fix a physical solution. According to physicists, a candidate for additional conditions is to require that u is real-valued, has no singularity on the real axis and, further, satisfies the boundary condition

$$u \sim \text{const.}x^{1/(k+1)} \quad \text{as } x \to -\infty \tag{51}$$

for the same constant as in (50). It is conjectured, by numerical analysis, that such a solution exists if and only if k is even [BMP]; in particular, the first Painlevé equation fails to have such a solution. Mathematically, this issue is obviously related to a global connection problem of the above nonlinear equations.

6.3. Possible link with Ecalle theory

Is such a global connection problem feasible from the point of view of the Ecalle theory or anything along that line? A final answer is far beyond our present scope, but let us show two observations suggesting a possible link with the Ecalle theory. For simplicity, we consider the first Painlevé equation alone.

One observation is due to physicists [DS], [GM]: Suppose that u is given as a linear combination of the form

$$u = x^{1/2} + v. \tag{52}$$

the second term v being, of course, expected to describe subleading contributions on the right hand side of (45). In terms v, the first Painlevé equation can be rewritten into a kind of nonlinear Sturm-Liouville eqation,

$$v_{xx} - 2x^{1/2}v = v^2 + \frac{1}{4}x^{-3/2}, \tag{53}$$

with an inhomogeneous term and a nonlinear term on the right hand side. Physicists infer from this fact that there should be "instanton effects" (i.e., subdominant effects like quantum tunneling) governed by the WKB solutions

$$v_0^{\pm} \sim x^{-1/8} \exp(\pm\frac{\sqrt{32}}{5}x^{5/4}) \tag{54}$$

of the linearized and homogenized equation. Mathematically, this simply means that (53) has an irregular singularity of the Poincaré rank one with respect the new variable

$$s = \frac{\sqrt{32}}{5}x^{5/4}. \tag{55}$$

As far as only a neighborhood of $s = \infty$ is concerned (i.e., within a *local* theory), this already takes a "prepared form" for the Ecalle theory. The boundary problem with two boundary condition at $x = \pm\infty$ is obviously related a the connection problem around $s = \infty$, hence with some *nonlinear Stokes phenomena.* Ecalle seems to suggest a general framework for studying such issues within his "alien calculus."

Another observation is due to Pham in the "Epilogue" of his forthcoming book [CNP]. As a typical example, Pham considers therein the Riccati equation

$$u_x = u^2 - x. \tag{56}$$

Obviously, this equation allows basically the same treatment as we have mentioned for the first Painlevé equation. By the substitution

$$u = x^{1/2} + v, \tag{57}$$

one indeed obtains the equation

$$v_x - 2x^{1/2}v = v^2 - \frac{1}{2}x^{1/2}, \tag{58}$$

which has a prepared form of the Ecalle theory.

Of course, the Riccati equation is far simpler than the first Painlevé equation. For example, the Riccati equation can be converted to the Airy equation

$$p_x - xp = 0 \tag{59}$$

by the well known transformation

$$u = -p_x/p. \tag{60}$$

This is never the case for the Painlevé equation; even a stronger property ("irreducibility") is known [N], [U], which implies its highly transcendental nature. Nevertheless, the Riccati equation may be viewed as a nice example to get some insight into the issues on the Painlevé equation mentioned above.

Let us show a result of numerical computation to the Riccati equation (Fig. 7). This is, actually, just a redrawing of a figure quoted in the book of Pham. [The caption therein says that it is reproduced from: Artigue and Gautheron, *Systèmes différentiels, étude graphique* (Cedic, Paris 1983).] The curves other than the axis in the figure are graphs, in the (x, u)-plane, of solutions of the Riccati equation with various initial conditions. On the right half plane, one will be able to see a parabola, which is exactly the curve $u^2 - x = 0$. The upper half of this parabola is "repulsive" in the sense that every point close to this part soon get far away as x increases; the lower half is "attractive" in the opposite sense. (A rigorous analysis of this phenomena can be found in Hille's book [H].) Note that drawing this kind of figure in the Painlevé case requires a fine tuning of two arbitrary constants in a general solutions; one has to find a nice one-parameter family of solutions, which is, even numerically, a hard problem.

6.4. New resurgence from τ function?

It seems thus very plausible that the Ecalle theory will be a useful tool for the study of the string equations at infinity. There is, however, another issue related to possible poles of solutions. At least for the case of the Painlevé equation, it is known that poles are all of second order, and come from zeros of the so called "τ function," and the τ function is proven to be an entire function. Actually, the notion of the "τ function" is

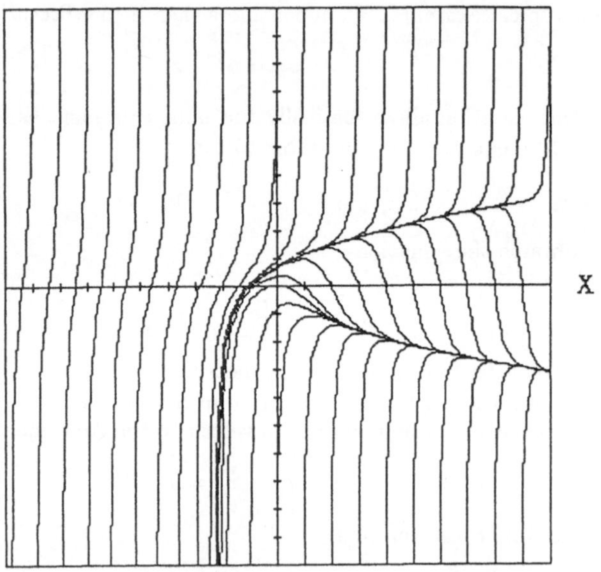

Fig. 7. Graphs of solutions to $u_x = u^2 - x$

also introduced in the KdV and higher KdV equations. For the study of poles of u, the τ function should play a key role. Its relation to the Ecalle theory is now far from our understanding; this might lead to a new aspect of resurgence. At this stage, we already have two variables, s in (55) and λ in (48), which will be responsible for two distinct resurgence phenomena ("equational" and "quantum" in the terminology of Ecalle). The small parameter G, if incorporated as in (44′), might yield a third resurgence that will be specific to the τ function.

References

[AKNS] Ablowitz, M.J., Kaup, D.J., Newell, A.C., and Segur, H., The inverse scattering transform — Fourier analysis for nonlinear problems, Stud. Appl. Math. vol. 53 (1974), 249-315.

[B] Boutroux, P., Recherches sur les transcendantes de M. Painlevé et l'etude asymptotique des équations différentielles du second ordre, Ann. Sci. Ecole Norm. Sup. (3) 30 (1913), 255-375; (3) 31 (1914), 99-159.

[BK] Brézin, E., and Kazakov, V.A., Exactly solvable field theories of closed strings, Phys. Lett. B236 (1990) 144-150.

[BMP] Brézin, E., Marinari, E.,and Parisi, G., A non-perturbative ambiguity free solution of a string model, Lett. Math. Phys. B242 (1990), 35-38.

[CNP] Candelpergher, B., Nosmas, J.C., and Pham, F., Approche de la résurgence (to appear).

[D] David, F., Loop equations and non-perturbative effects in two-dimensional quantum gravity, Mod. Phys. Lett. A 5 (1990), 1019-1029.

[DDP] Delabaere, E., Dillinger, H., and Pham, F., Développements semi-classiques exacts des niveau d'énergie d'un oscillateur à une dimension, C. R. Acad. Sci. Série I, 310 (1990), 141-146.

[DS] Douglas, M.R., and Shenker, H., Strings in less than one-dimension, Nucl. Phys. B335 (1990), 635-654.

[E] Ecalle, J., Les fonctions résurgentes, I-III, Publ. Math. Orsay 1981, 1985.

[GD] Gelfand, I.M., and Dikii, L.A., Asymptotic behavior of the resolvent of Sturm-Liouville equations and the algebra of the Korteweg-de Vries equations, Russian Math. Surveys 30:5 (1975), 77-113.

[GG] Gérard, C., and Grigis, A., Precise estimates of tunneling and eigenvalues near a potential barrier, J. Diff. Eq. 72 (1985), 149-177; Grigis, G., Sur l'équation de Hill analytique, Sem. Bony-Sjöstrand-Meyer 1984/85.

[GM] Gross, D.J., and Migdal, A.A., Nonperturbative two-dimensional quantum gravity, Phys. Rev. Lett. 64 (1990), 127-130.

[H] Hille, E., Ordinary differential equations in the complex domain (John Wiley and Sons, New York, 1976).

[N] Nishioka, K., A note on the transcendency of Painlevé's first transcendent, Nagoya Math. J. 109 (1988), 63-67.

[O] Olver, F.W.J., Asymptotics and special functions, Academic Press 1974.

[U] Umemura, H., On the irreducibility of the first differential equation of Painlevé, in *Algebraic geometry and commutative algebra, in honor of Masayoshi Nagata*, pp. 101-119 (Kinokuniya, Tokyo, 1988).

[V] Voros, A., The return of quartic oscillator, Ann. Inst. H. Poincaré, Section A, 39 (1983), 211-338.

[W] Wasow, W., Linear turning point problem, Springer-Verlag 1985.

[ZS] Zakharov, V.E., and Shabat, A.B., A scheme for integrating the nonlinear equations of mathematical physics by the method of the inverse scattering problem, Funct. Anal. and Appl. 8 (1974), 226-235.

Hayashibara Forum 1990
International Symposium on Special Functions
August 16 – 20

PROGRAMME

August 16

10:00–11:00	Mikio Sato (RIMS, Kyoto University, Japan) "Algebraic analysis of singular perturbation"
12:40–13:40	Frederic Pham (University of Nice, France) "Airy and cylindro-parabolic functions from a "resurgent" viewpoint"
14:00–15:00	B. M. McCoy (State University of New York at Stony Brook, U.S.A.) "The Chiral Potts Model: from Physics to Mathematics and back"
15:30–16:30	Ivan Cherednik (Moscow State University) "On generalizations of Kniznik-Zamolodchikov equations"
20:00–22:00	POSTER SESSION

August 17

9:20–10:20	Henry McKean (New York University, U.S.A.) "Theta functions and scattering"
10:40–11:40	V. Lakshmibai (Northeastern University, U.S.A.) "Generalized Schubert varieties"
12:40–13:40	Andrey V. Zelevinsky (Scientific council for Cybernetics, U.S.S.R.) "Hypergeometric functions, toric varieties and Newton polyhedra"
14:00–15:00	N. Reshetikhin (Harvard University) "q-Weyl group and multiplicative formula for the R-matrix"
15:30–16:30	G. J. Heckman (Math. Inst en Inst., The Netherlands) "Special functions associated with root systems"
20:00-22:00	POSTER SESSION

August 18

9:20l–10:20	Paul Malliavin (University Paris VI, France) "Integration on loop group"
10:40–11:40	B. Feigin (Inst. Solid State Physics, U.S.S.R.) "Differential Operators on the Moduli Space of G-Bundles on Algebraic Curve and Lie Algebra Cohomologies"

August 19

9:20–10:20	Kazuhiko Aomoto (Nagoya University, Japan) "q-analogue of de Rham cohomology associated with Jackson integrals"
10:40–11:40	Kyoji Saito (RIMS, Kyoto University, Japan) "Duality in the Teichmüller-Fuchs modular function"
12:40–13:40	Leonard Lewin (University of Colorado, U.S.A.) "Supernumary Polylogarithmic and Clausen-Function Ladders"
14:00–15:00	A. N. Kirillov (LOMI, Fontanka, U.S.S.R.) "Representation of quantum groups and Special functions"
15:30–16:30	Masatoshi Noumi (University of Tokyo, Japan) "Quantum groups and q-orthogonal polynomials"

August 20

9:20–10:20	Leon Ehrenpreis (Temple University, U.S.A.) "The Hypergeometric Function"
10:40–11:40	Takayuki Oda (RIMS, Kyoto University, Japan) "An interpretation of Lewin's ladder relation of polylogarithms"
12:40–13:40	Don B. Zagier (Max Planck Institute, F.R.G.) "Polylogarithms in geometry and arithmetics"
14:00–15:00	Don Blasius (The Institute for Advanced Study, U.S.A.) "Motivic and automorphic L-functions"

LIST OF PARTICIPANTS

AKAHORI, T. (Niigata U., Japan)

ANDRONIKOF, E. (U. de Paris-Nord XIII, France)

AOKI, T. (Kinki U., Japan)

AOMOTO, K. (Nagoya U., Japan)

BLASIUS, D. (UCLA, USA)

CHEREDNIK, I. V. (Moscow State U., USSR)

DATE, E. (Osaka U., Japan)

EHRENPREIS, L. (Temple U., USA)

FEIGIN, B. (Inst. Solid State Phys., USSR)

HASEGAWA, K. (Tohoku U., Japan)

HECKMAN, G. J. (Katholieke U., Netherlands)

HORIKAWA, E. (U. of Tokyo, Japan)

IKEDA, N. (Osaka U., Japan)

JIMBO, M. (Kyoto U., Japan)

KANEKO, J. (Kyushu U., Japan)

KASHIWARA, M. (Kyoto U., Japan)

KAWAI, T. (Kyoto U., Japan)

KIRILLOV, A. N. (LOMI, Fontanka, USSR)

KOBAYASHI, T. (U. of Tokyo, Japan)

LAKSHMIBAI, V. (Northeastern U., USA)

LEWIN, L. (U. of Colorado, USA)

MAJIMA, H. (Ochanomizu U., Japan)

MALLIAVIN, P. (U. de Paris VI, France)

MATSUO, A. (Kyoto U., Japan)

MCCOY, B. M. (State U. NY at Stony Brook, USA)

MCKEAN, H. P. (NY U., USA)

MIKI, T. (Kyoto U., Japan)

MIMACHI, K. (Nagoya U., Japan)

MIWA, T. (Kyoto U., Japan)

NETO, O. (Kyoto U., Japan)

NOUMI, M. (U. of Tokyo, Japan)

ODA, T. (Kyoto U., Japan)

OHYAMA, Y. (Osaka U., Japan)

OKADO, M. (Osaka U., Japan)

PASTRO, P. I. (Kyushu U., Japan)

PHAM, F. (U. Nice, France)

RESHETIKHIN, N. (Harvard U., USA)

SAITO, K. (Kyoto U., Japan)

SATO, M. (Kyoto U., Japan)

SATO Y. (Kyoto U., Japan)

SHIMIZU, Y. (Tohoku U., Japan)

TAKASAKI, K. (Kyoto U., Japan)

TAKEI, Y. (Kyoto U., Japan)

TANISAKI, T. (Osaka U., Japan)

USUI, S. (Osaka U., Japan)

YAMADA, H. (Tokyo Metropolitan U., Japan)

YAMANE, H. (Osaka U., Japan)

YOSHIDA, J. (Kyoto U., Japan)

YOSHIZAWA, H. (Okayama U. of Science, Japan)

ZAGIER, D. B. (Max-Planck-Inst. Bonn, Germany)

ZELEVINSKY, A. V. (Scientific Council Cybernetics, USSR)